THE SAUROPOD
DINOSAURS

THE SAUROPOD DINOSAURS

LIFE IN THE AGE OF GIANTS

MARK HALLETT

AND

MATHEW J. WEDEL

Johns Hopkins University Press

Baltimore

© 2016 Johns Hopkins University Press
All rights reserved. Published 2016
Printed in Canada on acid-free paper
9 8 7 6 5 4 3 2 1

Johns Hopkins University Press
2715 North Charles Street
Baltimore, Maryland 21218-4363
www.press.jhu.edu

Library of Congress Control
Number: 2015956754

ISBN 978-1-4214-2028-8
(hardcover: acid-free)
ISBN 978-4214-2029-5
(electronic)

A catalog record for this book is available from the British Library

*Special discounts are available for bulk purchases of this book. For more information, please
contact Special Sales at 410-516-6936 or specialsales@press.jhu.edu.*

Johns Hopkins University Press uses environmentally friendly book materials, including recycled
text paper that is composed of at least 30 percent post-consumer waste, whenever possible.

To the three women in my life who helped to set and keep me on my course as an artist, naturalist and writer. First, my mother, Joan, who encouraged me to draw and dream. Second, my Aunt Merlyn, who promised me that wonders awaited at the top of the museum steps that day, if I could overcome the ordeal of my prosthesis. Lastly, my wife, Turi, whose patience and love has eased the occasional loneliness of art and writing.

MARK HALLETT

To my parents, John and Norma Wedel, for taking me to countless museums as a kid, for never getting tired of hearing about dinosaurs, and for raising their kids to love animals, the outdoors, and reading. Mom and Dad, here's your book.

MATHEW J. WEDEL

The seasons came and went, and seemingly everything changed and nothing changed. There were images of sauropod necks and backs glistening in monsoonal rains. Of lush savannahs, clouds of insects, and streams of migrating giants. Of great, broad swaths of red and black mud cut by game trails . . . of masses of fibrous dung, of blood-soaked earth, and of chunky hatchlings arriving with the rains, representing the hope for generations and generations to come.

DALE A. RUSSELL
The Dinosaurs of North America: An Odyssey in Time

CONSULTANTS

MARK HALLETT
MATHEW J. WEDEL

In the writing of this book, the two coauthors express their most sincere appreciation and thanks for the knowledge and advice provided by consulting paleontologists pictured below.

KRISTI CURRY ROGERS, PhD
Macalester College

Curry Rogers is a leading authority on the anatomy, evolution and systematics of titanosaurian sauropods and has participated in several field seasons in Madagascar, one of which led to the discovery of the nemegtosaurid titanosaur *Rapetosaurus krausei*, which she described in 2001. Her work has produced major insights into the paleobiology of titanosaurs, and she is an editor and contributing author of *The Sauropods: Evolution and Biology,* an anthology of sauropod evolution, biomechanics and systematics.

CAROLE T. GEE, PhD
University of Bonn

Gee is a senior research scientist in paleobotany, Division of Paleontology at the Steinmann Institute, University of Bonn, Germany, where she studies the morphology, taxonomy and taphonomy of Mesozoic and Tertiary plants. The editor of the book *Plants in Mesozoic Time* and an author and editor for *Biology of the Sauropod Dinosaurs*, she currently studies the genus *Araucaria*, the Late Jurassic flora of the Morrison Formation and the relationship between Mesozoic plants and herbivorous dinosaurs.

JOHN R. HUTCHINSON, PhD
**Royal Veterinary College,
University of London**

Hutchinson is a professor of evolutionary biomechanics and studies how animals move and how the progressive adaptations of their movements evolved, especially in larger land animals. He has studied elephants, birds and diverse dinosaurs such as *Tyrannosaurus*, among other species. His work in the science of biomechanics has clarified how giant size limits the various abilities of land animals—including sauropods—to move quickly or support their weight. He is an associate editor of *Proceedings of the Royal Society*.

PAUL UPCHURCH, PhD
University College London

Upchurch has published 70 research articles and 12 book chapters, mainly focusing on the systematics and evolution of sauropod dinosaurs. Currently his work centers on large-scale evolutionary patterns that occurred during the Mesozoic, including the impact of plate tectonics, changes in sea level and climate change on the geographic distributions of dinosaurs and other land vertebrates. He is a major authority on sauropod feeding ecology, in which his studies have investigated the relationship of jaw and tooth morphology to feeding patterns and resource partitioning.

CONTENTS

Seconds later the mists are swept aside by tall, muscular columns like moving tree trunks, colossal bodies and towering necks.

OUT OF THE MISTS

If she or he could go back in time, what might a journal entry from a sauropod field biologist be like? Perhaps the following:

GEOCHRON—Aptian-Mid-Albian, **112 MYA**
Oklahoma, North America

5:26 a.m.: The milky ground fog, barely noticeable a few hours ago, has now covered everything except the trees. With the cooling temperature of last night, my recorder picks up a few last piercing trills and screeches of the marsh frogs, which have now dwindled to an occasional booming honk from the larger species that haven't chilled to inactivity. As late as yesterday, the baked, red-gray surface of the hardpan near my camp held no life. When the fury of the storm broke over the dry *Araucarites* woodlands after days of buildup, however, the rain must have awakened the sleeping lives entombed in the concrete clay below. Vast sheets of water collected within 20 minutes, and frogs in the thousands fought their way up to the sweet, percolating moisture, transforming the hardpan and other areas like it into frenzied mating areas.

5:38 a.m.: Chilled and shivering, I wait. As the sky lightens, there are crunches and snaps out in the fog toward the plain. Distant, but coming closer every few minutes, deep, throbbing rumbles begin to fill the air, and I can actually feel the ground vibrate. Seconds later the mists are swept aside by tall, muscular columns like moving tree trunks, colossal bodies and towering necks. Splashing across the pond, huge clawed and padded feet are propelling the enormous shapes smoothly, effortlessly toward the *Araucarites* woodland. They're *Sauroposeidon proteles*, the most gigantic sauropods of the Antlers Formation, Early Cretaceous Oklahoma.

5:53 a.m.: My legs ache from cold and stiffness, but I have to scramble over to a copse of tree ferns where the herd won't notice me, about 75' away. As massive as they are, the *Sauroposei-*

dons' skins hang loosely on their tall frames, evidence of the poor foraging opportunities of the Dry Season's last month. The herd, 17 in all, is made up of subadults spanning 10–25 years old, a handful of mature adults between 25–50 years old and one old individual, a 73-year-old survivor. The sauropods concentrate on the tender, protein-rich new growth at the ends of the branches, cropping with chisel-like, sharp-edged teeth. They soon have enough fodder in the wide jaws to sluice down their gullets. The biggest sauropods feed ravenously, slicing off huge, three-bushel bites of *Araucarites* leaves, swallowing and then reaching for more. Without the need to chew, their jaws can crop a tremendous quantity of vegetation, and within a few days it will be time to move on to the next patch of woodland. For now, however, there's enough, and while the morning's drizzle begins to tap and trickle down the jutting nasal crests, watchful amber-brown eyes blink the drops away. The Wet Season has begun.

Dinosaurs are staggeringly distant from us in time, as far away from our own moment as distant galaxies are from Earth. This remoteness extends far beyond any time frame of which we as humans can readily conceive, beyond the mastery of fire and the invention of stone tools, beyond the time when our primate ancestors came down from the trees and much farther into deep time. It's also hard to think of them as real animals that once populated our world because they are so different from the modern creatures with which we are familiar. In terms of appearance, behavior and sheer size they are unique to us, and like the Pyramids, they carry with them a sense of profound mystery. The word "dinosaur" itself conjures up many images in the human mind, and one of the most enduring and familiar is that of the huge, long-necked, long-tailed "brontosaurus" type of dinosaur. Correctly known as **sauropods**, they are the very symbol of dinosaurs for most people.

In this book, we'll explore the newest ideas—and sometimes the oldest—of what many paleontologists think these awesome creatures were like. Even among dinosaurs sauropods were different, not only for their size but also for their possible ways of eating, defending themselves from predators and reproducing their young. The skyscrapers of the animal world, the largest of them may have carried their heads 50' or more in the air and outweighed a small herd of elephants by many tons. They were also one of the longest-lived groups of dinosaurs, originating near the very beginning of the dinosaurian age and surviving until its close, a span of almost 150 million years.

Some of this information has come to us in the past several years by taking a fresh, new look at dinosaurs, by original approaches in the interpretation of anatomy and geology and by recent technological developments in the study of ancient animals and their worlds. Most of it, however, has only come through decades of old-fashioned fieldwork, patient study and research by legions of hardworking, insightful paleontologists from all over the world. These women and men are almost unknown to the general public, but what we know of the past has come from their efforts. Anything as large as a sauropod should be easy to find and study, but sauropods' very size can make them hard for paleontologists to tackle. When found, the huge bones take a lot of time and money to excavate, remove from the field and prepare for study and exhibition. As massive as they often are, they are often tantalizingly incomplete; the enormous hip, shoulder and limb bones can be well preserved, but the fragile, lightly constructed skulls are usually missing. The long chains of vertebrae are almost never found complete. Great size and weight also make the fossilized bones difficult to handle and have sometimes kept students and researchers from easily manipulating actual specimens in a way that would be less of a problem with smaller dinosaurs.

In spite of challenges like these, an ever-increasing roster of new species has been discovered in almost every part of the world, from Argentina to Mongolia and from Montana to Portugal. Although we think of all sauropods as being enormous, some had bodies about as big as a modern elephant, and a few were actually dwarfs, island outcasts that adapted to the limited resources of their restricted habitats by evolving a small body size. A few sauropod families developed bizarre, almost dragonesque extensions from their necks and backs, the function of which is unknown. Others had spiked tails and clubs for warding off their main predators, the flesh-eating theropods. Even as we write this book, both well-known and recently discovered species are providing us with exciting new glimpses into sauropod diversity and paleobiology, and more are certain to come.

As a group of animals, the sauropod dinosaurs push our understanding of how big animals can become, and how terrestrial creatures of this size, in some cases rivaling large whales in tonnage, may have evolved techniques of feeding, courtship, and reproduction. How did a 75- to 80-ton *Argentinosaurus* take in enough food to keep itself from starving? What were the consequences to a *Mamenchisaurus*'s blood pressure if it raised its 55' neck from horizontal to vertical in less than a minute? How did sauropod eggs and hatchlings survive without the presence of their parents, which were constantly moving in search of new forage? Much about how sauropods lived—and died—is still a mystery, but answers may come when the next field jacket is opened, the next slide is examined by a bone histologist or the next experiment is conducted by a biomechanist. In this volume we offer a holistic approach to sauropod paleobiology. Incorporated are recent concepts and theories from groundbreaking research efforts from members of the 533 Sauropod Research Biology Team, and from this and other recent studies the authors show what at present we confidently know about these dinosaurs' anatomy, paleobiology, phylogeny and evolution from leading researchers in many disciplines. At the same time we'll venture into some informed speculation when appropriate. We need to constantly look at sauropods as living animals within their constantly changing environments and shifting global geographies instead of as now extinct fossil forms, and for this reason much emphasis has been placed throughout the book on the botanical species these dinosaurs coevolved with instead of treating Mesozoic plants as merely stage props during the sauropods' time on Earth. Finally, as with any honest book that attempts to reconstruct the lives of extinct animals, we anticipate that with time new discoveries, and studies, will correct mistakes we've made and advance new ideas that we haven't thought of. Sauropods will continue to intrigue us because, of all the dinosaurs, they are the least well understood as a group; our planet has never had vertebrates like them before or since. They were and continue to be unique, and both the authors and consultants hope this book will contribute to everyone's knowledge of and fascination for these animals.

The rain is still pelting down on the woods where the Sauroposeidon herd is sheltering and feeding. We'll be starting our safari today, a safari of the mind's eye into the Mesozoic world of the sauropod dinosaurs—their evolution, daily lives and ways of survival. What will we see and learn? The alert, amber eyes of the huge female that accompanied the herd this morning are watching us, and in this book perhaps we'll come to understand some of the secrets they hold before we finish. One last look, and it's time to start off . . .

ACKNOWLEDGMENTS

Both authors would like to thank the following individuals and institutions for their generous contributions and assistance in making this book possible: Dennis Anderson, Laura and Bob Archer/Ash Creek Animal Clinic, Robert T. Bakker, Brant Bassam/Brantworks, Garrido A. Cerda, Luis Chiappe, Cleveland-Lloyd Quarry, John Collingwood, Ernie Cooper/Macrocritters, Jennifer A. Coulson/Coulson Harris Hawks, Ian Cross, Nathan Dahlstrom, Marty Daniel, Kieran Davis, Peter Falkingham, David C. Freitag, Kelly Gorham, Gerald Grellet-Tinner, Scott Hartman, Donald Henderson, Douglas Henderson, Mike Hettwer, Phil Hore, John Hutchinson/Royal Veterinary College, University of London, Tyler Keillor, Nils Knötschke/Dinosaurier Park, Eva Krocher, Susan Leibforth/Bastian Voice Institute, Frank Luerweg, Devon Lyon, Patricia and Philip Maher/Australian Ornithological Services, Heinrich Mallison, Andrew Milner, Minnesota Children's Museum, Museo Paleontologico Egidio Feruglio, Gregory S. Paul, Candace Paulos/Earth's Ancient Gifts, Phil Platt/Brontobuilder, George Poinar, H. C. Proctor, Hannah Rawe/AMNH, Robert Reisz, Kristi Curry Rogers, Kathy Rose/Imbala, Sam Noble Oklahoma Museum of Natural History, P. Martin Sander, Nima Sassani/Sassani Paleoart, Daniela Schwartz-Wings, William Sellers, Paul Sereno, Jeff Shaw/Bone Clones, Larry C. Simpson, Gesine Steiner, Thomas Steuber, Mike Taylor, Tony M. Thomas, Hall Train/Hall Train Studios, Turbosquid, Inc., Troy Weiss, John Whitlock, Lawrence Witmer/Witmer Studios, Grant Woods/Growing Deer, Mark Young and the Zigong Dinosaur Museum, for their kindness in allowing us to reproduce their photos and digital images free of charge.

We are grateful to Rachel A. Hallett, for her skills as layout and production artist; Karyn Servin, for her skills in vector art production; Michael M. Fredericks and Erik Fredericks/Prehistoric Times, for their tireless help in acquiring and editing photos; professors, academic advisors and research collaborators Paul Barrett, Matt Bonnan, Brooks Britt, Rich Cifelli, Leon Claessens, Bill Clemens, John Conway, Margie Dell, Brian Engh, John Foster, Don Henderson, Dave Hone, Jim Kirkland, Brian Kraatz, Darren Naish, Bob Nicholls, Pat O'Connor, Kevin Padian, Kent Sanders, Trish Schwagmeyer, Daniela Schwartz-Wings, Sarah Werning, Larry Witmer and Adam Yates, as well as others too numerous to mention, for the inspiration and help they provided to coauthor Wedel in launching and sustaining his career, as well as a special thanks to Mike Taylor for being a constant friend, coauthor, sounding board and fellow-traveler; Fiona, Daniel, Matthew and Johnno Taylor for their hospitality during coauthor Wedel's many visits.

Finally, the two authors thank Vicki and London Wedel, as well as Turi Hallett, for putting up with their often crippling bouts of dinomania.

THE SAUROPOD DINOSAURS

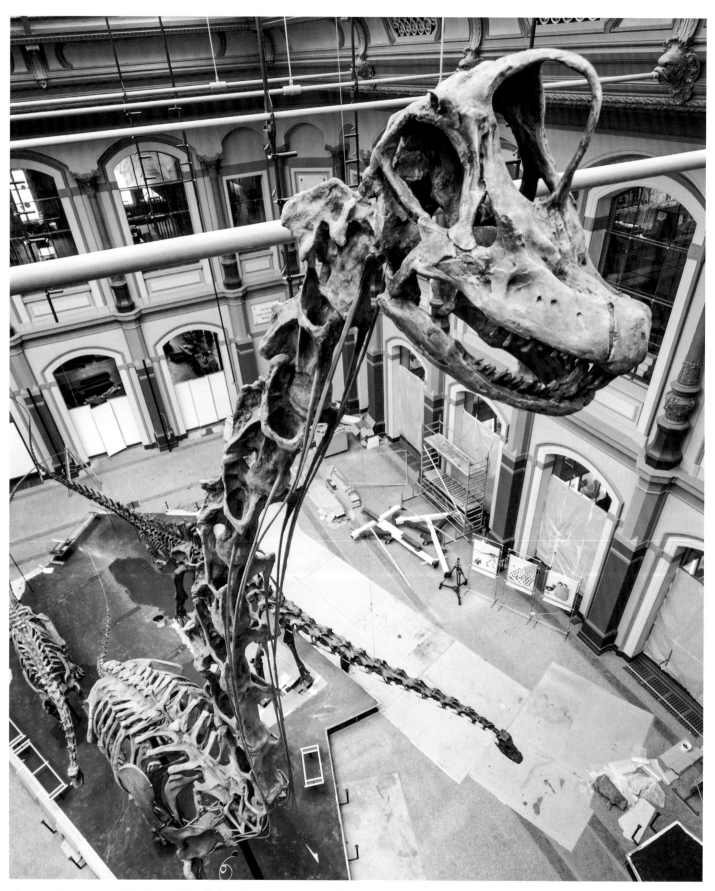

The recently remounted skeleton of *Giraffatitan brancai* towers over the main dinosaur hall at the Museum für Naturkunde in Berlin, Germany.

CHAPTER ONE
SIZING UP SAUROPODS

While gigantic vertebrae and other bones have been turning up for thousands of years to become woven into the legends of Native Americans and other cultures, the first scientifically recorded finds of what became known at first as "ceteosaurs"—and then sauropods—began in the 1830s. Like the evolution of the animals themselves, our human understanding of these dinosaurs took time to grow, often taking diversions until new finds and changes in scientific thinking happened to raise the curtain, each time a little bit more.

Georges Cuvier (1769–1832), because of his pioneering studies of Tertiary mammals from the Paris Basin and application of comparative anatomy to classification, is often considered to be the father of modern vertebrate paleontology.

RICHARD OWEN'S WHALE LIZARDS. As chilly as the Cabinet Room at Oxford University was, it was nothing compared to the coldly analytical eyes of the man whom Dean William Buckland ushered in to view his fossil collection. The visitor was Baron Georges Cuvier, the great French zoologist and anatomist whose opinion was taken as the last word when it came to assessing the identity of an ancient animal. Cuvier's anatomical knowledge was vast and his prowess in the new science of vertebrate paleontology legendary, and as he scanned the crowded tabletop of specimens, his gaze stopped at a group of enormous bones. Only the modern whale vertebrae and ribs he had become familiar with at the Jardin des Plantes in Paris equaled them in size. The Cabinet Room bones were petrified, and Buckland assured him that they had been found weathering out of the Mid-Jurassic Stonesfield Slates, near Woodstock. Apart from sheer size, their porous inner texture was very reminiscent of a whale's, but it was incredible to think that huge, mammalian whales had existed along with the giant terrestrial reptiles that William Buckland, Gideon Mantell and other British paleontologists had been studying and describing. Nonetheless, Cuvier's opinion was that these were prehistoric whale bones, and this verdict still held until sometime after the scientist's death in 1832.

Richard Owen (1804–1892) was the best-known British anatomist throughout much of the 19th century in Britain. Coining the term *Dinosauria*, he did much to establish the study of sauropods and other prehistoric animals. His description of a single tooth from *Cardiodon* was the first published study of a sauropod.

The bones had by now fallen under the eyes of the ambitious and brilliant young British anatomist Richard Owen, who had been pondering over a single, mysterious tooth that in 1841 he named *Cardiodon* because of the heartlike shape of its crown. Unknown to him, this would be the first fossil leading to what would later be called *sauropods*, but Owen was now intrigued by the whale-like, honeycombed inner texture of Buckland's bones. After rejecting the idea of an actual prehistoric whale, he began to toy with the notion of a marine, whale-like *saurian* with crocodilian features, since the bones possessed neural spines and chevron articulations that had some similarities to these reptiles. For Owen, the enormous vertebrae and long ribs seemed "to indicate that the present gigantic marine Saurian must have had a bulky and capacious trunk, but propelled by a longer and more Crocodilian tail than in the modern whales." He further hypothesized that the *Cete[i]osaurus*, or "whale-saurian" as he now called the animal, probably had a vertical fin on the dorsal part of the tail, similar to what he envisioned for the already-known, dolphin-like ichthyosaurs. *Cetiosaurus* (modern spelling), Owen believed, was even more specialized for an ocean existence than the ichthyosaurs, and he imagined it as a giant carnivore that preyed on smaller, true crocodiles and the long-necked plesiosaurs. In creating the order Dinosauria in 1842, based on the previous findings of *Megalosaurus*, *Iguanodon* and *Hylaeosaurus*, Owen excluded *Cetiosaurus*, which he believed, as a marine animal, to be in a totally different category than the terrestrial dinosaurs.

By 1848, however, the "whale saurian" model had started to crumble, owing to more discoveries near Oxford. These limb bones, attributed to *Cetiosaurus*, looked much more like those of a terrestrial rather than an ocean-dwelling animal. In 1850 Gideon Mantell described, on the basis of a humerus, *Pelorosaurus.* By the time sauropods were first formally recognized later in the century, it had become the first taxon of these named on the basis of appendicular skeletal elements and skin impressions. Even more importantly, *Pelorosaurus* was also the first "cetiosaur" (as the animals continued to be called at that time) bone to be recognized from its features as probably that of a land-dwelling animal. Other discoveries trickled in. In 1868 the Stonesfield Slates produced a gigantic femur, at 1.62 m (5′4″) bigger than any other skeletal element yet known. More were to follow, and along with the first femur, these were referred to Owen's *Cetiosaurus.* John Phillips, its discoverer and successor to Buckland at Oxford University, uncovered more bones after working the quarry for more than a year. Along with more material found later from Kirtlington, north of Oxford, these enabled Phillips to produce his groundbreaking description of *Cetiosaurus oxoniensis* in 1871. Based on material from several localities, the 50-page, illustrated work was a major step forward in bringing "cetiosaurs" closer toward inclusion into Owen's dinosaurian fold. As Phillips wrote, "The [femur] is nearly straight, in this respect differing much from a crocodilian, and approaching towards the d[e]inosaurian type." While no neck and skull material was known at the time, Phillips was one of the first workers to realize that the enigmatic giants had an erect posture and were capable of walking on land. In spite of increasing evidence of their probable identity, however, for Owen "cetiosaurs" were still not dinosaurs. Years before, the scholar had referred several more discoveries to *Cetiosaurus* but had rejected including them within Dinosauria because they lacked the five fused sacral vertebrae that he had determined characterized the other known forms. This, however, changed in 1874

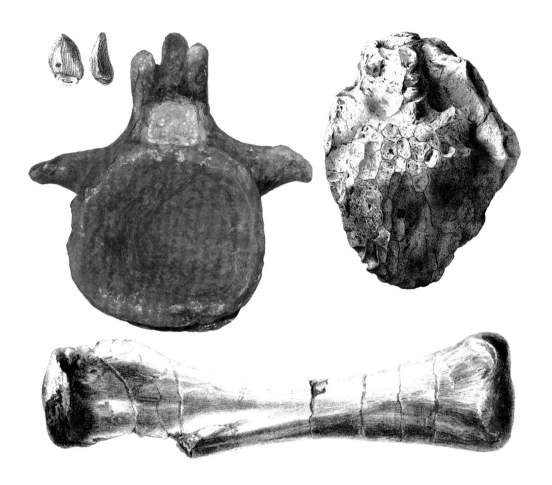

From a few strange yet tantalizingly familiar-looking specimens, early paleontologists like Cuvier and Owen made attempts to understand the enigmatic teeth and bones of animals that would later be first known as "ceteosaurs," then sauropods, as more material came to light. *Clockwise from left:* Cardiodon tooth, *Cetiosaurus* vertebra, indeterminate cancellous bone fragment, *Pelorosaurus* humerus.

with the discovery of a new partial skeleton. Although Owen's diagnosis of what he named *Omosaurus* as a "cetiosaur" was later shown to be incorrect, the well-preserved pelvis with its five fused sacrals of what was then thought to be one finally convinced Owen that these animals were actually dinosaurs, and the massive, elephantine proportions of the forelimb showed that they were capable of supporting themselves on solid earth. Other fragmentary but important discoveries had by then come in from outside of Britain, such as *Aepisaurus* (France, 1852), *Astrodon* (Maryland, 1859) and the first Gondwanan sauropod, "*Titanosaurus*" (now known as *Isisaurus*, India, 1877), showing that the "cetiosaurs" were a widespread type of dinosaur.

While the "cetiosaurs" had now been officially allowed admission into the "Dinosaur Club," Owen still felt that the giant saurians spent more time in water than on land. As the decades passed, his strict adherence to this idea, his rigidity toward new evolutionary concepts and a by now jealously maintained academic domination of British natural science increasingly antagonized younger scientists. One of these was Thomas Henry Huxley, an assertive researcher and evolutionary theorist. Earlier in 1867 he had focused his energies on dinosaurs, and after visiting the impressive collection of "cetiosaur" fossils in Oxford, he came away convinced that these saurians belonged in Dinosauria. Although he differed from the phylogenist Harry Govier Seeley about their exact classification, both he and Seeley agreed that "cetiosaurs" were a major dinosaurian order.

Thomas H. Huxley (1825–1895), an anatomist, evolutionary theorist and early supporter of Charles Darwin's principle of natural selection, was the first to make the dinosaur–bird connection that led to the recognition of avian-like respiration in sauropods.

Othniel Marsh (1831–1899) (*top*), **Edward D. Cope** (1840–1897) (*bottom*). As a result of their notorious rivalry, they described several of the best-known sauropods; their field crews exposed and removed a wealth of bones from the Badlands.

Huxley also held some remarkably farsighted views about dinosaurs' relationships to birds. One of these was the probability, based on anatomical similarities, that at least small bipedal dinosaurs, like birds, had a "warm-blooded," or *endothermic*, metabolism. He proposed that these forms were actually the ancestors of birds and, on the basis of the famous *Archaeopteryx* fossil discovered in 1861 and other studies, would have also had a bird-like heart and lungs. Although in his time Huxley didn't specifically suggest that the now dinosaurian "cetiosaurs" might have had these as well, the possession of an avian-type circulatory and respiration system for sauropodomorphs and other saurischian ("bird-hipped") dinosaurs is now largely accepted by today's workers, as we'll see in Chapter 5.

BONE WARS OF THE WEST

On the other side of the Atlantic in the USA, 1877 was a critical year, as much of a tipping point into the understanding of sauropods as it was in England and in Belgium for ornithopods, where a fabulous trove of complete *Iguanodon* skeletons had been found. The coming of the railroads and the opening of the American West following the Civil War had begun to produce scientific discoveries, and among these were the fossils eroding out of the Dakotas and other badlands. Arthur Lakes, an Oxford graduate who spent time prospecting for fossilized leaves in the sandstone beds near Morrison, Colorado, where he had taken a teaching post, in March came across a huge vertebra and other bones. Thanks to his time at Oxford, he was well aware of the "cetiosaur" bones in Phillips's collection, and he wrote and sent drawings of his finds to Othniel Charles Marsh, heir to a wealthy uncle who had set him up with money and a curatorial position at Yale University. The aloof, methodical and secretive Marsh had already established a reputation for his extensive field excavations and descriptions of the pterosaur and bird discoveries in the Kansas chalk beds, but amazingly Lakes's contacts produced no interest from the Eastern academic, even when Lakes shipped off 10 crates, weighing almost a ton, of "cetiosaur" and other bones to him. Frustrated, Lakes then turned to writing and sending material to Edward Drinker Cope, a highly strung and gifted son of Quaker parents who later became a paleontology professor at the University of Pennsylvania. Like Marsh, Cope had inherited a fortune and was as experienced in the field as he was in the museum. As young men the two had originally been friendly and exchanged cordial letters, but when their careers took off, what first emerged as competition became, after a series of academic insults and mutual incursions into what each considered his exclusive fossil-collecting territory, open hostility. By 1870 the two paleontologists were bitter rivals. When Marsh learned of his adversary's windfall, he reacted by directing his field man, Benjamin Mudge, to buy both Lakes's fossils and his further services. Cope, however, had by now received a box of even more massive and numerous fossils through another collector, O. W. Lucas, from a site near Canyon City. Soon after, each paleontologist's field team was swarming over the Morrison badlands in efforts to find and quarry the rich fossil deposits before the other could. Under the harsh conditions and spurred by competition, the collecting techniques the teams used were often hasty and poor, and sometimes the bones were excavated without the benefit of protection from the newly devised technique of using rice-glue-and-burlap coverings. If a team ran out of time by the end of the field season, occasionally the men would actually smash fossils in the quarry to prevent collection by their rivals. In spite of this, tons of bones were packed, crated and rumbled east on the new railroads to be received by the expectant Marsh and Cope. Following receipt of them, each then shot off dozens of equally hasty, poor descriptions and papers in an attempt to maintain priority as the preeminent American vertebrate paleontologist.

SAUROPODS TAKE NAME AND SHAPE

The Bone Wars brought spectacular results. In addition to the theropod *Allosaurus* and ornithischian *Stegosaurus*, Marsh, on the basis of a hindlimb, pelvis and caudal vertebrae with their "peculiar chevrons," described the first known diplodocid sauropod *Diplodocus*. This was soon followed by Cope's macronarian *Camarasaurus*, and although characteristically rushed, the description was published with a figured drawing by John Ryder, the very first scientific reconstruction of any sauropod skeleton. While it contained many errors by modern standards, the drawing depicts the animal's characteristic features of a large body, small skull, long neck and tail. In addition to his drawing, Ryder created a 15 m (50′), three-dimensional (3D) skeletal reconstruction based on material from several individuals, another first of its kind. One early outcome of the Bone Wars was the discovery of the iconic *Brontosaurus, Apatosaurus, Diplodocus, Camarasaurus* and other now well-known species. Besides these discoveries, the wealth of new material from the West also spurred a renewed attempt to make sense of the new and old "cetiosaur" species' *phylogenies*, or systematic relationships to one another. Owen, based on *Cetiosaurus* and *Streptospondylus,* had in 1859 proposed the order Opisthocoelia for these and the other known "cetiosaurs," while Seeley by 1874 favored Cetiosauria. Ignoring both these names, in 1878 Marsh coined *Sauropoda*. Although most workers didn't warm up to this, based as it was on a strangely claimed resemblance of "cetiosaur" feet to those of known saurians, by 1903 Sauropoda (*sauros*, "saurian or lizard"; *podus*, "foot") had become the accepted name for the dinosaurian order and stuck.

An outcome of the publicity and notoriety following Cope and Marsh's "Bone Wars" was that the public had now acquired a fascination for dinosaurs, and eastern museums like Pittsburgh's Carnegie and New York's American Museum of Natural History began to display awesome sauropod and other dinosaur skeletal mounts. In 1910 the tycoon and philanthropist Andrew Carnegie had ensured that a complete skeletal mount of *Diplodocus* could be seen at his museum in Pittsburgh, and he had casts of the skeleton sent to more than a dozen other museums around the world. More sauropod discoveries from the West and East Africa poured in through the teens, and the paleontologist and American Museum president Henry Fairfield Osborn sensed an opportunity to educate. He developed wide-ranging exhibit programs, one goal of which was to portray dinosaurs and prehistoric animals as vividly as possible, and for this he gathered a remarkable array of talents. One included Charles R. Knight, a shy, visually impaired artist who learned animal anatomy quickly from his taxidermist associates, had access to Central Park's zoo for sketching and possessed a genius for infusing his prehistoric creatures with a sense of believability and drama. Sauropods were among his favorite subjects, and during the 20th century's teens, 1920s and 1930s he produced classic renditions of giants like *Brontosaurus* and *Diplodocus*

Following the early discoveries of sauropod bones in Britain and Europe, other finds began to accumulate as a result of gradual western exploration and colonization around the world. One of these was the amazingly rich deposit of Late Jurassic dinosaur fossils at Tendaguru, Tanzania, then German East Africa. Prior to World War I this produced many specimens of the giant brachiosaurid *Giraffatitan*.

An enormous scapula, or shoulder blade, still encased within its sandstone matrix, awaits removal. It was bones like this that made early paleontologists realize the possibility that giant saurians existed on Earth well before the Age of Mammals.

As a result of the collecting efforts of Cope, Marsh and field directors like Charles Sternberg, Earl Douglass and others, by the first decades of the 20th century spectacular skeletal mounts began to appear in large western museums like that of *Apatosaurus* at the American Museum of Natural History in New York. Mounts like these gave the public their first impression of the awesome size of these dinosaurs.

in mural form. Knight's murals and sculptures accompanied the giant mounted skeletons, and these gave visitors their first understanding of how sauropods actually looked, which included an upright, erect posture like an elephant's. Old ideas sometimes continued to cast a shadow over these efforts, however. A bizarre episode using skeletal posture to counter the idea of upright sauropods occurred following the installations of the first *Diplodocus* mounts, when Oliver P. Hay and Gustav Tornier claimed that *Diplodocus* would have had a sprawling, croc-like posture. The hideously distorted reconstructions on paper made this idea so obviously improbable that the debate over sauropods' upright posture was quickly over. Another Osborn protégé, the sculptor Erwin Christman, in 1921 created a complete, accurate, to-scale skeletal and flesh restoration of *Camarasaurus*, aided by Osborn and Mook's classic osteology paper produced the same year.

HOME, WET HOME

As paleontologists came to get a better idea of sauropods' true appearance, they also began to come to grips with their *paleobiology*—how they lived. Because of their early associations with aquatic environments hypothesized by Owen and others, sauropods were cast in the role of being amphibious, mostly swamp or lake dwellers that eased their great weight with the support of water. This resulted in the first depictions by the young Knight of species like *Amphicoelias* well below the surface, feeding on lake vegetation and using their long necks as snorkels to reach the air. Although Osborn felt that sauropods were capable of feeding from a tripodal position on land and persuaded Knight to show a rearing *Diplodocus* in 1907, he too was part of the consensus that sauropods' preferred habitats were wetlands. Some workers actually believed that while lighter forms such as *Diplodocus* might sometimes have left the water to occasionally browse

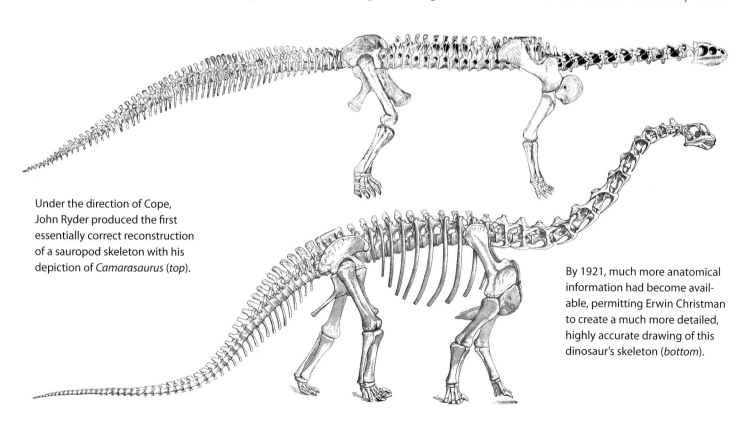

Under the direction of Cope, John Ryder produced the first essentially correct reconstruction of a sauropod skeleton with his depiction of *Camarasaurus* (*top*).

By 1921, much more anatomical information had become available, permitting Erwin Christman to create a much more detailed, highly accurate drawing of this dinosaur's skeleton (*bottom*).

on shore, heavier species like *Apatosaurus* weren't able to walk far on firm environments because their enormous weight would have actually crushed the elastic pads underneath their feet. Among the few who held out for a fully terrestrial lifestyle was Elmer S. Riggs, the discoverer and describer of *Brachiosaurus* in 1900. He argued, "There is no evidence among [sauropods] of that shortening or angulation [bending] of limb, or the broadening of foot, which is common to amphibious animals. . . . The straight hind leg occurs in quadrupeds only in those forms which inhabit the uplands. . . . The short, stout metapodials and blunted phalanges . . . would be as ill adapted for propulsion in water or upon marsh lands as those of an elephant. . . . In short, if the foot structure of these animals indicates anything, it indicates specialization for terrestrial locomotion." Few agreed with this, however, and the prevailing opinion was that sauropods didn't stray far from water. What about reproduction? As a saurian, *Apatosaurus* and other species would presumably need to emerge on land to lay eggs, but even here there was disagreement. William D. Matthew thought that, like ichthyosaurs, "brontosaurs" (as they were now popularly referred to) might have been *viviparous*, or capable of bearing live young.

Further presumed confirmation of "brontosaurs" as amphibious dinosaurs came in 1938 with field paleontologist Roland T. Bird's discovery and careful documentation of well-preserved trackways along the Paluxy and Purgatoire Rivers near Glen Rose, Texas. The very light forefoot impressions, only occasional hindfoot marks and lack of a "tail drag" (sauropods and other dinosaurs were assumed to have dragged their tails, instead of keeping them held up) all continued to indicate for most that sauropods were waders and swimmers. This discovery appeared to be convincing evidence of the animals' typical rather than occasional behavior and seemed to clench the aquatic habitat concept. In addition, no one questioned the assumption that sauropods' long necks, while certainly evolved to reach food, were also snorkels. This ensured their safety from predators, which were thought to be poor swimmers, by keeping as much of their bodies under water as possible—a submerged sauropod was a safe sauropod.

Speculations as to sauropods' levels of activity and probable diet also took a wrong turn. Huxley's ideas of possibly endothermic, bird or mammal-like physiologies for dinosaurs were shelved, especially when in 1927 Gerhard Heilmann claimed that the apparent lack of *clavicles*, or collar bones, disqualified saurischian dinosaurs as bird ancestors. Because of this, bird-like advanced lungs and hearts were no longer considered, and sauropods and other dinosaurs were assumed to have saurian or reptile-like metabolisms in spite of the fact that, unlike reptiles, their upright skeletal postures would have required a metabolism similar to a bird's or mammal's. Following on this idea was the perception that if they'd actually been warm-blooded like birds and mammals, the huge creatures with their small heads could never have eaten enough to stoke the internal fires required for such high activity levels. All this flawed reasoning was then used to support the conclusion that they then had to be *ectothermic*, or "cold-blooded," like modern reptiles. A swamp, marsh or other wetland was ideal for sauropods, since here there was probably enough relatively soft, easily plucked vegetation that could be harvested by the presumably feeble, simple-looking (by mammalian standards) teeth. Realistically executed artist's images can take hold and remain in the mind even when new evidence comes along to refute them, and Knight's beautiful 1897 painting of *Brontosaurus* in a weed-choked waterway was the symbol of

After the first enormous skeletons appeared in museum exhibit halls and the public became aware of their existence, sauropods quickly became the iconic image of a dinosaur, being popularized in the animated cartoon "Gertie the Dinosaur" by Winsor McKay in 1913.

Rebelling against entrenched ideas, **Robert T. Bakker**'s fresh examinations of dinosaur anatomy and behavior produced new evaluations of where and how sauropods lived. He incorporated these with earlier insights from older paleontologists to demonstrate that instead of being swamp dwellers, sauropods had fully terrestrial adaptations to walking and living on land.

Charles R. Knight's iconic painting of the diplodocids *Apatosaurus* (*foreground*) and *Diplodocus* (*background*) remained the most graphic, enduring image of sauropods for decades in hundreds of publications until the late 1960s.

Because of blindness in one eye, artist **Charles R. Knight** (1874–1953) often created 3D models of dinosaurs and other prehistoric animals to understand anatomy, volume and lighting before attempting a painted restoration when he worked at New York's American Museum and Chicago's Field Museum during the 1920s and 1930s.

sauropods in the popular mind for decades. It all seemed to fit together. While important skeletal discoveries and studies continued throughout the mid-20th century, sauropods and other dinosaurs, as spectacular as they were, were increasingly thought of as a sideshow to the main events in paleontology. Though now as popular as ever with the general public and a staple subject for children, paleontologists came to be preoccupied with the mammalian and early human evolutionary issues that by now were considered much more important.

GOODBYE TO THE SWAMP

As the 1960s ended, this quickly began to change with a popular article by a flamboyant student in a tattered cowboy hat, Robert T. Bakker, then at Yale and Harvard Universities and now affiliated with the Houston Museum of Natural History. Entitled "The Superiority of the Dinosaurs" and published in the 1968 issue of *Discovery*, it directly challenged the notion that sauropods and others were basically upscaled, non-adaptive giants, and it made the compelling case that dinosaurs were biologically advanced animals. Bakker's ideas came not only out of his own fresh observations but also from a new consideration of Huxley's, Riggs's and others' forgotten insights. Soon after, he focused directly on sauropods with his 1971 "Ecology of the Brontosaurs" in *Nature*. In this formal paper, Bakker methodically undermined the traditional, water-dwelling image of the sauropods and made the case for their being land dwellers, citing such features as the elephant-like long limbs, compact feet and slab-sided rib cage, rather than the typically short, spreading feet and round rib cage of an amphibious animal like a hippo. The author also resuscitated an early concept of Cope, that sauropods were mainly giraffe-like arboreal feeders, and in 1975 Walter C. Coombs (Pratt Museum, Amherst College) expanded on this concept and other possible adaptations in a major study. Bakker envisioned sauropods as major players in the dinosaurian communities and ecosystems of the Mesozoic world, stating, "The impact of brontosaurs on the Mesozoic terrestrial ecology must have been enormous.

The herds of different sympatric sauropods species must have opened up thick forest and kept undergrowth from becoming dense, much as elephants do in Africa. Smaller herbivorous dinosaurs adapted to open woodlands and plains must have depended on sauropods for keeping such habitats from being overrun by thick jungle." The "Dinosaur Renaissance" had begun.

Along with the realization that sauropods and other dinosaurs were anatomically and biologically as sophisticated as birds and mammals came a consideration of their possible gregariousness and other behaviors. As it became clear that sauropods were adapted for terrestrial environments and habitats, many new discoveries were made of trackways that added to Bird's original finds. Studies by Martin G. Lockley (University of Colorado at Denver), independent paleontologist Giuseppe Leonardi, Tony Thulborn (University of Queensland) and others enormously expanded the potential knowledge that these could provide, making the science of *ichnology*, the study of tracks and traces, as important a source of information as *osteology*, the study of bones. The 1960s and 1970s also ushered in a period of increasingly sophisticated laboratory studies that are still going on. Estimates of dinosaur mass, or weight, at first highly crude and subjective, began to become more refined. Edwin H. Colbert (American Museum of Natural History), a longtime major student of dinosaurs in general and one who continued his work when interest in the animals was at its lowest ebb, produced pioneering, systematic studies of dinosaur body masses in the 1960s using scale models. These were followed by those of R. McNeill Alexander (University of Leeds) and others in the 1980s.

As the end of the century approached, new applied technologies also improved the work of finding and excavating fossils in the field. David Gillette (New Mexico Museum of Natural History) in 1987 was faced with the need to locate the enormous skeleton of *Diplodocus* (*Seismosaurus*) *halli*. He had to do this before the designation of the locality site as part of a national wilderness area would make the extensive excavation needed to find the fossils off-limits, and he turned to the scientists at Los Alamos National Laboratory and associated institutions for help. One solution among several was the use of *ground-penetrating radar*, which uses differences in readings from radio frequency wave impulses projected into the ground to plot electronic disturbance profiles at different levels in rock. Unusual shifts in density may indicate the presence of fossil bone. A second solution was the use of *proton free-precession magnetometry*. This uses a pole and sensors to pick up and measure variations or divergences in Earth's normal magnetic lines of force or projection at a locality. Buried fossils can cause convergences in these lines, producing concentrations of high readings at a particular depth and place. Still another technique was *acoustic diffraction tomography*, which records and measures the differential rates of sound waves when projected through various rocks. Low-density rock (for example, sandstone) produces a different velocity of feedback than high-density objects (often fossil bone). These can be measured by creating sonic blasts and then recording the shock waves on high-sensitivity hydrophones, placed in evenly spaced, fluid-filled holes drilled at selected levels. All these techniques enabled Gillette to find bone concentrations much more quickly and efficiently than he would have by using traditional bone prospecting methods.

The alternative possibilities of sauropod lifestyles also forced a reexamination of known skeletal material—could this shed light on fossils already in museum collections and exhibits? The

The Long March, 1994 (*detail*). The column-like, massive limbs and powerful, clawed feet of sauropods gave them the means to travel great distances in search of food and water.

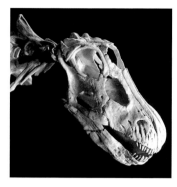

After studying the known skull of *Apatosaurus/Brontosaurus*, Australian paleosculptor **Brant Bassam** (Brantworks) has produced a new reconstruction that indicates that the jaws, in addition to their width, were probably deeper and more robust than previously thought, and that the orbital area was higher, more closely resembling that of *Diplodocus*.

first problem to be studied was neck posture and flexibility, an issue that is still being debated. As soon as they became known, the sauropods' long necks were, like those of birds and plesiosaurs, assumed to be highly flexible. While on paper and in skeletal mounts a variety of poses from horizontal to vertical had been portrayed, no actual scientific study took place until 1987, when John Martin (Leicester City Museum) made an analysis of *Cetiosaurus* cervical vertebrae. His findings and those of other researchers have widely differed, an example of the way paleontologists and other scientists can come up with different conclusions even when using exactly the same specimens. A new look at skeletal morphology also led to the correction of long-standing mistakes. One was the realization that *Brontosaurus* (until 2015 known only as *Apatosaurus*), as a diplodocid, had a skull like that of the closely related *Diplodocus* rather than the macronarian *Camarasaurus*. A subjective decision by Osborn in the 1930s condemned *Brontosaurus* and *Apatosaurus* skeletons to wear the wrong heads for decades, and it wasn't until David S. Berman (Carnegie Museum of Natural History) and John S. McIntosh's (Wesleyan University) 1978 paper that the actual skull was correctly identified and described.

The refutation of the swamp-dweller concept, of course, demolished the idea of the long neck as a snorkel. Studies had actually been conducted early in the century that related to how well humans could withstand water pressure and still be able to breathe, and in 1951 these were applied by Kenneth Kermack (University College, London) to sauropods. His findings showed that at the submerged depths proposed for *Brachiosaurus* and others, the water pressure would have been so great that it would have prevented the dinosaur's respiratory system from expanding to draw in air, resulting in suffocation and fatal hemorrhaging. Kermack's calculations, however, like those of Huxley, Cope and Riggs, were swept under the carpet of scientific conservatism and were finally recognized only after Bakker's articles forced a reexamination. Another myth that came under attack was the scenario of the water as refuge from predators. As graphically depicted in an illustration by independent artist-paleontologist Gregory S. Paul, allosaurs and other theropods had long, powerful hindlimbs with spreading toes, well suited for swimming if needed to follow huge prey into the water.

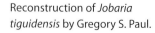

Reconstruction of *Jobaria tiguidensis* by Gregory S. Paul.

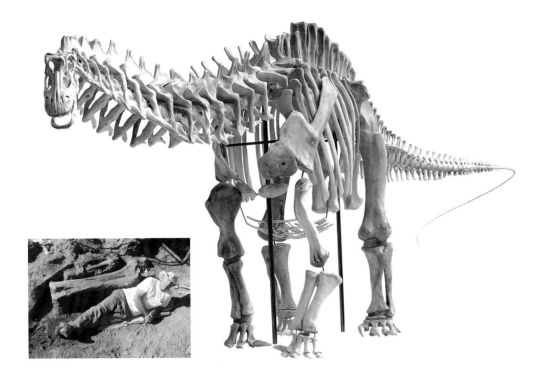

Focusing on *Apatosaurus/Bronto-saurus*, **Phil Platt** (Brontobuilder) (*far left*) has created new and accurate model skeletons of this diplodocid (*left*).

In spite of new discoveries throughout the decades, the evolutionary lineages of sauropods were until the mid-1990s poorly understood, largely owing to a lack of research in this area. A number of studies, beginning with Dale A. Russell (North Carolina State University) and Zhong Zheng's (Institute of Vertebrate Paleontology and Paleoanthropology) 1993 analysis of *Mamenchisaurus*, have led to a highly successful approach that uses the science of *cladistics*, rather than the traditional and more subjective traditional methods based exclusively on key characters, to determine sauropod relationships (see Chapter 2). The great usefulness of cladistic analysis, together with the ease and speed of computer technology, provides a means of sorting out and organizing the increasingly enormous numbers of known skeletal characteristics, making it possible to trace the evolution of certain body features and putting sauropod classification on a sound theoretical basis. Although there are sometimes differences among workers in the findings they derive from this, the new type of research has made it possible to carry out much more precise studies of sauropod diversity, paleobiology and ecology than otherwise would have been possible.

Artist-paleontologist **Gregory S. Paul** (*top*) and **Scott Hartman** (*bottom*) take a rigorous and logical approach to determining the anatomy and appearance of sauropods. Along with those of coauthor Mark Hallett, their step-by-step techniques of skeletal reconstruction, followed by muscle and integument, have resulted in highly probable depictions of these dinosaurs.

SAUROPODS IN THE 21st CENTURY

Today the ever-increasing sophistication of computers, software programs and their applications has resulted in the use of these technologies becoming standard practice, making paleontology a much more quantifiable science than it once was. As with analyzing measurements and other data, the ability to program, re-create and restore skeletal parts and other body features as virtual, 3D images has revolutionized the study of comparative anatomy and *biomechanics*, or body motions (see Chapter 4). While their findings have been disputed, one team, Kent Stevens (University of Oregon) and Mike Parrish (Northern Illinois University) of Dinomorph, was the first in the late 1990s to take full advantage of the possibilities offered by computer imaging technology. They scanned orthographic images of sauropod osteology and experimented with virtual 3D images instead of resorting to the daunting task of moving life-size bones to understand neck articulations, and now a fully articulated, massive sauropod skeleton can be put through its paces on-screen just like the most obedient horse in a stable. Virtual imaging has also increased our understanding of *pneumaticity*, or the air spaces within sauropod and other dinosaur bones. As we'll learn in Chapter 4, the determination of pneumaticity is vital in accurately determining *mass*, or body weight, and this, in turn, can offer valuable perspectives on questions like a given sauropod species' probable food and water needs, habitat requirements and reproductive potential.

Like phylogeny, research into sauropod diets and feeding was until recently a neglected field. Although this field is no longer dominated by the myth that the giants had weak jaws and teeth only suited for harvesting soft plants, until relatively recently there were no rigorous studies of tooth and jaw morphology. This research, of course, needed to go hand in hand with analyses of neck structure to produce a realistic theory of feeding ecology. Beginning in the 1990s, Anthony Fiorillo (Dallas Museum of Natural History), Paul Upchurch (University of Bristol), Jorge Calvo (Museo de Ciencias Naturales, Universidad Nacional de Comahue) and others undertook studies that abandoned more traditional analogies with living, large mammalian herbivores in favor of comparisons based on functional anatomy and actual dental microwear. These resulted in the consensus that sauropods did not chew their food and possessed dentitions suited to a distinctive feeding strategy—to eat as much as possible without stopping to chew, as mammalian herbivores must.

Sauropods also started to come under the microscope. Huge advances have been made in the science of *osteohistology*, or the microscopic study of bone tissues, in a search to find out how the

Few modern technologies have had as direct an impact in understanding the internal skeletal anatomy of extinct vertebrates as that of computerized axial tomography (CAT) scans. **Larry Witmer** and his lab team have produced extraordinary insights into the form and function of both prehistoric and living forms.

In addition to indoor dinosaur exhibits, the Dinosaurier Park (*left*) at Münchehagen, Germany, features accurate, life-sized restorations of sauropods and other dinosaurs that give visitors a true idea of their size. **Nils Knötschke** (*above*), a park curator, has extensively researched the osteology of the German dwarf brachiosaurid *Europasaurus* and its environment.

enormous animals grew and what kind of metabolism they had. In applying these techniques for the first time in 1991 to the juvenile, subadult and adult bones of *Apatosaurus*, Kristina Curry Rogers (Macalester College) discovered that dramatically rapid growth took place in these and (as discovered later on) in the bones of other sauropods, with near-adult size being attained in about 10 years. P. Martin Sander (Steinmann-Institut für Geologie, Mineralogie und Palaontologie, Rheinische Friedrich-Wilhelms-Universitat Bonn), in turn, analyzed the microstructure from a wide spectrum of sauropod species and was able to show that different taxa can actually now be distinguished on the basis of bone histology alone. The growth and reproduction of sauropods from egg laying, or *ovoviparity*, long suspected from much older finds but never proven, were dramatically confirmed by the discoveries at Auca Mahuevo, Patagonia, by Luis Chiappe (Los Angeles County Museum of Natural History) and his associates, where they uncovered vast nesting grounds of titanosaurs. The ongoing studies continue, and these and other nest sites around the world promise to continue yielding their secrets of baby sauropod growth.

Finally, there has been a flood of newly discovered sauropod species from many places around the world. Although often frustratingly incomplete, when found, the bones usually have enough details to make it at least possible to diagnose the basic type of sauropod to which they once belonged. Many of these have been *titanosaurs* (see Chapters 10 and 11), the last clade of sauropods to flourish, but at present the least understood group of all. Instead of the handful of families known at the beginning of the 1970s, there are now many families, genera and species, almost a third of them titanosaurs. Not only have the sheer numbers of sauropod forms increased but also knowledge of their *disparity*, or morphological variation. Starting with a similar basic body plan, some sauropods over their 160-million-year time on Earth evolved spikes, clubs, storage chambers, weird vacuum-cleaner heads and shearing jaws, while others actually became

Digital imagery has made a great impact in the design of museum exhibits. In this American Museum of Natural History traveling display, a rear projected image shows the cervical muscle system of *Apatosaurus*, based on original research by Mark Hallett.

dwarves instead of giants. The increasing number of the newly discovered species helps to refute a long-standing myth regarding sauropods in dinosaur paleontology. This is the idea that sauropods were, as the Mesozoic era transitioned from the Jurassic period into that of the Cretaceous, increasingly ill-adapted to world ecosystems that, in some cases, were becoming dominated by the flowering plants or angiosperms. Although it seems likely that the great radiation of ornithischians such as hadrosaurs and ceratopsians resulted from the increased prevalence of these plants, there is evidence that at least some Indian titanosaurs incorporated graminoid (grass) angiosperms into their diets. In addition, conifers, the sauropods' probable main food source, still remained the dominant plants in many Cretaceous ecosystems, particularly in the separating continents of Southern Gondwana (see Chapter 11).

Sauropod paleontology has come far since the first bones were spread out onto a museum table in 1823. As the 21st century progresses, we have the exciting prospect of seeing new finds from both traditional field work and insights into paleobiology from cutting-edge technologies—these promise to reveal more and more about the lives of sauropods as time goes on. The following chapters will show what we've definitely learned so far, what we can only speculate about and what we still don't know—small as it might be, each discovery helps in turning a collection of ancient, often incomplete bones into a virtual living animal, another thread in the vast, rich tapestry of prehistoric life. Here maybe the British paleontologist Michael Taylor (University of Bristol) has said it best: "Even though they've been gone for 65 million years, history still continues for sauropods."

A full skeletal mount of *Barosaurus lentus* defending a juvenile from a predatory *Allosaurus fragilis* awes visitors who enter the Theodore Roosevelt Memorial Wing of the American Museum of Natural History, New York. Although some researchers disagree with the sauropod's upright rearing stance, mounting evidence shows that many, if not most, of these dinosaurs were habitually capable of this posture (*opposite*).

Unsuccessful at catching insects, a *Panphagia* considers a mouthful of ferns.

CHAPTER TWO
PARTING OF THE WAYS

It might have started quite casually. Having failed to catch any of the insects and other small animals that formed its accustomed diet, the turkey-sized biped with the supple neck, grasping hands, long hindlimbs and tail looked at the spray of green growth next to it for a moment and then cropped off a mouthful. It wasn't as protein-rich as a roach or spider, but for the time being it took care of the desperate, gnawing hunger until something else could be found. This single act and, aided by natural selection, millions of others like it set the eaters on paths that resulted in a variety of experiments in large-scale herbivory. Ultimately these became sauropods.

BRANCHING INTO DINOSAURS. That biped was one of the first of the true dinosaurs and, along with some of its close relatives, was a minor player in a cast of much larger and more impressive vertebrates. These included bizarre, armadillo-like *aetosaurids*, whose sides bore curved sabers; lumbering, beaked, cow-sized *therapsids*, or "mammal-like" reptiles; and fierce *rauisuchians*, bear-sized crocodile relatives with slashing teeth that were the top predators of this time, the Late Triassic. Darting among all these heavyweights were the ***dinosaurs***. The first actual dinosaurs for us would have been undistinguishable from some other, very similar small ***archosaurs***, a class or group of biologically advanced reptiles or saurians. By this time in the Triassic the archosaurs had split into two main groups, the ***Crurotarsi***, or croc-like archosaurs, and the ***Ornithodira***. Besides pterosaurs (or "pterodactyls"), ornithodirans included *dinosauromorphs*, featuring, among others, quick little rabbit-sized bipeds. *Dinosaurs, right?* No, but getting warmer. As time went on, dinosauromorphs split further into the *dinosauriforms*.

NOW dinosaurs? No, but almost there. Finally, from the dinosauriforms we now branch off onto the subclass **Dinosauria**—the true dinosaurs. These include two groups. One is the **Saurischia**, the so-called saurian (lizard)-hipped dinosaurs that encompass both the carnivorous **theropods** (and their descendants, the birds) and the **sauropodomorphs** (sauropod-like dinosaurs and their descendents, the true **sauropods**, the subjects of our book). The other is the **Ornithischia**, or "bird-hipped" forms like *ornithopods* (iguanodonts, hadrosaurs, etc.), *thyreophoreans* (stegosaurs, ankylosaurs), *ceratopsians* (horned and shielded) and more.

That's a lot of branching, and if we're a little dizzy from this, occasionally so are some dinosaur paleontologists. We've just been on the roller-coaster ride of **phylogeny**, the study of how life-forms relate to one another. The earliest actual dinosaurs such as the cat- and dog-sized bipeds *Eoraptor, Herrerasaurus* and *Staurikosaurus* differ so little from their closest non-true dinosaur relatives that there aren't any really obvious body features, known to phylogenists as **characters**, to distinguish them. To do this, we've got to look closely at anatomical details such as the muscle scars on limb bones and the way the bones of braincases fit together. Maybe not very dramatic stuff, but it allows us to trace out the many small evolutionary changes that resulted in dinosaurs. At one fossil locality, the Hayden Quarry in New Mexico, we can look down and see the actual point in time when true dinosaurs and their dinosaur-like close relatives, the dinosaurimorphs and dinosauriforms, each began to take a separate evolutionary path from one another. Here we have not only a true but *basal*, or primitive, dinosaur in the form of the herrerasaurid *Chindesaurus* but by now an actual, more *derived* or specialized theropod dinosaur very similar to *Coelophysis*. Along with these are a dinosauromorph, *Dromomeron*, and the dinosauriform *Eocoelophysis*. Poised to go their three separate ways, these would look pretty much alike to us, and without prior knowledge we'd never predict that it was the *dinosaurs*, not the other two relatives, that would become the biggest success story of Mesozoic vertebrates.

As we learned, dinosauriforms are the types closest to the dinosaurs themselves, and of these the closest of all are *Silesaurus* and other *silesaurids*—this group throws a new wrinkle into how we think about the evolutionary origins of dinosaurs. For a long time, scientists inferred that the earliest dinosaurs were all carnivores. Why? From their beginning as land dwellers, small and later large reptiles probably preyed on the hordes of evolving invertebrates they encountered, graduating to bigger things until they started eating their fellow reptiles. Everyone couldn't succeed in being a carnivore, however, and some forms began to warm up to the nutritional possibilities of plants, a less concentrated but more easily obtained and very abundant source of food. With time all kinds of specialized herbivores evolved to take advantage of the increasing variety of plants, and this, in turn, broadened the playing field for carnivores. The body plan of a small, agile running dinosauromorph, dinosauriform or dinosaur, with its flexible neck and grasping hands to grab and manipulate food, was assumed to be an overall adaptation for running down and snapping up small prey—a carnivore. One branch of these small, presumably carnivorous bipeds became the above-mentioned Saurischia, which, although it included conservative, meat-eating theropods, also became the *sauropodomorphs*, giant plant eaters. So, as the logic ran, carnivory preceded herbivory in dinosaur evolution. But did it? As we've seen, silesaurids had just these kinds of bodies, but instead of a carnivore's typical sharp teeth, these animals had somewhat leaf-shaped dentition, the hallmark of an herbivore. This means that *if* the earliest true dinosaurs (still poorly known, with probably undiscovered forms) were like their silesaurid close cousins, maybe they were originally *plant eaters*, only later switching to insects and meat.

ENTER THE SAUROPODOMORPHS

There are other wrinkles as well. As we've said, the relationships of many of the earliest dinosaurs are not well understood, and such currently recognized true forms as *Herrerasaurus* and *Staurikosaurus* might turn out to be a **clade**, or phylogenetic group, of early saurischians that lie not in but just outside of the theropod-sauropodomorph group. Since these two bipeds were definitely carnivores based on the shape of their teeth, they may represent the way the earliest saurischians (and therefore sauropodomorph descendants) actually were, and if this is true, then carnivory still holds good as the basal condition of this group. Thanks to a flood of new discoveries from North and South America, Europe, Africa and Madagascar, our ideas about early saurischians and stem sauropodomorphs are constantly changing. These are both a blessing and a curse: a blessing because they're helping to clear up the question of sauropodomorph and other dinosaur origins, but also a curse in that our view of dinosaur origins is constantly being revised. At the present time there are several competing views of how the first sauropodomorphs relate to one another, not necessarily because of professional egotism or entrenched opinions but simply because the data can be confusing, a problem that also comes up in studying later, *advanced or derived* sauropods. For example, there seem to be not just one but several different lineages of basal sauropodomorphs. All these were acquiring sauropod-like characters but not all in the same way or at the same time, a condition called *mosaic evolution* because it happens piece by piece, like a faulty, pixelated image trying to come up on a screen. We now have so many forms of near and early sauropodomorphs that the evolutionary picture has become a lot more complex; as Darwin once said, it's easy to divide animals into different groups when extinction has knocked out the intermediate forms. Birds once seemed very different from dinosaurs until the discovery of *Archaeopteryx* and later feathered theropods like *Sapeornis* linked these two forms that were once seemingly remote from each other. So it goes with sauropodomorph beginnings—it was once simpler to distinguish them and to come up with hypothetical relationships until transitional forms began to turn up and erase the differences between species.

Let's try to pick up the thread of sauropodomorph evolution by focusing on what we have at the moment. By the Late Triassic period, there were at least four groups of interest. The first is typified by *Panphagia protos* (= *Pampadromaeus barberenai*) and also *Eoraptor lunensis*, now considered the most *basal* sauropodomorph. We can call these **basal sauropodomorphs**, although they didn't have external characters we'd think of as even remotely sauropodomorph-

Close to but just outside the ancestry of the sauropodomorphs, the swift and agile *Staurikosaurus pricei* gives us an idea of how much the sauropods' ancestors resembled theropods.

Formerly considered to be a basal theropod, the skull of turkey-sized *Eoraptor lunensis* from Argentina has now actually been shown to belong to a very early ancestral sauropodomorph. Along with *Panphagia* and *Pampadromaeus*, it was still probably carnivorous but may have started to include plants in its diet.

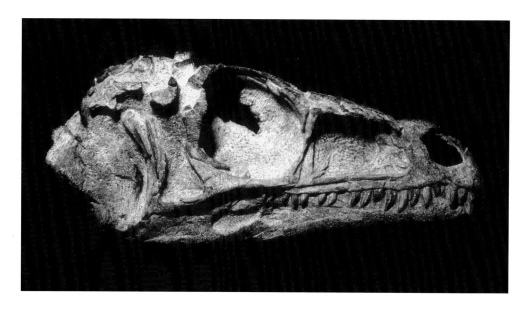

like (big bodies, long necks and small heads) and were still dog-sized, agile bipeds. While *Eoraptor* and *Panphagia* at first glance seem to have pointed teeth like a carnivore's, in both the teeth toward the back of the jaws (more so in *Panphagia*) are shorter than the longer, sharper ones in front, are leaf-shaped and have serrations, all herbivore dental characters. These showed that *Panphagia and Eoraptor*, although probably not committed herbivores, were capable of processing plant matter, but just how much plant material is an open question. In mammals, dental *morphology*, or shape, is closely related to the job it has to do. Generally, pointed, conical teeth like *canines* are evolved to pierce tough surfaces, *premolars* (in humans, *bicuspids*) to crush and *molars* to crush and grind. Extinct and modern saurians, however, usually don't have teeth that are as complex as those of mammals, and their functions aren't always so clear-cut. Studies by Paul Barrett (Natural History Museum, London) of iguanas and other modern lizards show that it isn't easy to predict diet from tooth shape, a fact that influenced his findings in hypothesizing basal sauropodomorph diets. He concluded that the "basals," like modern animals, could have consumed a potentially wide variety of plant and animal foods, with herbivores sometimes devouring animal material and carnivores occasionally snacking on plants. Even in some modern vertebrates there are shocking observed dietary behaviors, with deer known to eat baby bird nestlings (possibly for the mineral content rather than protein) and crocs eating fruit. So with the "probable sauropodomorphs" we just can't tell for sure, just that the teeth of some types are showing the potential for dealing with plants as well as flesh.

The next category can be labeled **definitive sauropodomorphs**, since although still rather small (2 m [6.5′] adult lengths), their necks were proportionately longer and the head smaller than in their ancestors, packing teeth that were more leaf shaped and serrated. Also resembling trends that the later sauropods would take to extremes, the rib cage was also longer and wider, adapted to hold the longer guts required to process plants (see Chapter 6). Definitive sauropodomorphs include *Thecodontosaurus antiquus* and *Pantydraco caducus* from Britain, as well as *Saturnalia tupiniquim* from Brazil. The features mentioned above, as much as they seem to foreshadow the sauropods themselves, should be seen in a broad perspective and not in a simple line leading to the later animals. As Thomas Holtz (University of Maryland) has pointed out, theropod evolution wasn't just a "bird factory," and in the same way "definitive sauropodomomphs" and the larger clade typically referred to as "prosauropods" weren't a dress rehearsal for the sauropods that came later. In its earliest days the evolution of life-forms, such as that of horses and humans, was often incorrectly portrayed as a simplistic, linear progress of increasingly advanced creatures moving toward a final, "finished" form. We now know that it isn't like that. Instead of treelike, simple lines of ascent, evolution takes the form of a *bush*—there's no predetermined "goal," and clades often radiate in several directions at once when the group's more typical animal species' *bauplan,* or body plan, is sufficiently flexible. So it wasn't by any means a straight march toward sauropods, and while basal sauropodomorphs as a group didn't last as long as other dinosaurs, they produced some very successful forms that differed from sauropods in surprising ways.

THE PROSAUROPODS

This brings us to the third group, the ***core prosauropods***. This one, apart from the sauropods themselves, was the most numerous and morphologically diverse, or *disparate*, group of sauropodomorphs and features not only better-known species like *Plateosaurus engelhardti* but ones like *Riojasaurus incertus, Massospondylus carinatus* and *Lufengosaurus hueni*. It isn't clear if all the known species in this informal group are more closely related to each other than other "basals," but they're all very similar and have a body form and way of life that were probably quite successful, since they had a worldwide distribution that stretched all the way from the Late Triassic to the Middle Jurassic. The "core prosauropods" have several important characters that set them apart from earlier forms. They continue the tendency to develop a long neck, which as herbivores maximizes their *feeding envelope*, the size of the area within which they can move their heads up, down and all around to harvest foliage. Certain "core prosauropods," especially massospondylids like Argentinean *Adeopapposaurus mognai, Coloradosaurus brevis* and the Chinese *Lufengosaurus*, accomplished this by evolving long *cervical*, or neck, vertebrae, some of which have a length 4 times the width. In *Leyesaurus maryensis*, this goes up to 5. In the later sauropods such as *Mamenchisaurus, Sauroposeidon, Erketu* and others this evolutionary adaptation would become even more extreme, but the "cores" show us that this was a feature that evolved repeatedly and independently in different sauropodomorph lineages.

"Cores" also have deeper snouts, larger, somewhat retracted *nares* (bony nose openings) and proportionately smaller heads. The function of the large nares isn't clear, but they possibly housed soft tissues to warm and moisten inhaled air or to conserve moisture. The joint attaching the lower jaw to the skull is now also more *ventral,* or farther down on the skull, as well as being *offset* (at a short distance below the line of the teeth when they *occlude,* or come together). This makes the upper and lower rows of teeth all close at the same time like a pair of pliers, instead of a pair of scissors, in which the blades first cross near the pivot point and then gradually keep crossing the farther away they get. The second condition is great if you're a carnivore and you want to slice through meat, but if you're a herbivore, having your teeth occlude all at the same time is an advantage—grabbing, seizing and chewing a lot of little, flexible leaves or shoots is much easier when your jaws are equipped like this. This makes sense, but why proportionately

Like some modern iguanas, the teeth of "definitive" sauropodomorphs and some other dinosaurian herbivores have denticles, or tiny pointed structures, along the edges that aided in cutting through vegetation. Instead of sticking out to the side like the serrations on some theropod teeth to slice flesh, these point toward the apex, better suited to cropping plants when the jaws close.

smaller heads? Wouldn't it be even better to have a relatively big head, so you could pack in more teeth to bite off and chew bigger quantities of plants? Maybe it would, but there's a problem if you want that head to be at the end of a long, slender neck: weight. To do the jobs they're supposed to, teeth have to be made of hard and durable materials like dentine and enamel, and this makes them heavy. Making them more numerous and smaller only works up to a certain point. "Cores" and other "prosauropods" faced a trade-off between heads and jaws that were big enough to harvest the amount of food the animals needed and becoming so big that the increased weight would be hard to support on those long, slender (for their body size) necks. Mammalian herbivores chew their food and, unlike long-necked sauropodomorphs and birds, usually have extensive sets of big, heavy molar teeth that have to be supported on fairly robust, comparatively thick necks. Giraffes are an exception, but even these big-toothed feeders have necks that never became as long and slender as those of sauropodomorphs. The solution if you need a longer and longer neck? Don't let your head get too big, and this means *give up chewing*. The "cores" and the ***near sauropods*** were well on their way to achieving this and became huge in doing so, and we'll learn how in Chapters 3 and 5. Core sauropods were generally big. Whereas the earlier "probables" and "basals" hadn't surpassed the dimensions of a big dog, "cores" had gotten up to the size of cows and often much bigger—the largest individuals of *Plateosaurus* were 10 m (34′) and probably weighed 4 metric tons (the form of tonnage used throughout the rest of this book), the size of an adult female elephant. This was far bigger than either any of the other contemporary non-dinosaurian herbivores or any animals that had ever lived on Earth up until this point, and they needed a lot of chow.

How did "cores" support those bigger bodies? Perhaps surprisingly, not on four legs, a very recent discovery. Traditionally, both paleontologists and artists have reconstructed all "prosauropods" as being equally happy either standing on their hind legs or on all fours, but in 2007 Matt Bonnan (Western Illinois University) and Phil Senter (Fayetteville State University) used computerized digital modeling to convincingly demonstrate that *Plateosaurus* and *Massospondylus* were *obligate bipeds*, meaning they could walk only on their hindfeet. This finding rests on several lines of evidence, starting with the bones of the forelimb, the *radius* and *ulna*. In humans, the radius can rotate around the ulna, which allows us to turn our wrists. From a neutral position (with the palm facing toward our body) we can turn our palms up towards the ceiling, an action called *supination*, or downward facing the floor, the action known as *pronation*. Pronating your hand onto the ground is obviously what you want if you need to walk on all fours, but core sauropods couldn't do it. Whereas our radii and ulnae have almost perfectly circular wheel-and-socket joints, those of prosauropods have only shallow grooves—these allow the adjacent bone to articulate but not rotate. Both radius and ulna are too short and blocky to move past each other even if the joint allowed this.

Another indication of four-footedness, or *quadrupedality*, lies in the great difference between the lengths of the forelimb and hindlimb. In *Plateosaurus* the hindlimbs are more than twice as long as the forelimbs, proportions much more typical of a biped than a quadruped. By laser-scanning a complete skeleton of *Plateosaurus*, Heinrich Mallison (Museum für Naturkunde, Berlin) was able to build an accurate, 3D digital model of the whole animal, and onto this he

programmed guts, muscles and other soft tissue. This, in turn, enabled him to put his virtual *Plateosaurus* "through its paces" and to investigate the probable locomotion of the living dinosaur. From this he determined that *Plateosaurus* was perfectly stable on its two hind legs and could walk more efficiently than on four. Like Bonnan and Senter, he also found that the forelimb bones of this prosauropod would have prevented pronation and therefore the ability to walk on its forefoot, but in the program he gave his model *Plateosaurus* the benefit of the doubt to see how it would look if it had tried: it would have been very uncomfortable. The fossil record also sheds light on "core prosauropod" bipedality. Plateosaurids and massospondylids are now known from hundreds of partial to complete skeletons, and so far not one has its wrist pronated. There is also no convincing trackway evidence for these animals coming down on all fours. Although they may have been able to briefly use their forelimbs to help in getting to their feet or lying down to rest, they would have to have done this by either turning their forefeet out to the side or supporting themselves on their knuckles. In short, *Plateosaurus* could've given you a handshake but not a high five. Not that you'd want a high five from *Plateosaurus* or any other basal sauropodomorph. They had sharp claws on the first three fingers, and the thumb claw, or digit I, was especially large and viciously hooked. We really don't yet definitely know what they were doing with those claws, but digging, pulling up plants and defense against predators are all very likely.

DOWN ON ALL FOURS

The last group of basal sauropodomorphs we can call ***near sauropods***, animals more closely related to the true sauropods than core sauropods, but not quite yet sauropods. While the "cores" are a relatively uniform group all roughly doing the same thing ecologically (being big-bodied,

Rather than simply being regarded as the ancestors of sauropods, early sauropodomorphs radiated widely throughout the Late Triassic and Early Jurassic world into a variety of forms. *Counterclockwise from top*: the basal sauropodomorph *Panphagia protos,* the "core sauropod" *Plateosaurus engelhardti* and the "definitive sauropodomorph" *Thecodontosaurus antiquus.*

Plateosaurus and other core sauropods had intimidating fore-claws (*right*) that could not only dig and pull on leafy branches but also deal effectively with predatory theropods. *Plateosaurus*'s hindfoot (*below*) shows the short, blocky fifth toe that anticipated the transition of some core sauropods' feet to those of the earliest true sauropods.

bipedal herbivores), the near sauropods represent a diversity of adaptations. Some, such as *Anchisaurus polyzelous*, are small and at first glance not very sauropod-like, much more closely resembling early sauropodomorphs like *Saturnalia* in body size and proportions. But "under the hood" there are anatomical details—the bones of their braincases and faces, the way their vertebrae fit together and their pelvic morphology—that show that they're in fact closely related to sauropods. At the other extreme are large, now probably quadrupedal forms such as *Camelotia borealis* and *Lessemsaurus sauropoides* that might actually be early basal, true sauropods. In between these are species like *Antetonitrus ingenipes* that superficially resemble *Plateosaurus* but are even bigger and more robust. As with the origins of basal sauropodomorphs among the early saurischians we encountered at the beginning of this chapter, many aspects of how the first true sauropods evolved aren't clear right now.

Three known species, however, may help us understand near sauropods and how some of these might have transitioned to the sauropods themselves. All are from the Upper Triassic of South Africa, a geochron that appears to be as much of a hotbed of sauropodomorph evolution as the Early Cretaceous of China is for the understanding of bird origins. The first of these is *Aardonyx celestae*, and at first glance this animal could be mistaken for just a fat "core sauropod" like *Plateosaurus*. It still has short forelimbs and a forefoot that won't allow it to pronate to walk on all fours, but the large muscle attachments on the femora, or thigh bones, are located farther from the pelvis than in "cores." This tells us that *Aardonyx* was trading speed for power, a common feature among animals that are evolving toward large body size. Another sauropod-like character is a wall of bone, the *lateral plate*, that braces its teeth to strengthen them against stress while feeding. Last, it has a shorter, broader hindfoot. What does this mean?

The hindfoot is particularly important in the transition from basal sauropodomorphs to actual sauropods. In the first true sauropods, the *metatarsals* (the bones in the middle of the foot, to which the *phalanges*, or toes, attach) are all short, blocky and similar in length, and although the foot in these early types has only *four* metatarsals and phalanges, this goes back up to *five* in advanced or derived forms. In most "basals," however, the fifth metatarsal was shorter than the rest and supported a very small toe, suggesting at first that these bones were in the process of disappearing. If this had actually happened as some workers think, then "basals" couldn't have been the ancestors of sauropods, because for this to be true sauropods would have had to re-evolve the fifth toe—most evolutionary theorists believe that once a structure has disappeared or been lost from disuse, it can never be re-evolved. In response to this anatomical dilemma, however, Adam Yates (Museum of Central Australia) has pointed out that the fifth toe in sauropods consists of actually just a single, blocky button of bone, not very different from that of a "basal," in which the actual toe is likewise quite reduced. As for the differences in the relative lengths of the metatarsals, it's probable that the longer, more slender metatarsals of early sauropodomorphs had to become shorter and thicker to support more weight as some became sauropods. So transforming a "basal" foot into a sauropod's didn't require enlarging the fifth metatarsal to match the rest because the fifth toe was *already* short and thick. Instead, it was the *other metatarsals* that did the shortening to match the already chunky fifth metatarsal—this can be seen in *Aardonyx*.

The second near sauropod, *Melanorosaurus readi*, if not actually the most primitive sauropod, at least gives us our current best glimpse of how one might look. Most crucially, the forelimb's radius and ulna show the ability to pronate, indicating that the forelimb, which is also now longer, is at least some of the time playing a role in helping to support the body. The sacrum, instead of a basal's three fused vertebrae, is now composed of four, typical of sauropods and a way of strengthening the pelvis to bear more stress and weight. In spite of these adaptations toward weight bearing, *Melanorosaurus* still retains a biped's curved femora and its vertebrae continue to be more like a "basal." These conditions, although they might seem contradictory, are evidence of mosaic evolution. Not all body structures marched equally in step with each other along a road to sauropods, but were chosen by natural selection as the needs arose.

With *Antetonitrus* we finally come to a true **sauropod**, the earliest known member of this clade. It's now got forelimbs almost as long as the hindlimbs, and while its relative *Melanorosaurus* occasionally might have used these to walk, *Antetonitrus* is an obligate quadruped when it comes to walking. Both its fore- and hindfeet are short, blocky and clearly adapted for weight bearing, while its vertebrae are taller than those of "basals" and for the first time show thin struts of bone called *laminae*, a hallmark of true sauropods. From a distance this dinosaur might have looked like a prosauropod walking on all fours, but it nevertheless had the classic sauropod profile: a large body supported on four stout legs, a small head on a long neck and a long tail. In some ways *Antenonitrus* was rather primitive—it still had a mobile thumb that could have been used like a prosauropod's to grasp food—but it probably was one of the first sauropodomorphs to look and act like a sauropod. Here it's important to remember that even though at this point the earliest true sauropods had become quadrupedal as a way of supporting their greater weight while they walked, they never lost their ancestors' ability to rear bipedally, if needed, to reach vegetation.

Matt Bonnan (Western Illinois University) has conducted elegant computer model studies that have cast much light on the evolution of prosauropod locomotion and its implication for their derived descendants, the sauropods.

The skull of the "core" prosauropod *Plateosaurus engelhardti* displays a key feature of herbivores: jaws with tooth rows that all close together at the same time, making them more efficient than a carnivore's scissor-like jaws in cropping plants.

KEY ADAPTATIONS

Looking back on all these forms, we can see them as evolutionary experiments in building large-bodied herbivores. *Eoraptor* and *Panphagia* show the first steps away from the uniformity of the dinosaurs. *Saturnalia* and *Thecodontosaurus* consolidated the characters of sauropodomorphs in a distinct clade from those of other saurischians, a first "parting of the ways" as increasingly committed herbivores that set them on a path away from their theropod cousins and more omnivorous close relatives. After this came the second important split. While other lines of sauropodomorphs tended to later go quadrupedal, the "core prosauropods" were obligate bipeds and stayed this way throughout the rest of their evolution. *Plateosaurus*, *Massospondylus* and their "core" relatives became large, even huge, by retaining bipedal locomotion and the versatility of grasping hands. It was a successful strategy because these animals became global success stories for 25 million years, with only the *facultative* (having the capability but not the obligation) bipedal Upper Cretaceous iguanodontids and hadrosaurs adopting these same features. The near sauropods went another way. Although remaining bipedal at least through *Aardonyx*, they afterward shifted down onto all fours, straightened their limbs and shortened their fore- and hindfeet, all to bear up their more extensive digestive systems and heavier bodies. That transition from bipedal to quadrupedal locomotion (although never losing the option of being bipedal while feeding) and support was one of the keys to the evolution of large size in sauropods and led to their eventual success. A fascinating thing is that these evolutionary experiments played out alongside each other. Core sauropods didn't evolve before the earliest true sauropods; they evolved at the *same time*. If we'd been there without being able to see into the future, we couldn't have predicted which strategy would eventually prove to be the more successful.

OUT OF A JOB

From the vantage point of our own human perspective we know that some of the "near sauropods" evolved to become true sauropods, the subject of the rest of our book. But why did the "core sauropod" strategy fail, when they were so successful? Holtz has suggested that, counterintuitively, it was actually *because* they were such successful, "jack-of-all-trades" animals. Since these kinds of species often have a survival edge, unlike overly specialized forms, their generalized versatility means they can turn around and adapt to other lifestyles when the evolutionary going gets tough. They can squeak by when other species have no place to go but extinction. This is usually an excellent survival strategy, but not always. By the Middle Jurassic "core prosauropods" were facing competition from other dinosaurs that, like a candle burning at both ends toward the center, were both figuratively and literally consuming the core prosauropods' lifestyles. Early ornithischians such as basal iguanodontids and heterodontosaurids had developed sophisticated ways of chewing that enabled them to devour a wide variety of plants, coming to dominate the low- to middle-level feeding niches. At the other end the core prosauropods' relatives, the true sauropods, had gotten very big, very fast—a sauropod humerus from the Upper Triassic of Thailand is 1 m (3.5′) long, indicating an animal the size of a modern bull elephant. The sauropods were better low nutritional, non-specialist bulk feeders, while the ornithischians were better high-quality, specialist feeders—in terms of ecology, this simply left the core prosauropods out of a job.

Down on All Fours

1. *Eoraptor lunensis*
   Basal sauropodomorph, obligate biped

2. *Massospondylus carinatus*
   Basal prosauropod, obligate biped

3. *Gongxianosaurus shibeiensis*
   Vulcanodontid basal sauropod, habitual quadruped/facultative biped

4. *Barapasaurus tagorei*
   Cetiosaur basal eusauropod, habitual quadruped/facultative biped

As the prosauropods evolved away from a theropod-like carnivorous/omnivorous lifestyle (shown by *Eoraptor*) toward herbivory, their hindlimbs became more robust and bones of their feet shorter and blockier for supporting the greater size and weight imposed by their increasing larger gut systems (*Massospondylus*). While some prosauropods remained bipedal and grew bigger, others like basal sauropods (*Gongxianosaurus*) evolved quadrupedality as a solution. While they never lost the ability to rear up to feed, these and later true sauropods (*Baraposaurus*) developed longer forelimbs and feet that were more evenly proportioned, as well as even shorter and more compact foot bones for quadrupedal support. (Skeletons and pedal bones not to same scale.)

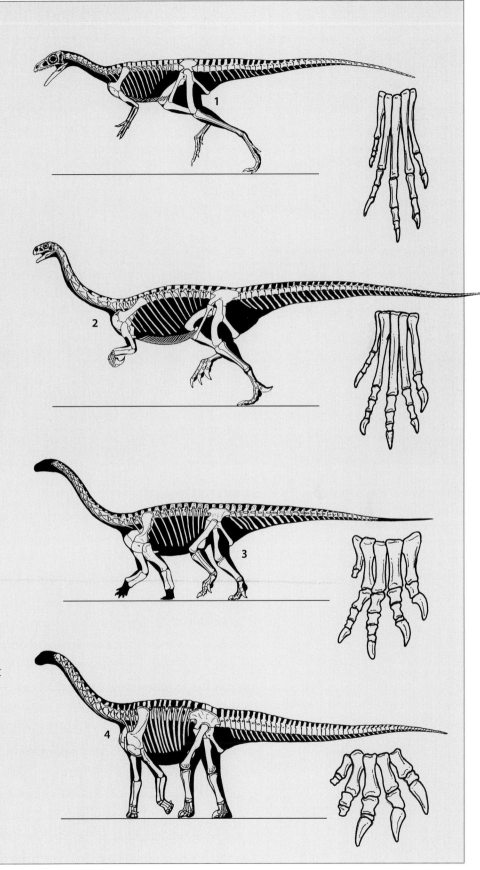

Although missing its skull, *Vulcanodon*, restored here as a life-sized model at the Dinosaurier Park, Münchehagen, is one of the most basal, or earliest known, sauropods.

SURPRISES INSIDE EGGS

"Core sauropods," although they ultimately lost out to the competition, have given us some surprises that shed light on both general sauropodomorph evolution and particularly the way in which their sauropod cousins evolved. Nests and eggs of *Massospondylus* have been known since the late 1970s, but the first babies, described in 2005, were almost ready to hatch (one may even have been in the act of hatching when it died), telling us what the chicks looked like at this stage in life. *And what they looked like were miniature sauropods.* The necks were short and the heads proportionately huge—not any different from most newborn animals. But the forelimbs and hindlimbs were almost equal in length and short compared to the body, so the hatchlings, just like sauropods, would've had to have walked quadrupedally, at least at first. As with all newborns, they had to "grow into" their heads, but to become typically "prosauropod" in their body proportions the neck had to become longer and the hindlimbs also had to lengthen, by 2 or 3 times. *Massospondylus* chicks must have looked very odd as they went through these radical changes. When they lost the ability to walk quadrupedally and became obligate bipeds isn't known, but it might have been triggered by the progressive ossification of their forelimb bones and joints as they became adults.

Another surprise is that these near-hatchling embryos lacked teeth. Teeth don't just appear overnight, so that means that chicks were also toothless, and we know from the details of their jawbones that basal sauropodomorphs didn't have any kind of beak. Then how did they feed themselves? The answer is that they couldn't and would have needed their parents to bring them food, as with modern baby birds and mammals. Indirect evidence for this comes from the fact that juveniles up to twice hatchling size have been found in *Massospondylus* nests, indicating that at projected growth rates the young were probably staying in the nest for at least several weeks. Among living animals, babies only stay in the nest if they're being fed and protected by their parents. In addition to indirect fossil evidence, the behaviors of living archosaurs like crocs and birds suggest that parental care was likely in some types of dinosaurs such as theropods and hadrosaurs. But parental care in sauropodomorphs is something we wouldn't expect, since there's no evidence for it in sauropods themselves. Because it was still present in *Massospondylus* and other sauropodomorphs, the loss of this behavior must have occurred later and was a critical step in allowing sauropods to become successful giants. As mammals we're used to the idea that parental care is an advantage in the survival of young, but as we'll discover in Chapter 8, for sauropods it was actually the opposite.

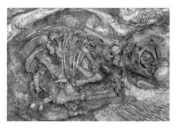

Exquisitely preserved, the skeletal development of this South African *Massospondylus carinatus* chick indicates that it was almost ready to hatch. Embryological studies by **Gerald Grellet-Tinner** of these core prosauropods offer clues as to the evolution of their giant descendants, the sauropods.

GENETIC SHORTCUTS

There's one more thing the *Massospondylus* embryos can tell us, and that's how sauropods got to be the way they were. These babies, instead of resembling the adults of their own kind, much more closely resemble the later sauropods in their limb proportions of equal length. If "near sauropod" ancestors of sauropods also had babies like this, it would have been an evolutionary advantage or "shortcut" for early sauropods to have retained these features as positive adaptations to the requirements of getting bigger and bearing more weight. It's here that the discipline of *evolutionary developmental biology*, sometimes called "evo-devo" for short, can come to the aid of more traditional paleontology to help explain how evolutionary trends can occur. "Evo-devo" focuses on the way genes and proteins interact to affect the structure of a developing embryo. It also explains the mechanisms for how an organism may display certain characters and not necessarily others that exist in its genetic repertoire of possibilities. One important discovery is the existence of *Hox genes*, "master control" genes that coordinate to a large extent not only *when* but also *where* key areas of the embryo's organs will form. Even though an organism's characters are previously encoded into its genetic makeup, certain genes can be *expressed* (allowed to happen) or *suppressed* (not allowed) through the interactions of the Hox gene's proteins. Here we can think of an animal's encoded DNA as being like an enormous recipe book with thousands of chapters and subchapters containing variations on how to make a certain kind of cake. You can follow the recipe just the way it's written, but you can also make changes along the way if you want the outcome to be slightly different. Hox genes come in different types, and by influencing gene expression and suppression to form an embryonic structure, they act almost like a sculptor, adding or carving off pieces of clay to make a particular shape. In this case it was the shape and length of a near sauropod embryo's limb bones and surrounding soft tissues. Another probably key process for near sauropods taking the evolutionary "shortcut" to evolving into sauropods was *paedomorphism* ("*paedo*," youthful; "*morphos*," shape). Here juvenile characteristics are retained throughout an animal's life instead of giving way to those of an adult, an outcome made possible by the suppression of some adult genes and retention of certain juvenile ones. This, enabled by the mechanism of the Hox genes, is what kept the hindlimbs from attaining the lengths

In looking at the adult *Massospondylus carinatus* and half-grown young (*above*), one would never imagine that prosauropods similar to these would someday become the massive, 50-ton *Astrodon johnstoni* (*right*), seen here as a life-sized model at the North Carolina Museum of Natural History.

of an adult "prosauropod" and made the forelimb's length and its morphology stay at the same proportions as the animal grew up. As a result, sauropods can be thought of in a way as giant babies. In spite of such "genetic shortcuts," however, we should remember that these dinosaurs, like all other organisms, during their existence were still subject to the principle of *natural selection* as the main force that acted on them and drove them in the direction in which they evolved. An individual creature's final morphology and potential for behavior, although determined by its inherited genes and their expression during its early development, are still subject to natural selection the moment the animal begins life on its own. When an individual inherits its combined genetic recipes from its parents, these may sometimes include *mutations*, or randomly occurring slight occasional variations in the genetic code. When they do pop up, they are what natural selection acts upon. An animal that's born (or hatched) with a minor but detrimental mutation has just that much less of a chance to live long enough to reproduce, while one that inherits some small (but helpful) feature has a slightly *better* chance to live long enough to pass on its favorable mutation—in this way organic evolution often gradually (but sometimes very rapidly) forms new species over time.

The earliest true sauropods, having emerged in the Late Triassic, had now finally evolved into the shape and body plan we'd recognize, but they were still very primitive or basal compared to later types. Over the ages, these dinosaurs would take many forms, and like the now relict, almost extinct "prosauropods," were poised to spread out over almost every corner of vast, arid Pangaea. The sauropods now had big bodies and a promising way to feed and support themselves, but where would they go next?

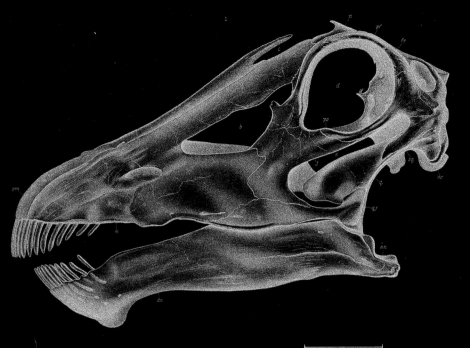

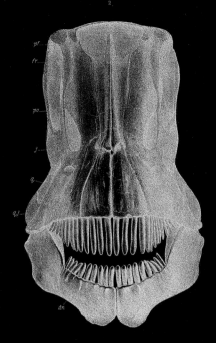

10 cm

Inside the Fossil Prep Room

The sun-drenched drudgery of the outdoor field season is in direct contrast to the museum's fossil vertebrate prep lab, where under the whine of air drills a sauropod skeleton like *Diplodocus* (skull, *above*) may take months, even years, of liberation from its rocky matrix before becoming available for study and exhibits. If to be exhibited, the bones are usually reproduced in foam-fiberglas-resin to save mounting weight and to make the bones available for study. The pose of the skeleton may be planned as a scale model (*below, left*) before its final, full-scale mounting (*below, right*).

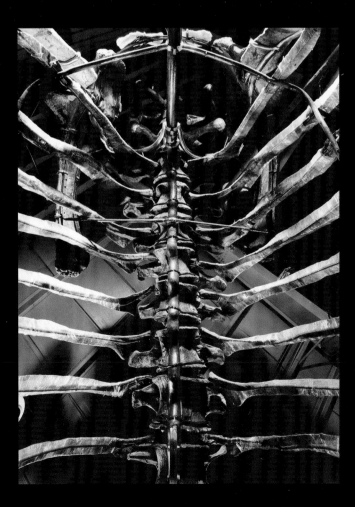

A pod of *Dicraeosaurus hansemanni* enjoys a watering hole at Tendaguru, Africa, during Tithonian times.

CHAPTER THREE
A SAUROPOD FIELD GUIDE

At first glance, it might seem that sauropods were all pretty much alike. Compared to the variety of horns on the ceratopsians and strange head crests of the hadrosaurs, sauropods didn't stray from the basic plan of small head, long neck, big body, long tail. To some extent this was true because their basic *bauplan* (a German word originally used to refer to architectural design, but also here to body plan) was highly successful and didn't need to radically change, but each major type of sauropod had very distinctive, sometimes bizarre-looking, features that evolved to enable them to adjust to different ecological niches. Modern refinements in the science of classification have enabled us to begin seeing the big picture of sauropod lineages and their relationships to each other.

ALL IN THE FAMILY. Sauropods, like all other living things, are classified and formally named according to a system devised in the mid-1700s by the Swedish naturalist Karl von Linné, or Linnaeus, the latinized name by which he's usually known. Linnaeus realized that the many popular names given to a modern animal often created confusion (for example, Europeans use the name "elk" for what North Americans would call a "moose," while "elk" to Canadians and those in the USA is the name for the large deer closely related to the red deer of Europe). He conceived of a *binomial*, or two-part, name for each living and extinct lifeform, usually based on their more distinctive features, called **traits** or **characters**. The first word

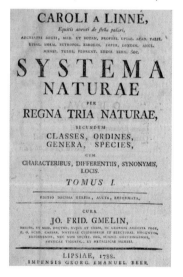

Karl von Linné (1707(10?)–1778), or **Linnaeus**, as he is better known, was a Swedish botanist who devised the system of *binomial nomenclature*, or double naming, still used today and which is the foundation of the modern discipline of taxonomy.

in the name (the **genus,** *plural* **genera**) is the equivalent of a human last or surname, while the next (the **species**, never capitalized) is like a first name, and together they define that organism in a precise way. Although other creatures can have the same specific name, none can have this *and* the same genus name. A lion, for example, is *Panthera leo*—(*Panthera*, "big cat"; *leo*, "lion"). Just as a human family with several members shares a last name but each has his or her own first name, this genus includes other big cats like *P. tigris* (tiger), *P. pardus* (leopard) and *P. onca* (jaguar). All are "sister species" because of their very close relationship in genus *Panthera,* and their unique two-part names eliminate confusion. Sometimes a third, the **trivial name**, is added to indicate a distinctive **subspecies**, or race, among others, that a species may include: *Panthera tigris altaica* refers to the Amur ("Siberian") tiger, while *P. tigris sumatrae* is the very rare Sumatran tiger. Biologists sometimes differ as to how a species should be defined: if you plucked the feathers from many small birds like finches or warblers that some ornithologists consider totally separate species, it would be almost impossible to tell them apart. Found only as fossils, a paleontologist would conclude that they all were just one species, since their bones would appear almost identical. In spite of this, there's a growing consensus that species can be defined as groups of actually or potentially interbreeding natural populations that are *reproductively isolated* (by either geography or behavior) from one another.

Taxonomy, the discipline of naming life-forms, also applies to extinct species like dinosaurs. It goes further than genera and species, however, and these, in turn, are usually grouped into **families** (sometimes subfamilies), followed by **orders**, **classes** (sometimes subclasses) and even broader catagories. Now we're talking about **phylogeny** (*phyllos*, "tribe"; *genesis*, "origin"), the paths by which living things evolved from one another and are classified. If we think of phylogeny as being like the growth on a tree or bush, our lion species, *leo*, would be a bud, along with other big cat buds, on the end twig of genus *Panthera*. Panthera, or "big cats," is, in turn, a twig close to that of a variety of other genera, unofficially known as the "small cats" (ocelots, pumas, etc.) in the subfamily *Felinae*. Both of these differ from their other cat cousins, subfamily *Machairodontinae*, or sabertooths. *Felinae* (note the "**n**") would, in turn, be a bigger stem of the family *Felidae* (all modern and extinct cats; note the "**d**"). The Felidae arise from order *Carnivora* (all mammalian carnivores), and Carnivora is just one branch of progressively larger ones arising from the subclass *Eutheria* (mammals that reproductively differ from marsupials like kangaroos and monotremes like a platypus). Eutheria is still another, larger branch coming from a very big main limb, the class *Mammalia*, or mammals. Scientists in fact refer to these kinds of systems as *phylogenetic trees*, and they're based on a comparison of the anatomical similarities and differences between one creature and another. A common ancestor and all of its descendants form a **monophyletic** ("one tribe") group known as a **clade** (*clados*, "branch"). Depending on their own inclinations, taxonomists sometimes have a tendency to be "lumpers" (lumping or rolling several taxa into one clade) or "splitters" (splitting or assigning different species status based on minor differences). This doesn't mean that dinosaur identification is just arbitrary, however, and we've now learned much about sauropod and other dinosaur lineages. This can change in the light of new discoveries, and paleontology, rather than being just a body of accumulated information, revises its hypotheses, or tentative ideas, when new data show that these may be in error.

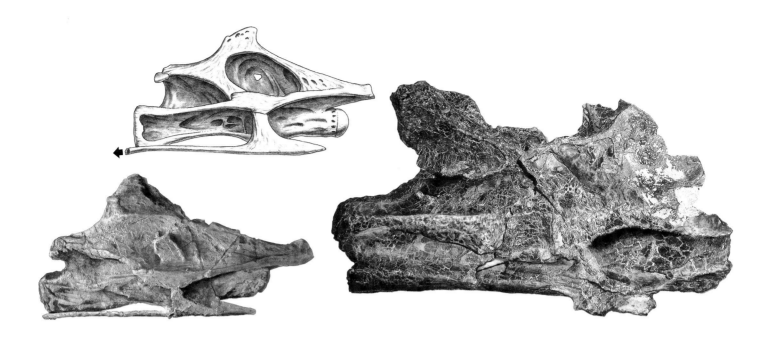

SORTING OUT SAUROPODS

Since bones, and only rarely soft tissues, are usually what's preserved in the case of dinosaurs and other prehistoric animals, **comparative anatomy** is usually all we have as a way of considering whether a particular species of animal might have been an ancestor, descendent or other close relative of another. The skeletons of sauropods, like many fossil vertebrates, are often very incomplete and fragmentary when found, and sometimes the bones don't provide much information about what kind it was. It's like alien scientists going to a human's aircraft scrap yard millions of years in the future: in this case their goal might be to find out how the Boeing 700 series jumbo jets originated and were developed, but all they would probably recover would be a scrap here or there of an engine, wing or tail. Unless they were lucky enough to find enough of a plane intact, they'd have to painstakingly accumulate and study the parts they had in order to learn how the giant aircraft were built, where they came from and how they flew. This is often the task facing sauropod and other dinosaur specialists. Certain parts of a sauropod's skeleton, like the vertebrae, are highly distinctive, and if found, these bones allow a worker to diagnose what type of sauropod it is. This is the goal of comparative anatomy, and it comes down to what looks similar to or different from what. With these and other dinosaurs the most common units of classification tend to be genera, since it's on this level that workers find that most skeletal differences or similarities are easily observed and can usually be agreed upon. For example, it's not always clear whether a specimen of the "brontosaur" *Apatosaurus* (and/or *Brontosaurus*) belongs to *A. ajax*, *A. excelsus* or *A. louisae*, but no one has any difficulty telling *Apatosaurus* or *Brontosaurus* from the genus *Diplodocus*. This is why it's desirable to have as large a **data** (information) **set** as possible based on bones, since this larger sampling makes it possible to start seeing patterns of similarity. If you have only one bone specimen that just slightly differs from another, it could be a new species, but it could also just reflect the natural variation that occurs in each individual animal.

Most discoveries of new sauropods are based on incomplete remains, like this partial vertebra from England (*bottom, right*). The new specimen must be compared to more complete material from a better-known taxon (*bottom, left*) to assess its identity and discover its relationships to other sauropods. This is the basis of comparative anatomy. In this case, the new animal is most similar to *Sauroposeidon* (*top, left*).

Because of the reasonably large number of well-preserved bones, the skeletal characters of lightly built *Diplodocus* (*left*) and contrastingly massive *Apatosaurus* (*right*) are now well enough known to eliminate any confusion as to their identity even when only fragments are found.

In the case of sauropod taxa and clades, by now we know that there are a lot of them. By the beginning of the 21st century's second decade, well over 100 sauropod genera had been named, with more being discovered every year since. Every new dinosaur name is a **hypothesis**, a rational but unproven idea, and not all may stand the test of time. In 1991 the bones of a large *Diplodocus*-like sauropod from New Mexico were assigned to the name *Seismosaurus halli*, but over time the characters of the partial skeleton originally used to distinguish *Seismosaurus* from its close relative, *Diplodocus*, were demonstrated to be misinterpretations, possibly the effects of old age and extreme growth. The most recent studies suggest that "*Seismosaurus*" is simply an old, very large specimen of *Diplodocus*, although it isn't clear whether the species *halli* is still valid or whether it should instead be assigned to the well-known *Diplodocus* species *D. longus* or another species. If this is true, under the rules of *biological nomenclature*, or naming, the oldest name is the accepted one, which is why "*Seismosaurus*" was "sunk" into *Diplodocus* instead of remaining valid. Sometimes, however, an older, supposedly invalidated name may be legitimate. Following a new analysis by Emanuel Tschopp (University Nova) and his colleagues, there are possibly enough demonstrable morphological differences between the popular (but long-considered invalid) genus name *Brontosaurus* and the later-discovered, supposedly identical *Apatosaurus* to allow *Brontosaurus* to stand as valid, and not simply the junior synonym, for the earlier-discovered specimens—a happy possibility for those who love the old name.

In making comparisons of characters or traits to find out actual relationships and lineages, we have to be careful—just because two types of sauropods have bifurcated neural spines, for example, or tall shoulders and long necks, doesn't mean that they are closely related. In some sauropod lineages, these similar anatomical features evolved because two separate, distantly related clades took similar evolutionary pathways to become similar in shape and behavior, a condition known as **convergence**. This produced the similar, but not closely related, high-browsing brachiosaurids and omeisaurids. Also called a **homoplasy** (*homo*, "same"; *plessein*, "to mold or form"), it's sometimes misleading in determining true relationships. This situation is opposite of a **homology**, the quality of having characters that are truly shared through a common inheritance. By counting and **weighting** (evaluating) homologies, we can identify monophyletic clades based on actual, shared characters. This can be tricky, however—how do we know when characters are homologies or homoplasies? We need to know if we're going to determine actual sauropod relationships and lineages.

Since the 1990s, a relatively new science, **cladistics**, has gradually come to the aid of traditional comparative anatomy studies to help understand sauropods' (and other organisms') relationships. Cladistics is a discipline that, like comparative anatomy, deals with shared characters, but instead of attempting to build phylogenetic trees and track possible evolutionary routes starting on a macro level, it initially focuses on shared characters at a *micro level*. Here the approach focuses on grouping species together based on whether or not they share one or more unique characters that come from a *common ancestor* and that are not found in other species.

Let's say that we want to know whether the relationship between two sauropods, the contemporaneous, now well-known *Diplodocus* and the longer- and wider-necked *Barosaurus*, forms a natural group made up of a common ancestor and all its descendants, or **clade**, within the subfamily Diplodocinae (together with another subfamily, the robust Apatosaurinae, informally called the "Flagellicaudata"), the more slender, so-called brontosaurus-like sauropods with wide, wedge-shaped heads, short forelimbs, long, whiplash tails and other distinctive features. At what point did the two separate, recognized species share a common ancestor, based on their shared characters, and at what point did one of them cease to share a given character (or characters) with the other? We start by constructing a **cladogram** (a diagram of the two species' possible relationship) that establishes perimeters, or boundaries—these fence off or define the area of investigation. This is done by choosing representative animals, called **outgroups**, that are more distant relatives of the species whose relationship we want to establish, and act like "bookends" that keep us from going too far afield in our search. Since we're focused on diplodocids, we can use a species like *Kaatedocus*, an earlier, more **basal** (less advanced) member of the bigger clade, the Diplodocinae ("Flagellicaudata"), to bracket the later *Diplodocus* and *Barosaurus* earlier in time, while another, the contemporary apatosaurine *Apatosaurus*, can act as a bracket for a more **derived** (phylogenetically advanced) type that evolved in a different direction from our two sauropods. Now we can compare the known characters from each of the two species to create a hypothesis as to how they were related to each other. The more shared, derived characters they both have in common, and which other sauropods don't, the more likely that there is a close relationship, or **synapomorphy** (*syn*, "shared"; *morphos*, "shape"). If an earlier form shares only some of the more basic characters that other sauropods don't have and only our two species have in common, it's likely that this sauropod (or one similar) was either on or close to the ancestral line that resulted in *Diplodocus* and *Barosaurus*. Here on our cladogram we make a dot, or **node**, representing this presumed ancestor, with lines going from this to the names of our two later species. If they're two closely related species sharing all or most similar characters that branched off from their shared common or parent ancestor, they're **sister species**—larger, closely related associations are called

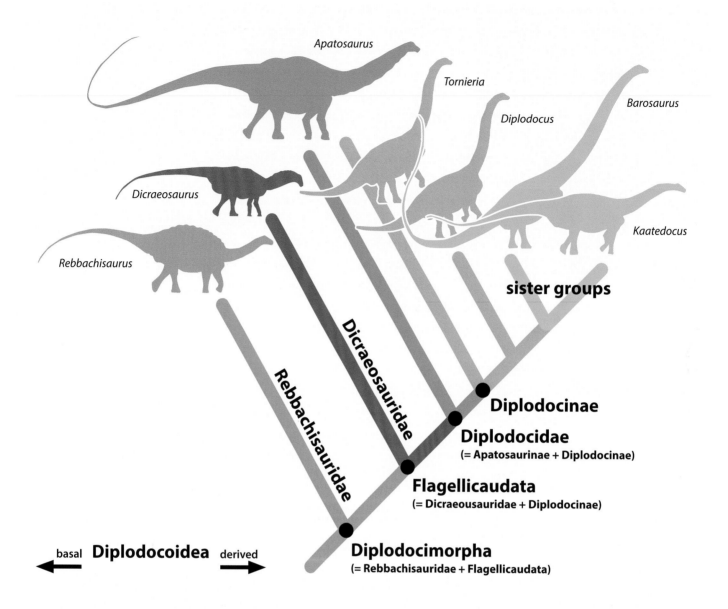

Apatosaurus

Tornieria

Diplodocus

Barosaurus

Dicraeosaurus

Kaatedocus

Rebbachisaurus

sister groups

Dicraeosauridae

Rebbachisauridae

● **Diplodocinae**

● **Diplodocidae**
(= Apatosaurinae + Diplodocinae)

● **Flagellicaudata**
(= Dicraeousauridae + Diplodocinae)

basal **Diplodocoidea** derived

● **Diplodocimorpha**
(= Rebbachisauridae + Flagellicaudata)

Building a Cladogram

This cladogram shows the possible relationship, based on an evaluation of shared skeletal characters, between *Barosaurus* and *Diplodocus*. Both of these are considered to be sister species that originated at a close point to each other in time, and are bracketed not only by more basal members of the subfamily Diplodocidae to which both these sauropods belong, like *Kaatedocus*, and less closely related diplodocoids like dicraeosaurids and rebbachisaurids, but also by other derived diplodocid species like *Tornieria*. These are known as outgroups and serve to bracket or limit the scope of the characters we are considering for the two species. Each circle (or node) represents the point at which a group separated from the others and is the last point in time when the two branches shared a common ancestor. These are called sister groups, whose closeness (suggested by the similarity of colors) is based on the number of characters they share.

sister groups. *Diplodocus* and *Barosaurus* are sister species (known from a later stratigraphic horizon where *Kaatedocus* doesn't occur) and share certain characters, or synapomorphies, that *Kaatedocus* and *Apatosaurus* don't have. In spite of their many similarities, *Diplodocus* and *Barosaurus* are assigned to different species because each has skeletal characters that are consistently distinctive from one another (the shapes and proportions of their vertebrae) and are known only in that species. These are called **autopomorphies** (*auto*, "self"). It's likely that both *Diplodocus* and *Barosaurus* either split off or **speciated** separately close to the same time from the *Kaatedocus* (or *Kaatodocus*-like) lineage, or that this sister group stayed together but then *later* split or speciated again into *Diplodocus* and *Barosaurus*. By constructing a cladogram, we can attempt to learn at what point two separate, recognized species shared a common ancestor, and at what point one of them ceased to share a given character (or characters) with the other. In this case *Kaatedocus* and *Apatosaurus* are successive outgroups to a clade containing *Diplodocus, Barosaurus* and a third sauropod, the African *Tornieria*. These sauropods and their relatives, in turn, form the family Diplodocidae, which is united with two other families, the Dicraeosauridae and the Rebbachisauridae, within the superfamily Diplodocoidea. All these groups, in spite of their many derived skull and other characters, share enough synapomorphies to be placed together in progressively larger clades. When applying this method to sauropod phylogenetics, the large number of rapidly growing known skeletal characters can be a serious challenge in evaluating these, but fortunately computers have now made it possible to process the large quantities of character data.

But even with the aid of computers, sometimes the sheer number of known characters can confuse cladistic studies. Some workers place very different interpretations on the same sets of data because they view the importance, or *weight*, of certain characters differently than others, which can result in very different cladograms. Here a scientific principle called **parsimony** can help out. Parsimony is a methodological preference for the most simple, straightforward explanation, based on observable facts, that best explains or provides an answer to a problem. When we look at certain closely related fossil and modern animal species in such widely separated parts of the world as South America and Africa, for example, one could offer the hypothesis that these areas were once connected by long, now non-existent land bridges over which the animals crossed. On the other hand, it's more parsimonious to hypothesize that these continents were once united, and that the similarity of forms is due to the fact that they evolved in what was once a single location, which then separated. In the case of constructing cladograms, the simplest, shortest version (one requiring the fewest amount of **steps**, or branches) is preferred over the most complicated one. This doesn't mean that the short version is always right, but it's more likely than the more elaborate one to be correct.

VARIATIONS ON A THEME

The above description of phylogeny, characters and cladistics, although abstract, has special relevance for sauropods. These dinosaurs were globally distributed from the Late Triassic to the very end of the Cretaceous period, and during that time they branched off into numerous lineages and probably hundreds of genera and species. During this great evolutionary radiation, sauropods kept their most defining features—large bodies, straight, columnar legs, long necks and tails and small heads. In other ways, however, they innovated wildly: the form of the jaws and types of teeth, the size and placement of nostrils, relative length of the neck, limbs, and tail, width and depth of body, numbers and shapes of vertebrae, muscle attachments on the spines and limb girdles, as well as skin ornaments like spikes, armor plates and tail clubs. Not surprisingly, many sauropod lineages came up with the same evolutionary solutions independently. Mamenchisaurs, diplodocoids, brachiosaurs and titanosaurs mostly evolved extremely long necks, but some had very short ones. Likewise, some unrelated clades independently developed narrow teeth. Let's look at where sauropod evolution went, focusing on some of the currently most well known families and genera (see picture essay at end of chapter).

In Chapter 2 we left off with the earliest known sauropods like *Melanorosaurus* and *Antetonitrus*. Other "earlies" include *Vulcanodon* from South Africa, *Kotasaurus* from India, *Tazoudasaurus* from Morocco, *Spinophorosaurus* from Niger and the small and very fragmentary *Ohmdenosaurus* from Germany. These animals were radiating in the Late Triassic and Early Jurassic, and they rapidly achieved a nearly global distribution. They typify the basic sauropod *bauplan* mentioned above. Beyond that, they're pretty unspecialized, although *Spinophorosaurus* is unusual in having a longer neck and paired spikes that might have gone to the end of its tail. Although known fossils shows that these were rather "small" for sauropods—roughly elephant sized—there are fragmentary remains of much bigger ones. One humerus from Thailand is a meter long, and a femur from the Early Jurassic of Morocco is more than 2 m (6′) long, indicating an animal the size of *Brachiosaurus* but living just after the dawn of sauropod history.

The next group of later and more derived sauropods is known as the ***Eusauropoda*** ("true" sauropods). Basal forms of these are *Shunosaurus lii* from China, *Barapasaurus tagorei* from India, *Cetiosaurus oxfordiensis* from England and *Patagosaurus fariasi* from Argentina. As with the "earlies," these were apparently worldwide in distribution. Most of them are from the Middle Jurassic, which is generally poor in fossil-bearing rocks, and as a result sauropod and other dinosaur evolution is not well known. This is unfortunate, because most of the later sauropod groups seem to have their origins at this point, so better finds from this time could be very important for understanding how and why sauropods went their separate ways. Eusauropods are characterized by having fairly large bony nares, or nostrils, recessed from the front of the muzzle, spoon-shaped teeth, large areas for muscle attachment on the pelvis, metacarpals (forefoot bones) that formed a shallow crescent shape and short toes with big claws. Eusauropoda includes three relatively *disparate*, or diverse, clades: the mostly Chinese, mostly long-necked **Mamenchisauridae** (a group in which some workers include **omeisaurids**), the **cetiosaurs** (not recognized by some workers) and the European **Turiasauria**.

Cetiosaurs, among the earliest sauropods to be historically described (see Chapter 1), are not recognized by some workers as members of a true, monophyletic family or clade and are sometimes considered a "catchall" or "ragbag" group to which it's convenient to assign poorly diagnosed eusauropods. In spite of this, they are probably the most basal of the eusauropods, and in addition to the above-mentioned characters, they show a final transition to other features that complete the basic external *bauplan* of sauropods: shorter, deeper, slab-sided torsos, longer upper and shorter lower limbs and a bigger pelvis for more muscle attachment.

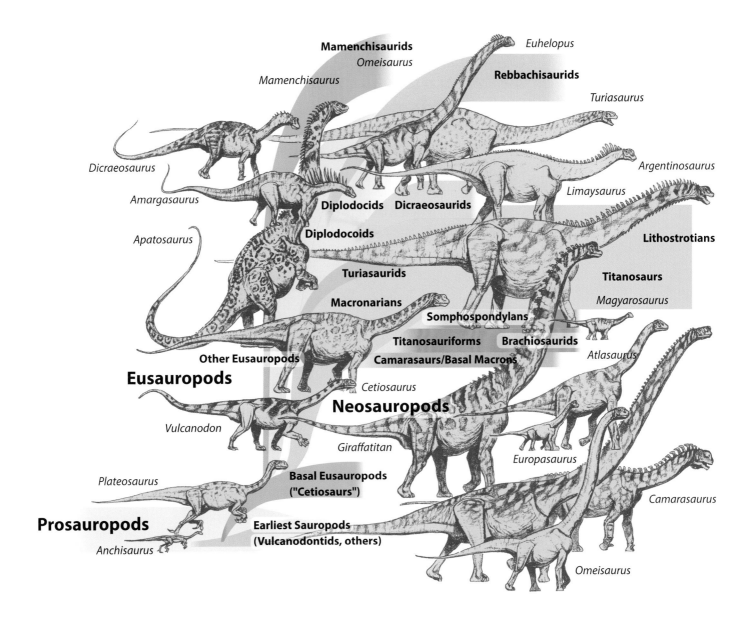

Mamenchisaurids
Omeisaurus
Mamenchisaurus

Euhelopus

Rebbachisaurids

Turiasaurus

Dicraeosaurus

Argentinosaurus

Amargasaurus

Diplodocids **Dicraeosaurids**

Limaysaurus

Apatosaurus

Diplodocoids

Lithostrotians

Turiasaurids

Titanosaurs

Macronarians

Magyarosaurus

Somphospondylans

Titanosauriforms **Brachiosaurids**

Atlasaurus

Other Eusauropods

Camarasaurs/Basal Macrons

Eusauropods

Cetiosaurus

Neosauropods

Vulcanodon

Giraffatitan

Europasaurus

Plateosaurus

**Basal Eusauropods
("Cetiosaurs")**

Camarasaurus

Prosauropods

**Earliest Sauropods
(Vulcanodontids, others)**

Anchisaurus

Omeisaurus

Sauropod Family Tree

Sauropod phylogeny as is currently known. During their geologic history, sauropods radiated into a remarkable array of huge-bodied, long-necked animals generally adapted to reach abundant, high-level browse. Some, however, actually reversed this trend to develop short necks for low-level browsing, while others became cow- and pony-sized dwarves in order to adapt to the limited resources of the islands on which they lived.

	Triassic				Jurassic											Cretaceous											
	Middle		Late			Early				Middle			Late					Early						Late			
	Anisian	Ladinian	Carnian	Norian	Rhaetian	Hettangian	Sinemurian	Pliensbachian	Toarcian	Aalenian	Bajocian	Bathonian	Callovian	Oxfordian	Kimmeridgian	Tithonian	Berriasian	Valanginian	Hauterivian	Barremian	Aptian	Albian	Cenomanian	Turonian	Coniacian Santonian	Campanian	Maastrichtian
	245	237	230	217	204	200	197	190	183	176	172	168	165	161	156	151	145	140	134	130	125	112	100	94	89 86 84	71	65.5

Mamenchisauridae, typified by fantastically long and long-necked forms like *Mamenchisaurus hochuanensis* and the even more extreme *M. sinocanadorum*, also includes recently described genera like *Eomamenchisaurus*, *Hudiesaurus*, *Tonganosaurus* and *Yuanmousaurus*, which have expanded this clade and deepened its fossil record. The small- to medium-sized **omeisaurids**, such as *Omeisaurus tianfunensis*, are considered by some workers as a separate clade and are distinctive from mamenchisaurs. These also have proportionately long necks, broad muzzles as opposed to the mamenchisaurs' narrow ones and forelimbs that are longer than the hindlimbs, paralleling the later brachiosaurs from Late Jurassic Africa, North America and possibly Europe. Mamenchisaurs, once thought to be diplodocoids, appear to have evolved in isolation in Proto-Asia, which then included much of China, Mongolia and Siberia, but there is a possibility that there were other forms that existed outside this area. Some omeisaurids may have had tiny clubs at the end of their tails.

Turiasauria is a newly recognized group of eusauropods from the Late Jurassic and very Early Cretaceous of Europe, and although known species are anatomically similar and from roughly the same geographic area, not all workers are convinced that this is a monophyletic clade. The known members are *Turiasaurus riodevensis*, *Galveosaurus herreroi*, and *Losillasaurus giganteus*, but teeth and other fragmentary remains from England, France and Portugal may also represent turiasaurs. The best known is currently *Turiasaurus* itself, a giant that almost reached the size of *Brachiosaurus* (23 m [>75′]). Along with an immense femur from Morocco, this shows that enormous size in sauropods wasn't restricted to other eusauropods and neosauropods (see next paragraph). Turiasaurs have spoon-shaped teeth that are pointed at the apices, suggesting a relationship with the earliest-described sauropod taxon *Cardiodon oxoniensis* (see Chapter 1). Though huge, they had relatively short necks, about 4–5 m (13′–16′) long compared to the longer necks (7 m [22′]) of the far less bulky *Diplodocus*.

Neosauropoda ("new" sauropods) is a very large clade that breaks down into two main branches, the ***Diplodocoidea***, including diplodocoids and their relatives, and the ***Macronaria***, which is made up of the **basal macronaria** (basal macronarians), ***Camarasauromorphs***, ***Titano-***

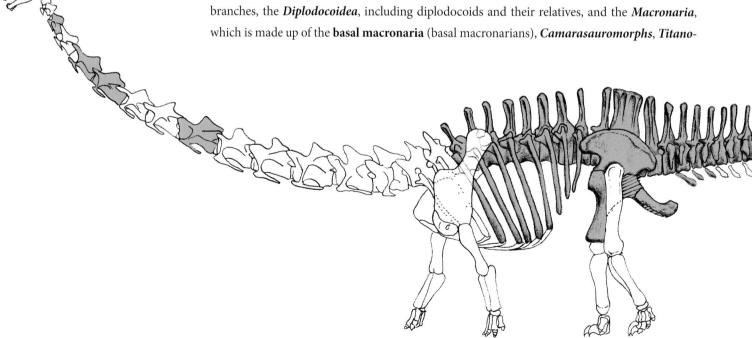

sauriforms (derived macronarians, including brachiosaurids) and ***Somphospondyli***, the last containing the ***euhelopodids*** and actual ***titanosaurs***. In life early neosauropods wouldn't have been noticeably different on the outside from most eusauropods, but their skeletons have subtle, important changes. One is a tendency to have larger and more complex air spaces inside their vertebrae (evolved independently in mamenchisaurs), pelvises that turned outward to support larger guts and, most importantly, metacarpal bones that were bent into a tight, horseshoe-shaped arc. This tubular shape of the forefeet made them into more solid supports for weight bearing than in the eusauropods.

Diplodocoidea is the branch of neosauropods that perhaps made the most radical departures in head morphology, neck anatomy and feeding techniques. The *nares* are retracted far back onto the top of the skull, sometimes as far as the *orbits*, or eye sockets, while the teeth, restricted to the front of the wide, somewhat wedge-shaped jaws, often become narrow and pencil-like for stripping off leaves in one clade, the **Diplodocidae**. In this subgroup the necks are sometimes extremely long, while two others, the **Dicraeosauridae** and **Rebbachisauridae**, feature some highly derived forms that have the shortest necks of all sauropods, an adaptation for low-to-the-ground feeding. As a further adaptation to intensive low browsing, the teeth in one short-necked, rebbachisaurid form have become tiny, closely packed "batteries." Diplodocoids have short cervical ribs, which may have improved neck flexibility, and shortened forelimbs, which lightened the anterior part of the body to aid bipedal rearing for arboreal browsing in some species. Finally, all diplodocoids have very long tails, which in "flagellicaudatan" diplodocids are developed to the most extreme, ending in thin "whips" possibly for defense and maintaining individual space. Some diplodocids (like the unrelated mamenchisaurids) also possessed double chevron bones underneath the mid-caudal vertebrae, which gave extra support while tripodally resting on their tails. The *diplodocids*, whose fossils are found in North America, Africa and Europe, contain some of the most famous and iconic sauropods, such as *Brontosaurus*, *Diplodocus* and *Apatosaurus*, and reached huge sizes. Their bifurcated neural spines helped give lateral stability while feeding vertically.

Dicraeosaurids are generally smallish, short necked (for sauropods) and currently found in the Late Jurassic to Early Cretaceous of Africa, South America and North America. Sharing with the diplodocids bifurcated neural spines, in the dicraeosaurids these were often extremely tall and deeply split, and in one case there evolved a bizarre, spike-necked species, *Amargasaurus cazui*. Another lineage resulted in *Brachytrachelopan mesai* (see Chapters 4 and 10), a very small, low-browsing animal with the shortest known neck (1 m [3.5′]) of any sauropod. Except for the basal North American form *Suuwassea emeliae*, dicraeosaurids have reduced spaces in

Seismosaurus halli was first described as a new genus and species because some characters of its known skeleton (*color*) were at first considered distinctive from those of other diplodocids. Now known to probably be an exceptionally large, old *Diplodocus*, it's a lesson to paleontologists: avoid the temptation to describe a new fossil species from only one specimen when osteological variability of the species isn't well known.

their vertebrae, typically a juvenile character in sauropods. The combination of small size and reduced vertebral pneumaticity suggests that the smaller, more derived dicraeosaurids evolved by paedomorphosis (see Chapter 2), an evolutionary process in which adults of descendent species resemble juveniles of their ancestors—in this case an ancestor much like *Suuwassea*.

Rebbachisaurids, like the turiasaurs, are a clade still relatively new to paleontologists, and they are divided into the **rebbachisaurines** and **limaysaurines**. Currently known only from the Early and Middle Cretaceous of South America, North Africa and Europe, they had single neural spines, unlike the bifurcated neural spines of diplodocids and dicraeosaurids. The limaysaurine group features some medium-sized species like *Limaysaurus tessonei,* while the rebbachisaurines are known from such forms as *Rebbachisaurus garasbae* and a very small, cow-sized form, *Nigersaurus taqueti*. Its short neck and skull, with bizarrely wide, flaring jaws (containing up to 600 minute teeth that together functioned as a single surface for nibbling short plants), was adapted for close-to-the-ground browsing. Some limaysaurine rebbachisaurids also show some unusual articulations in their dorsal or trunk vertebrae and their limb girdles, suggesting that they had evolved innovations in walking over rough, possibly mountainous terrain.

Macronaria is the other great clade that, along with the Diplodocoidea, makes up the Neosauropoda. It includes the tall, iconic **brachiosaurids**, the aptly named and often bizarre **titanosaurs** that were so successful during the Cretaceous period and a host of other interesting sauropods that don't fit into either of these groups. *Macronaria* means "big nostrils," and in general the very large, bony nares of earlier eusauropod taxa like *Mamenchisaurus* were retained in macronarians, sometimes getting even bigger. *Jobaria tiguidensis* from Morocco (considered by some workers to be actually a eusauropod) was a possible **basal macronarian** of average shape and size, with stout cropping teeth that could feed on most arboreal browse. A **camarasauromorph**, the well-known *Camarasaurus lentus* of western North America, takes the basic plan further, with the beginning of tall shoulders and bifurcated neural spines that helped it to stabilize better in high feeding and a deep, powerful skull and teeth that could wrench off coarse vegetation.

Brachiosaurids like the North American *Brachiosaurus altithorax* took the adaptation of high browsing much further, with tall shoulders, long forelimbs, a very long neck and generally enormous size. Some, however, such as *Europasaurus holgeri* from Germany, were island-bound dwarfs, among the smallest known sauropods (see Chapter 10). The giant species formerly known as *B. brancai* found in East Africa, for years thought to be longer bodied but essentially the same as the North American *Brachiosaurus*, was shown by paleontologist-artist Gregory S. Paul and later Michael P. Taylor in 2009 to be a separate genus, *Giraffatitan*, on the basis of a careful bone-by-bone study. Other brachiosaurids include *Lusotitan atalaiensis* from Portugal and *Abydosaurus mcintoshi*, *Cedarosaurus weiskopfae* and *Venenosaurus dicrocei* from North America. Brachiosaurids, known from Late Jurassic and Early Cretaceous, evolved their long necks not by adding more vertebrae, as some other sauropod clades did, but by increasing vertebral length. With its long neck and immense frame, *Giraffatitan*'s 13.27 m (43.5′) tall skeleton at the Museum für Naturkunde in Berlin is the largest skeleton mount in the world, and it is even more awesome when we learn that this was actually a subadult—larger specimens show that it could grow at least 13% bigger.

Somphospondyli is a newly named category of sauropods and includes the most basal titanosaurs and all the taxa closer to these than brachiosaurs. One of the latter is a poorly known group called **euhelopodids,** once thought to be basal eusauropods, typified by the Chinese *Euhelopus zdanskyi.* This and other members of this clade show an odd mix of characters: a skull superficially like that of *Camarasaurus,* tall, brachiosaurid-like shoulders, long forelimbs and a very long neck compared to the rest of the body. Unlike the 13 cervical bones of brachiosaurs, however, *Euhelopus* has 17 cervicals, and the individual bones are shorter. Despite its proportionately long neck, however, this sauropod was relatively small, perhaps only 10.5 m (35′) and weighing less than 5 tons. In contrast to small *Euhelopus* is the Oklahoma mystery giant *Sauroposeidon proteles,* originally known only from cervical vertebrae. These are enormous: the largest is 125 cm (4.1′), second only to a cervical of the diplodocid *Supersaurus* as the longest vertebra discovered for any animal; if there had been buildings around in the Early Cretaceous, an adult *Sauroposeidon* could have peered through a six-story window. One lineage, represented by *Atlasaurus imelekai,* is a twist on the usual sauropod adaptation of having a long neck for high browsing, instead having a moderate neck length but extremely long limbs, proportionately longer than any other known sauropod's. Other basal somphospondyls are known from Argentina (*Tehuelchesaurus benitezii*) and Mongolia (*Erketu ellisoni*), showing that these were quite broadly distributed across the globe during the Early Cretaceous.

Titanosauria was the last clade of sauropods to flourish. Although they were around by the Mid-Jurassic, titanosaurs didn't become abundant until the Early Cretaceous, and they were still going strong up to the very end of the Mesozoic, when all dinosaurs abruptly went extinct. Titanosauria was a very diverse clade encompassing everything from pony-sized *Magyarosaurus* to truly titanic animals like *Argentinosaurus huinculensis, Puertasaurus roulli* and *Futalognkosaurus dukei.* These may have outweighed the enormous *Supersaurus* and *Giraffatitan* by a factor of two and were probably the heaviest animals ever to walk on land. Titanosaurs also came in a variety of shapes, from long-necked *Rapetosaurus krausei* to the bizarrely short-necked *Isisaurus colberti,* and from the tall and almost brachiosaur-like *Alamosaurus sanjuanensis* to the wide, squat and chunky *Opisthocoelicaudia skarzynskii.* Some, like *Malawisaurus dixeyi,* had basal-looking skulls with big nasal openings, whereas *Nemegtosaurus mongoliensis* and its kin had skulls like those of diplodocids and small nostrils tucked up between their eyes. Titanosaurs are unique among sauropods in that many, but not all, bore armor, in the form of bony plates called *osteoderms* embedded in their skins.

Altogether, the sauropods were an amazingly diverse order, as variable in their sizes, shapes and probable behaviors as any other major group of dinosaurs. The following pages, with skeletal illustrations based on original reconstructions by paleontologist-artists Gregory S. Paul, Scott Hartman and coauthor Mark Hallett, show only a fraction of this incredible diversity but will help acquaint the reader with the main clades and their skeletal morphologies.

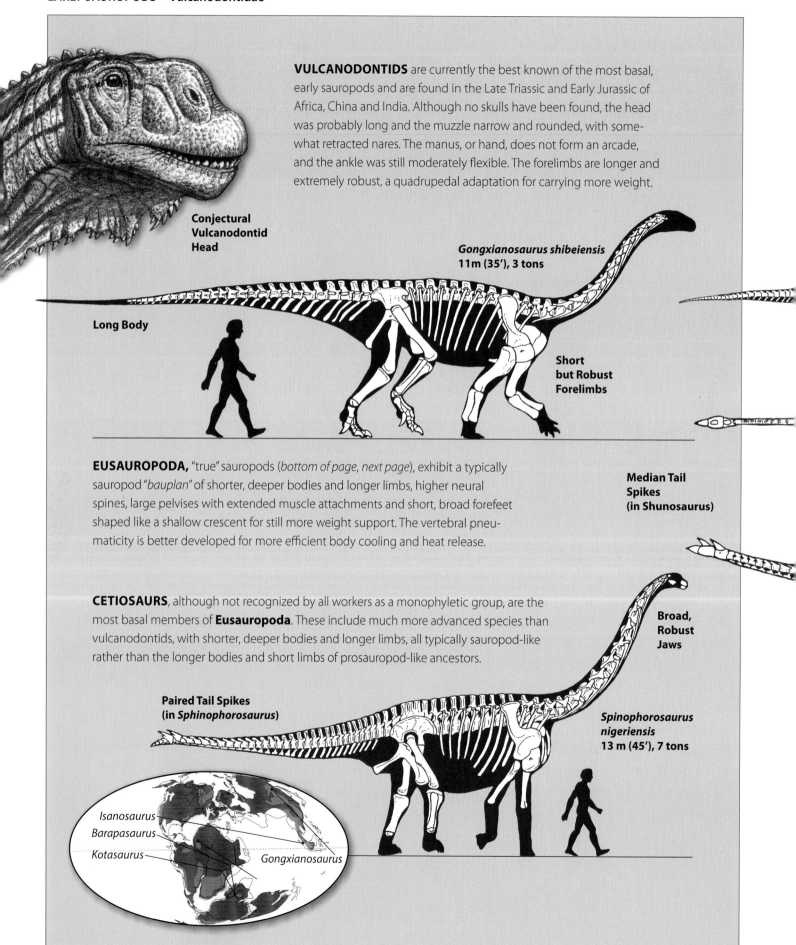

Conjectural Vulcanodontid Head

VULCANODONTIDS are currently the best known of the most basal, early sauropods and are found in the Late Triassic and Early Jurassic of Africa, China and India. Although no skulls have been found, the head was probably long and the muzzle narrow and rounded, with somewhat retracted nares. The manus, or hand, does not form an arcade, and the ankle was still moderately flexible. The forelimbs are longer and extremely robust, a quadrupedal adaptation for carrying more weight.

Gongxianosaurus shibeiensis
11m (35'), 3 tons

Long Body

Short but Robust Forelimbs

EUSAUROPODA, "true" sauropods (*bottom of page, next page*), exhibit a typically sauropod "*bauplan*" of shorter, deeper bodies and longer limbs, higher neural spines, large pelvises with extended muscle attachments and short, broad forefeet shaped like a shallow crescent for still more weight support. The vertebral pneumaticity is better developed for more efficient body cooling and heat release.

Median Tail Spikes (in Shunosaurus)

CETIOSAURS, although not recognized by all workers as a monophyletic group, are the most basal members of **Eusauropoda**. These include much more advanced species than vulcanodontids, with shorter, deeper bodies and longer limbs, all typically sauropod-like rather than the longer bodies and short limbs of prosauropod-like ancestors.

Broad, Robust Jaws

Paired Tail Spikes (in *Sphinophorosaurus*)

Spinophorosaurus nigeriensis
13 m (45'), 7 tons

Isanosaurus
Barapasaurus
Kotasaurus
Gongxianosaurus

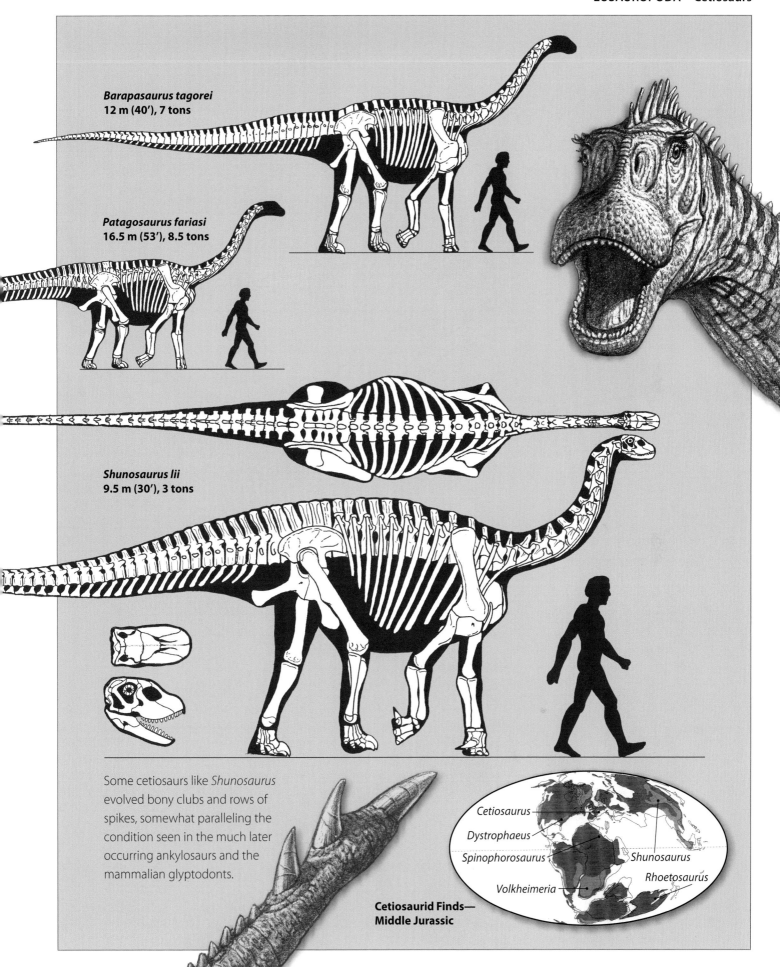

Barapasaurus tagorei
12 m (40′), 7 tons

Patagosaurus fariasi
16.5 m (53′), 8.5 tons

Shunosaurus lii
9.5 m (30′), 3 tons

Some cetiosaurs like *Shunosaurus* evolved bony clubs and rows of spikes, somewhat paralleling the condition seen in the much later occurring ankylosaurs and the mammalian glyptodonts.

Cetiosaurus

Dystrophaeus

Spinophorosaurus

Shunosaurus

Volkheimeria

Rhoetosaurús

**Cetiosaurid Finds—
Middle Jurassic**

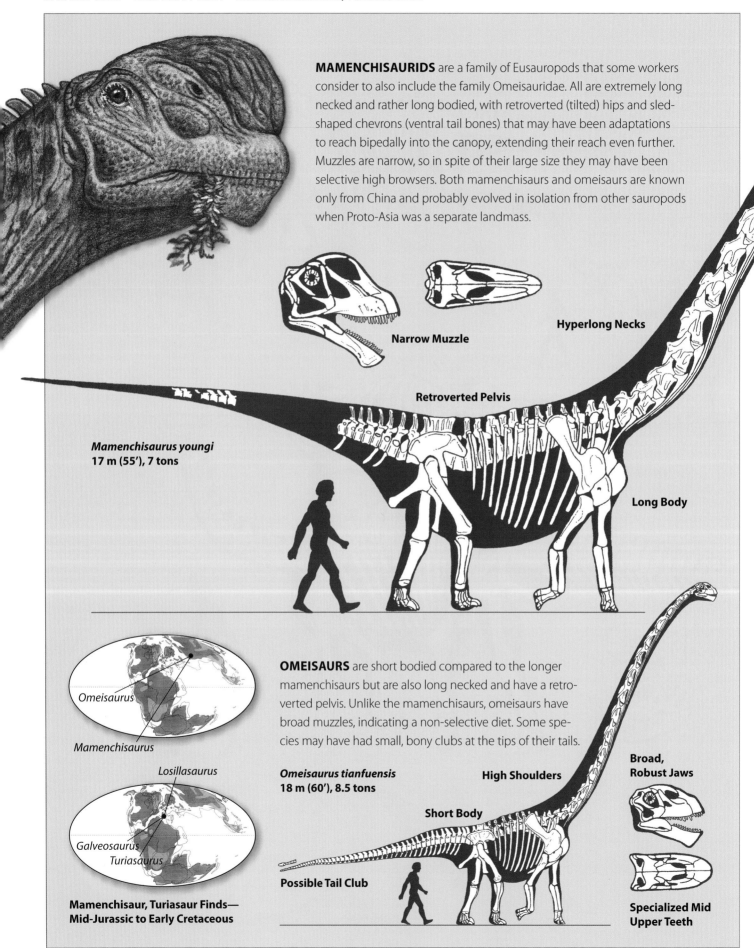

MAMENCHISAURIDS are a family of Eusauropods that some workers consider to also include the family Omeisauridae. All are extremely long necked and rather long bodied, with retroverted (tilted) hips and sled-shaped chevrons (ventral tail bones) that may have been adaptations to reach bipedally into the canopy, extending their reach even further. Muzzles are narrow, so in spite of their large size they may have been selective high browsers. Both mamenchisaurs and omeisaurs are known only from China and probably evolved in isolation from other sauropods when Proto-Asia was a separate landmass.

Narrow Muzzle

Hyperlong Necks

Retroverted Pelvis

Mamenchisaurus youngi
17 m (55'), 7 tons

Long Body

OMEISAURS are short bodied compared to the longer mamenchisaurs but are also long necked and have a retroverted pelvis. Unlike the mamenchisaurs, omeisaurs have broad muzzles, indicating a non-selective diet. Some species may have had small, bony clubs at the tips of their tails.

Omeisaurus tianfuensis
18 m (60'), 8.5 tons

High Shoulders

Broad, Robust Jaws

Short Body

Possible Tail Club

Specialized Mid Upper Teeth

Omeisaurus

Mamenchisaurus

Losillasaurus

Galveosaurus
Turiasaurus

Mamenchisaur, Turiasaur Finds—Mid-Jurassic to Early Cretaceous

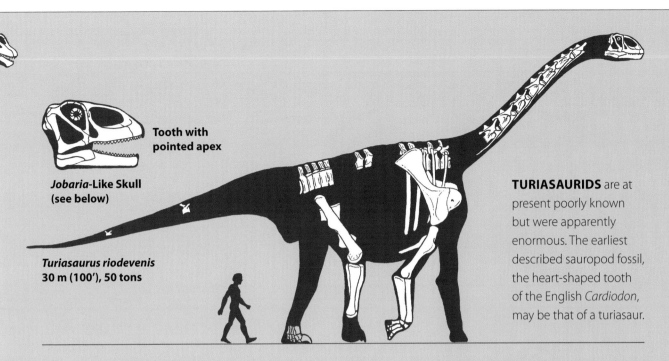

Tooth with pointed apex

Jobaria-Like Skull (see below)

Turiasaurus riodevenis
30 m (100'), 50 tons

TURIASAURIDS are at present poorly known but were apparently enormous. The earliest described sauropod fossil, the heart-shaped tooth of the English *Cardiodon*, may be that of a turiasaur.

NEOSAUROPODA ("new sauropods") is the major sauropod group that comprises two major clades, the *Diplodocoidea* and the *Macronaria*, both of which by the Mid-Jurassic produced highly diverse, often extremely enormous but sometimes dwarfed species that adapted to a variety of feeding niches.

Major characters are:
· Larger and more complex air spaces in vertebrae
· Illium bones of pelvis bend outward to better support abdomen
· Bones of forefeet form tight, horseshoe-shaped arch
· Bifurcated (split) neural spines commonly evolve in some clades

Jobaria tiguidensis
16 m (52'), 16 tons

Basal neosauropods have necks of moderate length, relatively short and robust skulls with spoon-shaped teeth and somewhat broad pelvises.

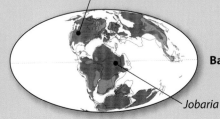

Abrosaurus

Basal Neosauropod Finds—Mid-Jurassic

Jobaria

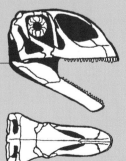

Skull of *Jobaria*

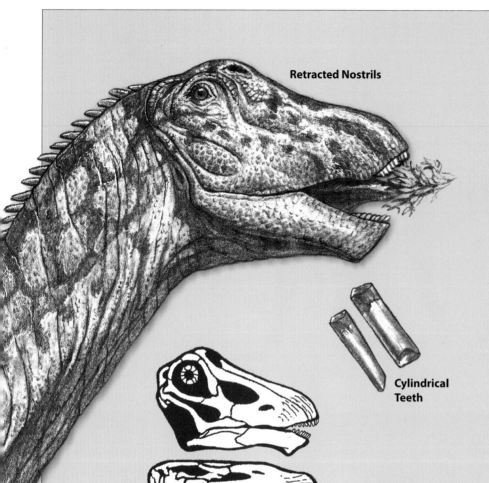

Retracted Nostrils

Cylindrical Teeth

Broad, Wedge-Shaped Muzzle

DIPLODOCOIDEA is the branch of Neosauropoda that includes all sauropods more closely related to *Diplodocus* than *Saltasaurus*, and all share obvious features.

Typical features are:
- Nostrils retracted up between (and even behind) the orbits
- Teeth often cylindrical (but not always)
- Pencil-like teeth restricted to anterior jaws
- Forelimbs shorter (up to 2/3) than the hindlimbs—helped rearing
- Long whiplash tails
- Cervical vertebrae with short ribs, may have improved neck flexibility

DIPLODOCIDS, like *Diplodocus*, used a leaf-stripping technique to feed and were highly specialized as bipedal or tripodal browsers. Usually enormous, with moderately long necks and long, wide heads, some were among the largest known sauropods.

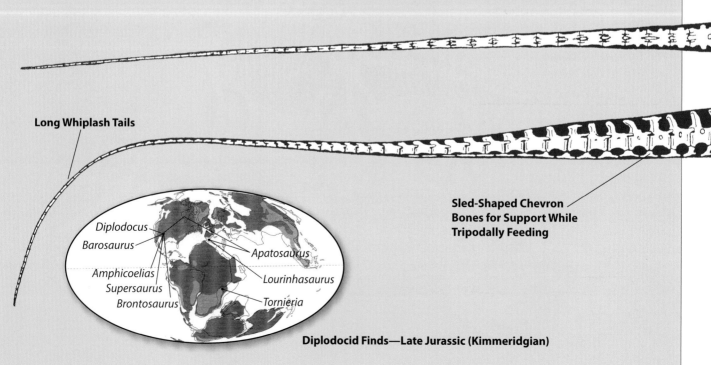

Long Whiplash Tails

Sled-Shaped Chevron Bones for Support While Tripodally Feeding

Diplodocus
Barosaurus
Amphicoelias
Supersaurus
Brontosaurus
Apatosaurus
Lourinhasaurus
Tornieria

Diplodocid Finds—Late Jurassic (Kimmeridgian)

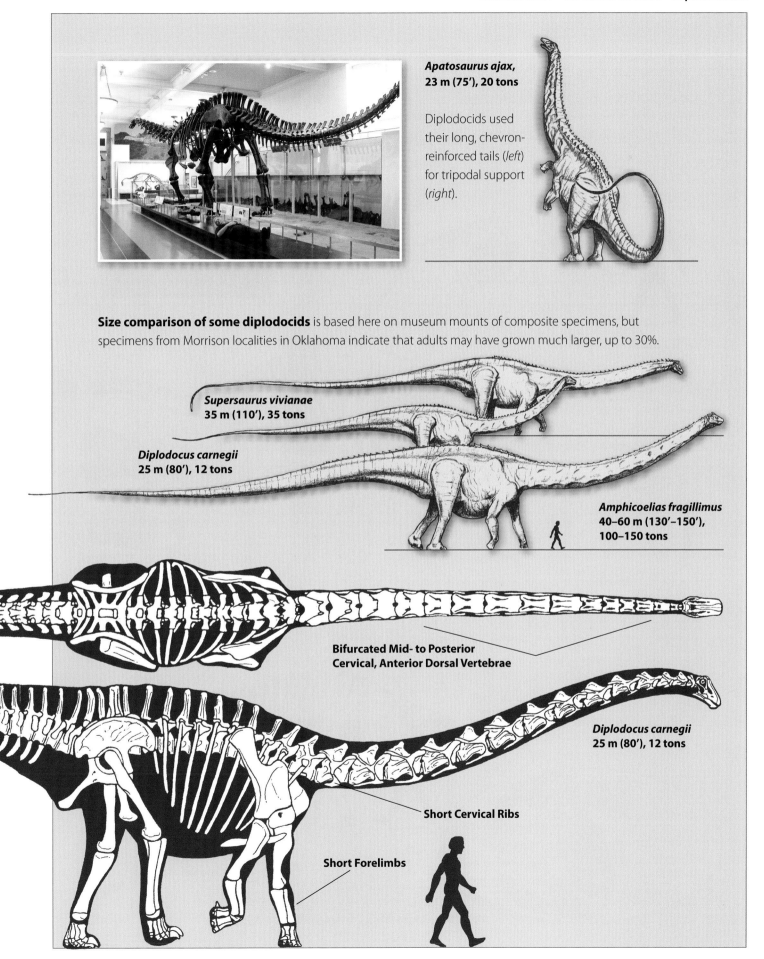

Apatosaurus ajax,
23 m (75′), 20 tons

Diplodocids used their long, chevron-reinforced tails (*left*) for tripodal support (*right*).

Size comparison of some diplodocids is based here on museum mounts of composite specimens, but specimens from Morrison localities in Oklahoma indicate that adults may have grown much larger, up to 30%.

Supersaurus vivianae
35 m (110′), 35 tons

Diplodocus carnegii
25 m (80′), 12 tons

Amphicoelias fragillimus
40–60 m (130′–150′),
100–150 tons

Bifurcated Mid- to Posterior Cervical, Anterior Dorsal Vertebrae

Diplodocus carnegii
25 m (80′), 12 tons

Short Cervical Ribs

Short Forelimbs

REBBACHISAURIDS were rather small bodied for sauropods, and known forms had unbifurcated neural spines, upturned transverse processes and very lightly constructed, short skulls. While **limaysaurines'** taller neural spines suggest that they may have been mid-canopy feeders, the **nigersaurines'** short neural spines and bizarrely wide skulls suggest that they were probably low browsers. One species, *Nigersaurus taqueti* (*right and below*), had vacuum cleaner–like jaws holding up to 600 tiny, closely packed teeth.

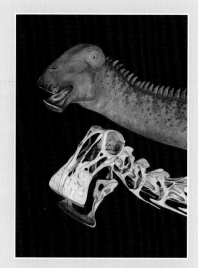

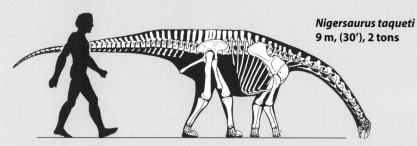

Nigersaurus taqueti
9 m, (30'), 2 tons

Limaysaurus tessonei
15 m (50'), 7 tons

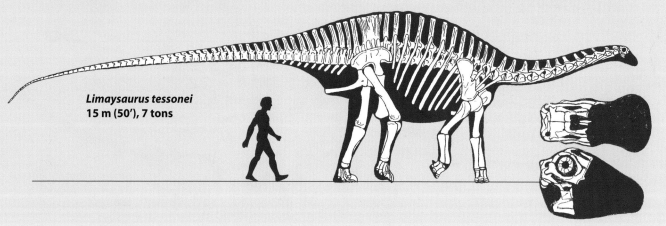

DICRAEOSAURIDS were also generally small for sauropods, with somewhat short necks, reduced air spaces in their vertebrae and deeply divided neural spines in their presacral vertebrae. In *Amargasaurus cazui* the neck spines were elongated into spikes, possibly for defense. Another, *Brachytrachelopan*, had the shortest known neck of any sauropod and may have, like *Nigersaurus*, been specialized for low browsing.

Amargasaurus cazui
13 m (43'), 4 tons

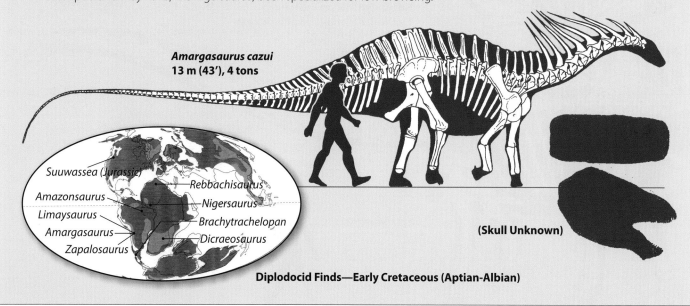

Suuwassea (Jurassic)
Amazonsaurus
Limaysaurus
Amargasaurus
Zapalosaurus
Rebbachisaurus
Nigersaurus
Brachytrachelopan
Dicraeosaurus

(Skull Unknown)

Diplodocid Finds—Early Cretaceous (Aptian-Albian)

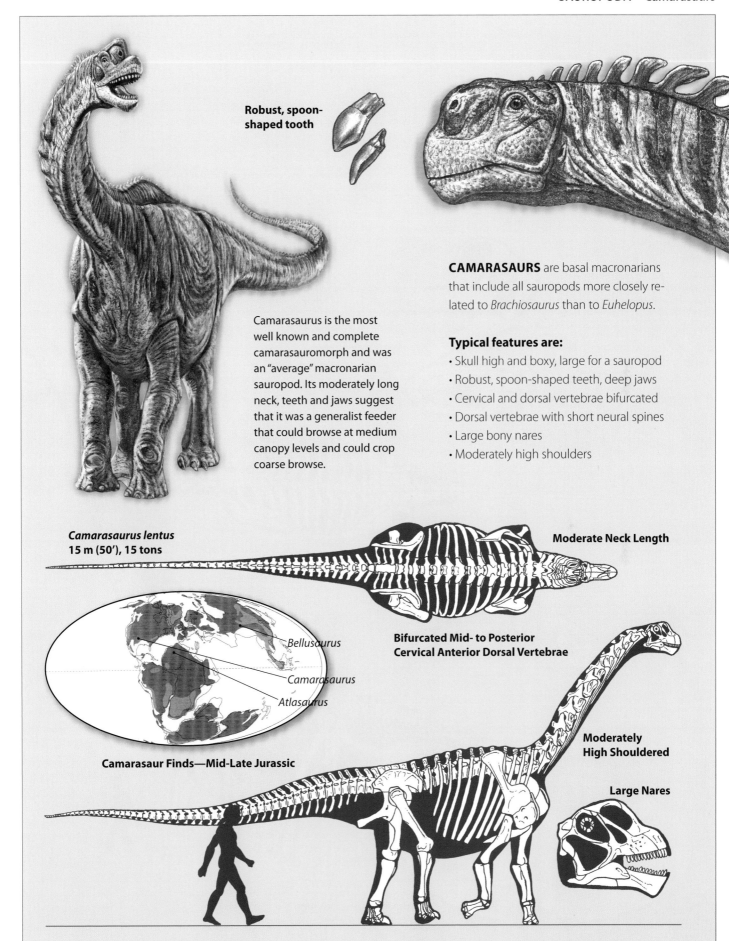

Robust, spoon-shaped tooth

CAMARASAURS are basal macronarians that include all sauropods more closely related to *Brachiosaurus* than to *Euhelopus*.

Typical features are:
- Skull high and boxy, large for a sauropod
- Robust, spoon-shaped teeth, deep jaws
- Cervical and dorsal vertebrae bifurcated
- Dorsal vertebrae with short neural spines
- Large bony nares
- Moderately high shoulders

Camarasaurus is the most well known and complete camarasauromorph and was an "average" macronarian sauropod. Its moderately long neck, teeth and jaws suggest that it was a generalist feeder that could browse at medium canopy levels and could crop coarse browse.

Camarasaurus lentus
15 m (50'), 15 tons

Moderate Neck Length

Bifurcated Mid- to Posterior Cervical Anterior Dorsal Vertebrae

Bellusaurus

Camarasaurus

Atlasaurus

Camarasaur Finds—Mid-Late Jurassic

Moderately High Shouldered

Large Nares

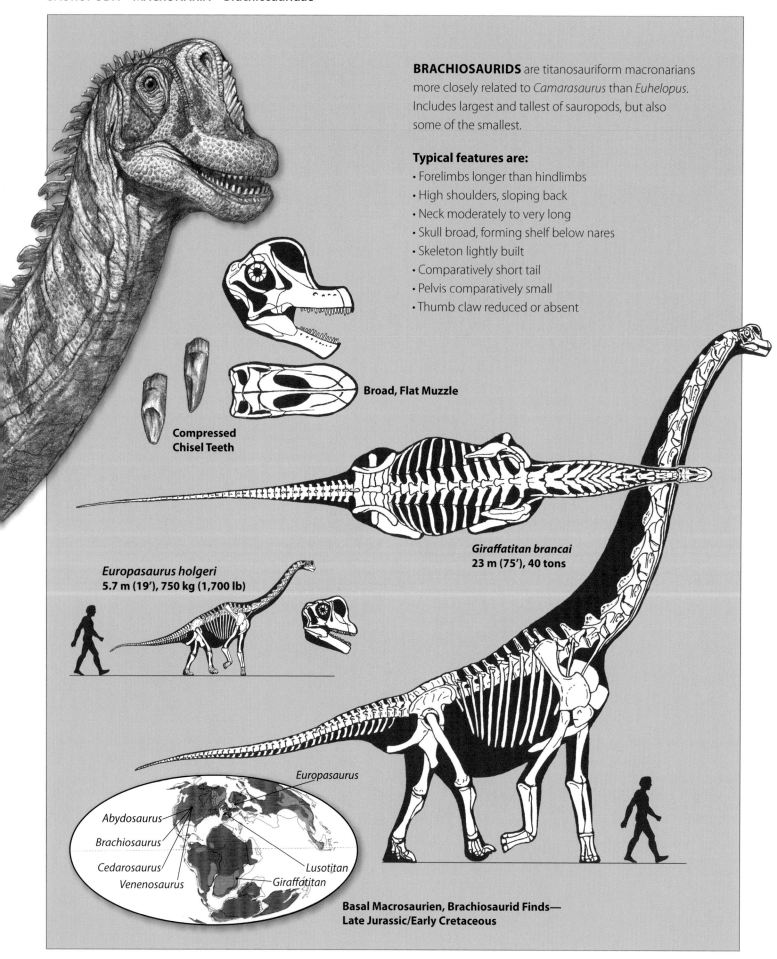

BRACHIOSAURIDS are titanosauriform macronarians more closely related to *Camarasaurus* than *Euhelopus*. Includes largest and tallest of sauropods, but also some of the smallest.

Typical features are:
- Forelimbs longer than hindlimbs
- High shoulders, sloping back
- Neck moderately to very long
- Skull broad, forming shelf below nares
- Skeleton lightly built
- Comparatively short tail
- Pelvis comparatively small
- Thumb claw reduced or absent

Broad, Flat Muzzle

Compressed Chisel Teeth

Giraffatitan brancai
23 m (75'), 40 tons

Europasaurus holgeri
5.7 m (19'), 750 kg (1,700 lb)

Europasaurus
Abydosaurus
Brachiosaurus
Cedarosaurus
Venenosaurus
Lusotitan
Giraffatitan

**Basal Macrosaurien, Brachiosaurid Finds—
Late Jurassic/Early Cretaceous**

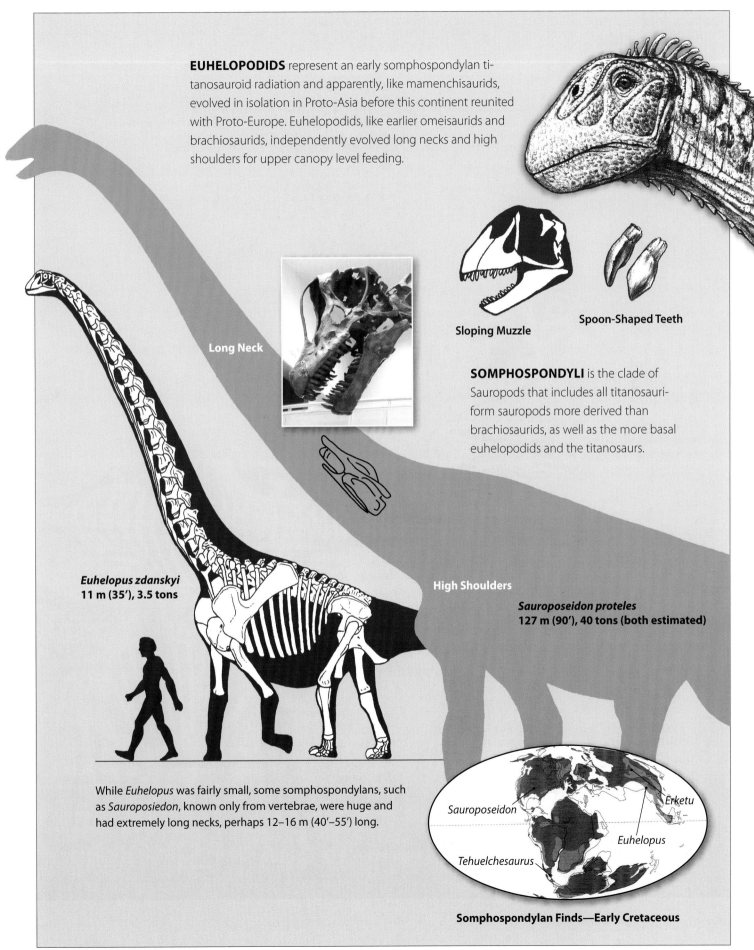

EUHELOPODIDS represent an early somphospondylan titanosauroid radiation and apparently, like mamenchisaurids, evolved in isolation in Proto-Asia before this continent reunited with Proto-Europe. Euhelopodids, like earlier omeisaurids and brachiosaurids, independently evolved long necks and high shoulders for upper canopy level feeding.

Long Neck

Sloping Muzzle

Spoon-Shaped Teeth

SOMPHOSPONDYLI is the clade of Sauropods that includes all titanosauriform sauropods more derived than brachiosaurids, as well as the more basal euhelopodids and the titanosaurs.

Euhelopus zdanskyi
11 m (35'), 3.5 tons

High Shoulders

Sauroposeidon proteles
127 m (90'), 40 tons (both estimated)

While *Euhelopus* was fairly small, some somphospondylans, such as *Sauroposiedon*, known only from vertebrae, were huge and had extremely long necks, perhaps 12–16 m (40'–55') long.

Sauroposeidon

Erketu

Euhelopus

Tehuelchesaurus

Somphospondylan Finds—Early Cretaceous

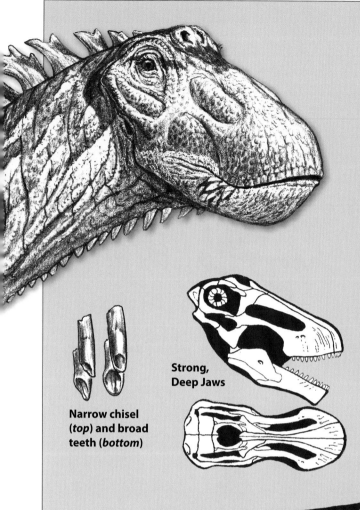

TITANOSAURS are a clade that arose from macronarians and were highly diverse in shape and size. Like brachiosaurids, they included not only some of the largest but also among the smallest known sauropods. Some known skulls resembled those of diplodocoids and may indicate an evolution toward similar feeding techniques after diplodocoids' extinction left this niche open.

Typical features are:
- Teeth often cylindrical but sometimes broad
- Diplodocoid-like skull in some known forms
- Trunk vertebrae moderately flexible
- Tail very flexible
- Osteoderms in some species
- Femoral head slants inward
- "Wide track" body stance
- Anterior vertebrae opisthocoelous (concave in front, convex in back)
- Manual phalange, thumb claw absent
- Extremely wide variation in length, width

Narrow chisel (*top*) **and broad teeth** (*bottom*)

Strong, Deep Jaws

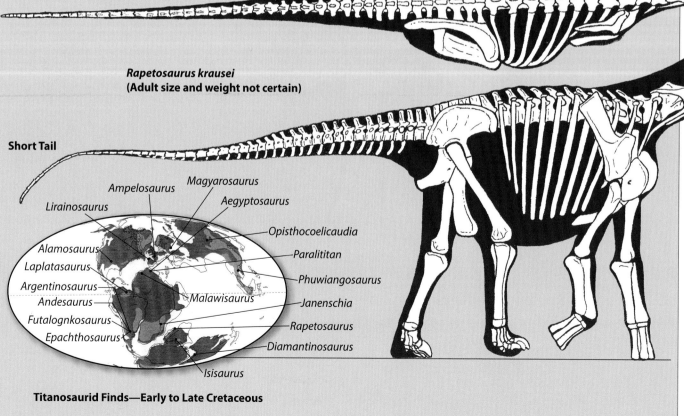

Rapetosaurus krausei
(Adult size and weight not certain)

Short Tail

Ampelosaurus
Magyarosaurus
Lirainosaurus
Aegyptosaurus
Alamosaurus
Opisthocoelicaudia
Laplatasaurus
Paralititan
Argentinosaurus
Phuwiangosaurus
Andesaurus
Malawisaurus
Janenschia
Futalognkosaurus
Rapetosaurus
Epachthosaurus
Diamantinosaurus
Isisaurus

Titanosaurid Finds—Early to Late Cretaceous

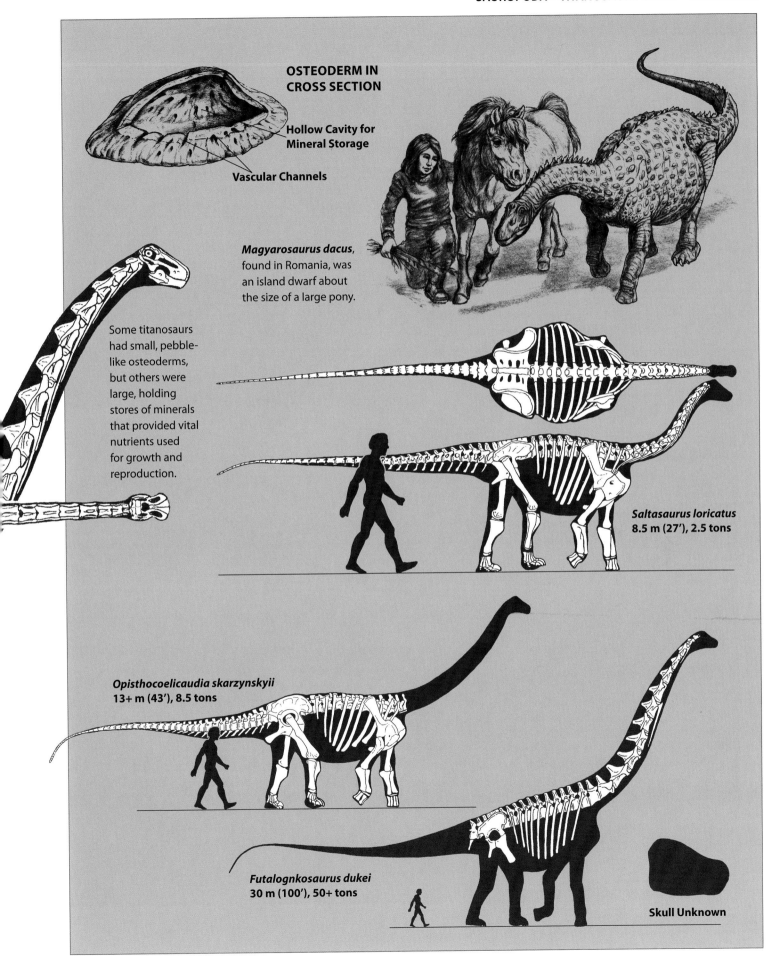

OSTEODERM IN CROSS SECTION

Hollow Cavity for Mineral Storage

Vascular Channels

Magyarosaurus dacus, found in Romania, was an island dwarf about the size of a large pony.

Some titanosaurs had small, pebble-like osteoderms, but others were large, holding stores of minerals that provided vital nutrients used for growth and reproduction.

Saltasaurus loricatus
8.5 m (27'), 2.5 tons

Opisthocoelicaudia skarzynskyii
13+ m (43'), 8.5 tons

Futalognkosaurus dukei
30 m (100'), 50+ tons

Skull Unknown

Mounted as they strode forward as in life, the massive hindlimb and pelvic bones of *Giraffatitan brancai* at the Museum für Naturkunde, Berlin, were once embedded in equally powerful muscles that were its main engines of propulsion.

CHAPTER FOUR

OF BONES AND BRIDGES

Even in our modern age of grand structural technology, the sight of a mounted sauropod skeleton is enough to make you gasp. It's hard to wrap your mind around any living thing having grown that big, but it did. The huge, bony framework of exquisite struts, spaces and columns is one of the most marvelously engineered of any organic form, and one leaves its presence haunted by what the animal was like in life. Assuming that the sauropods' gigantic size was the outcome of a need for support and movement on a huge scale to satisfy biological needs, how did this evolve?

A SKELETAL PRIMER. In discussing why the bodies of sauropods were the way they were and how they probably worked, it's helpful to first look at a sauropod skeleton and identify some of its characteristic features. Sauropods were such a diverse group that no one taxon is truly typical, but all shared certain key features that can be kept in mind as we explore the ways these dinosaurs changed through time. Anatomists start by dividing *tetrapod* (*tetra*, "four"; *podus*, "foot") skeletons into two parts: the ***axial skeleton*** (skull, vertebrae, rib cage) and ***appendicular skeleton*** (shoulders, hips, limbs and feet). The axial skeleton starts with the skull, which is composed of the ***cranium*** (braincase, facial bones and upper jaw) and the ***mandible,*** or lower jaw. The cranium is attached to a series of vertebrae, specialized as ***cervicals*** (neck), ***dorsals*** (chest and abdomen), ***sacrals*** (pelvis or hip) and ***caudals*** (tail). There are

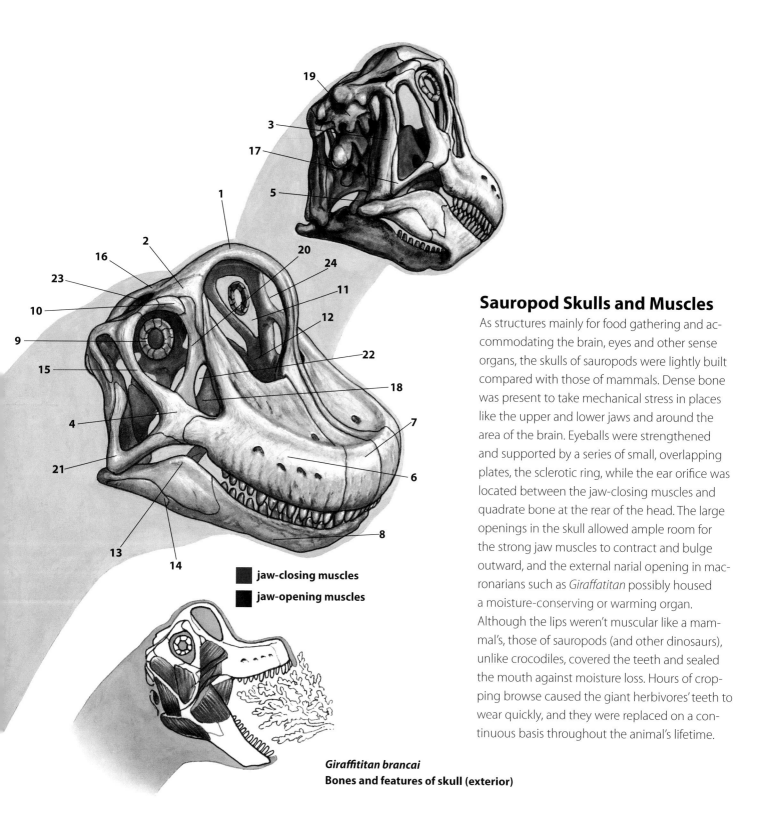

Sauropod Skulls and Muscles

As structures mainly for food gathering and accommodating the brain, eyes and other sense organs, the skulls of sauropods were lightly built compared with those of mammals. Dense bone was present to take mechanical stress in places like the upper and lower jaws and around the area of the brain. Eyeballs were strengthened and supported by a series of small, overlapping plates, the sclerotic ring, while the ear orifice was located between the jaw-closing muscles and quadrate bone at the rear of the head. The large openings in the skull allowed ample room for the strong jaw muscles to contract and bulge outward, and the external narial opening in macronarians such as *Giraffatitan* possibly housed a moisture-conserving or warming organ. Although the lips weren't muscular like a mammal's, those of sauropods (and other dinosaurs), unlike crocodiles, covered the teeth and sealed the mouth against moisture loss. Hours of cropping browse caused the giant herbivores' teeth to wear quickly, and they were replaced on a continuous basis throughout the animal's lifetime.

■ jaw-closing muscles
■ jaw-opening muscles

Giraffititan brancai
Bones and features of skull (exterior)

(1) nasal, (2) frontal, (3) quadratojugal, (4) jugal, (5) articular, (6) maxilla, (7) premaxilla, (8) dentary, (9) sclerotic ring, (10) prefrontal, (11) orbital fenestra, (12) external naris, (13) coronoid process, (14) angular, (15) postorbital, (16) parietal, (17) quadrate, (18) pterygoid, (19) basioccipital, (20) lacrimal, (21) surangular, (22) antorbital fenestra, (23) infratemporal fenestra, (24) narial fenestra

not only the expected long, curved and paired **dorsal ribs**, which correspond to the ribs in the thorax of a mammal, but also **cervical ribs** in the neck, as in birds. Whereas adult humans have a single breastbone, or sternum, sauropods were like most other non-avian dinosaurs in having paired **sternal plates**, which were connected to the dorsal ribs by **gastralia**, or ventral belly ribs. Together, the dorsal vertebrae, dorsal ribs, sternal plates and gastralia, as well as the two small paired **clavicles** (in humans, "collar bones"), formed a box of bone and cartilage around the heart and lungs and are the equivalent bones (except for the sternal plates, which are replaced by the sternum bone, and gastralia, which are instead cartilaginous) to those in our own bodies. The appendicular skeleton includes the limbs and limb girdles. At the front, and corresponding to our shoulders and arms, are the **scapula** and **coracoid**, which together form the shoulder girdle, a single bone in the upper part of the front limb (**humerus**), two lower bones below (**radius** and **ulna**), and finally the bones of the **manus**, or forefoot: the **carpals** (wrist), **metacarpals** (palm) and **manual phalanges** (digits or fingers). The hindlimb and its girdle include the **pelvis**, or hips (a three-part bone composed of the paired **ilia, pubes** and **ischia**), the bones of the thigh (**femur**) and calf (**tibia** and **fibula**) and finally the bones of the **pes**, or hindfoot: the **tarsals** (ankle), **metatarsals** (arch) and **pedal phalanges** (toes).

In anatomy keeping directions straight is also important, and when talking about skeletons and their various parts, we use the following basic terms (see also Glossary):

 Dorsal: toward the spine, which is up (relative to gravity) for a quadrupedal animal

 Ventral: toward the belly, or down for a quadruped

 Lateral: outer side, away from the body's midline

 Medial: inner side, toward the body's midline

 Anterior: front (or *cranial*, toward the head)

 Posterior: back (or *caudal*, toward the tail)

 Sagittal: anything within a midline vertical plane running from anterior to posterior

 Transverse: anything within a horizontal plane running from left to right, at right angle to sagittal plane

 Proximal: closer to main part of body

 Distal: farther away from main part of body

Few bones are as distinctive as the vertebrae of a sauropod. Each elegant, sculpted-looking shape played a part in providing the giant animals with the most support needed with the least amount of bone. All four series of vertebrae—cervicals, dorsals, sacrals and caudals—have distinctive shapes resulting from the way they provided support and attachment surfaces for muscles and ligaments. The differences in various sauropods' lifestyles give each of these bones a characteristic look. They also usually allow a paleontologist to identify the genera and species they belong to, as well as providing the ability to place them in a phylogenetic context and even make inferences as to how a sauropod lived. In spite of their differences, all had similar basic features. Like those of other vertebrates, each sauropod vertebra has a drum- or cylinder-shaped *centrum*, which articulated end to end with cartilage. Dorsal to the centrum, a *neural arch* protected the

Stresses and Strains

Like all other land vertebrates, sauropods' skeletons had to deal with two basic kinds of stress in supporting their bodies. Compressive loading (*green arrow*) from gravity pulled straight down and was best resisted by vertical limb bones whose shapes gave them a minimal tendency to bend, while tensional loading (*pink arrows*) pulled sideways and down and was dealt with by evolving air-filled, lightweight spines whose bony struts equalized and transferred weight with the help of tough, elastic ligaments. Because the skeleton did most of the work in supporting the weight, the muscles could be far less massive and bulky.

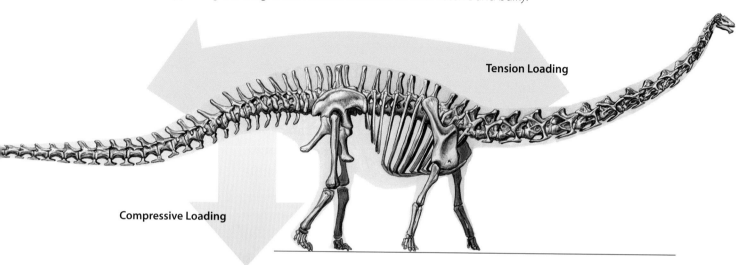

Tension Loading

Compressive Loading

Major Sauropod Muscles

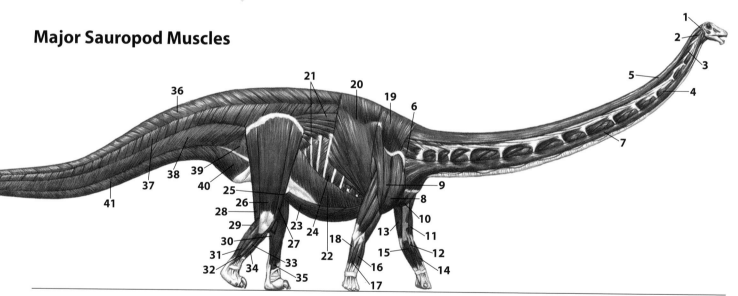

1. Complexus
2. Rectus capitis ventralis
3. Multifidis cervicis
4. Intertransversales
5. Semispinalis capitis
6. Levator anguli scapulae
7. Longus colli
8. Deltoideus
9. Scapularis
10. Pectoralis major
11. Biceps brachii

12. Brachialis anticus
13. Triceps caput brachii
14. Extensor carpi radialis
15. Flexor carpi radialis
16. Extensor digiti communis
17. Extensor digiti minimi
18. Flexor carpi ulnaris
19. Trapezius
20. Latissimus dorsi
21. Iliocostalis serratus
22. Obliquus externis

23. Rectus abdominis
24. Intercostales
25. Sartorius
26. Iliotibialis
27. Iliofemoralis
28. Semitendinosus
29. Gastrocnemius (lat. head)
30. Gastrocnemius (med. head)
31. Flexor digit. pedis sublimis
32. Peroneus longus
33. Extensor digit. pedis longus

34. Tibialis anterior
35. Soleus
36. Lumbodorsalis caudae
37. Lateralis caud. longissimus
38. Caudofemoralis longus
39. Caudofemoralis brevis
40. Ilio-ischiocaudalis
41. Depressor caud. longissimus

spinal cord. At the front and back these flare out into points called *zygapophyses*, which could slide past the ones on adjacent vertebrae to provide additional connection and movement and were securely fastened together by strong joint capsules. Sauropod vertebrae differ from those of many other vertebrates in having *laminae* (struts or braces) and *fossae* (spaces) that alternate with each other to provide just the right amount of strength, and at the angle needed, to take stress where it was required but eliminate bone and save weight where it wasn't. The fossae were filled with air spaces in life, which not only helped lighten the sauropods' bodies but also give us clues to how they breathed.

As distinctive as the vertebrae are, a sauropod skull is likewise a masterpiece of economy when it comes to bone versus space. Because they were mainly structures for food gathering and for housing the senses of sight, smell, hearing and taste, the thickest, strongest areas are concentrated in the jaws and around the vulnerable brain. Most of the cranium is a series of braces and spaces for framing the eyes, nose, ears and housing muscles for chewing. Like our own skulls, those of sauropods were composed of many separate bones; unlike ours, the bones of sauropod skulls were not always tightly interconnected, and so they tended to fall apart after death. This can cause sauropod specialists great frustration, since the skull's small size and fragile construction made its preservation less likely than the rest of the skeleton when the dinosaur died. Because of this, it's an exciting event when a complete sauropod skull is discovered, and by now enough are known to allow us to begin to understand what sauropods ate and how diverse they were. From the shapes of their skulls and teeth, as well as the rest of their skeletons, we know that there were several basic sauropod head morphologies, each implying a distinctive method of food gathering. Although there were some very unusual skull adaptations due to diet that we'll learn about later, most advanced sauropod skulls typically take four basic shapes: **Eusauropod skulls,** like those of *Jobaria* (possibly a basal neosauropod) and *Mamenchisaurus,* belong to a basal or primitive sauropod group from which others may have descended, and are short and somewhat deep, with broad to narrow muzzles and moderately large nasal openings. The large, spoon-shaped teeth suggest that they were probably feeding generalists. **Macronarian skulls** are wide, deep and chunky looking, with huge, forwardly located nares that sometimes form an arch. The reason the nares are so big in this clade is unknown, but the most promising idea is that they housed special structures to warm and moisten incoming air, just as in mammals and birds today. All sauropods also had big *orbits* (eye openings), suggesting that vision was very important. Macronarian teeth are robust and spoon shaped and occupy the margins of both upper and lower jaws in an arch when seen in dorsal and ventral view. **Diplodocoid skulls** are also wide (extremely so in some forms,

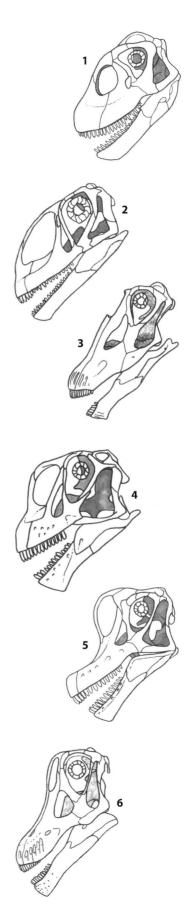

Sauropod skulls generally form six major morphotypes, seen here from top to bottom. Eusauropods are represented by the basal form **(1) Shunosaurus**, whose broad, robust construction and teeth indicate a basal, generalized condition, and **(2) Jobaria**, an early neosauropod. Diplodocoids like **(3) Diplodocus** made a radical departure in developing retracted nares and a specialized, leaf-stripping dentition. Macronarians like **(4) Camarasaurus** and **(5) Brachiosaurus** often had extremely large nares and spoon-shaped or chisel-like teeth for cropping large quantities of browse in single bites, while **(6) Nemegtosaurus** and other titanosaurs, although descended from macronarians, acquired some skull features of diplodocoids.

Forefeet into Pillars

To provide more stability, the manus, or forefoot, became reduced, with short, blocky bones to increase the area of support, like an elephant's forefoot or the base of a pillar. Unlike an elephant, however, sauropods traded many tightly fitting carpal bones in the wrist and tarsal bones of the ankle (*below*) for huge pads of fibrous cartilage that acted as shock absorbers fore the great weight.

Over time, the slightly arched shape of the manus's five toes became more of a circle (*right*) to give more support. Toes became smaller and in some titanosaur clades disappeared. (*Opposite, left*) Compared with elephants, sauropods had more extensive fibroid cartilage (*blue areas*) in their feet, which took the place of tightly fitting ankle and wrist bones.

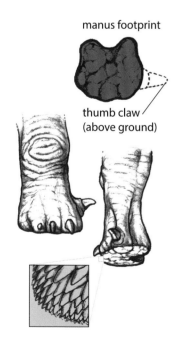

manus footprint

thumb claw
(above ground)

A cast of one well-preserved footprint shows that some sauropods had **keratin spikes** on the underside of the manus pad.

such as some rebbachisaurids) but wedge shaped in profile, and the nares are much smaller and retracted (located far back) in the skull between the orbits. In diplodocoids the teeth are slender and cylindrical or pencil shaped and restricted to the anterior areas of the jaws. There are often bony extensions or flanges protruding from the upper jaw posterior to the upper teeth that may have helped keep food in place just before swallowing. *Titanosaur skulls* belong to a group of derived macronarians. Now better known from recent discoveries, the skulls of some forms combined some of the features of basal macronarians (deep upper and lower jaws with medium-length tooth rows) with those of diplodocoids (wedge-shaped head, retracted nares) and had distinctive teeth that, although generally slender like a diplodocoid's, were sometimes broader and extended farther back along both upper and lower jaws. The known titanosaur skulls also tend to be somewhat hourglass shaped in dorsal view, with the tooth bearing, or *alveolar*, area being very wide and deep. Sauropods, as dinosaurs and therefore archosaurians, possessed most of the features of advanced reptile, as well as some bird and crocodilian (both dinosaur cousins), skeletal systems. As they evolved, the first sauropods kept some important innovations already possessed by their ancestors, but they also developed their own unique, and in some cases bird- and mammal-like, skeletal adaptations that enabled them to live on a huge scale.

HOW TO SUPPORT A SAUROPOD

By the Carnian stage of the Late Triassic period, prosauropods had evolved into a number of medium-sized to large, bipedal herbivores with forefeet that were suited not only for grasping but also for support when these dinosaurs came down on all four legs. This became a habit as the prosauropods developed larger body sizes to carry around their increasingly capacious, and more efficient, digestive systems. As adults some forms such as *Plateosaurus* and *Massospondylus* were *obligate*, or required, bipeds. Bigger, of course, also means heavier: you now have more weight to support, and if you're a land animal, gravity will place more stress in the form of both **compressive** and **tensional** force, or *loading*, on your skeleton. In *compressive force*, stress from

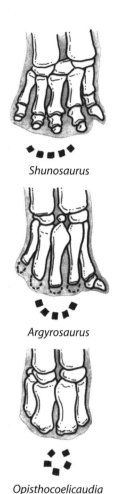

Shunosaurus

Argyrosaurus

Opisthocoelicaudia

Digging In

When a sauropod took a step forward, it dug its angled, laterally compressed claws into the ground. At the same time, these were pulled and twisted by the foot muscles to pull the entire foot sideways and back, beginning medially **(1)** and moving laterally **(2)**, aided by the *caudofemoralis muscle* (*below, right*), a power extender of the femur; this combined action helped propel the animal forward. Modern turtles like sliders (*photo, below, left*), although they have a sprawling instead of an erect leg posture, also have long hindfoot claws to push themselves forward.

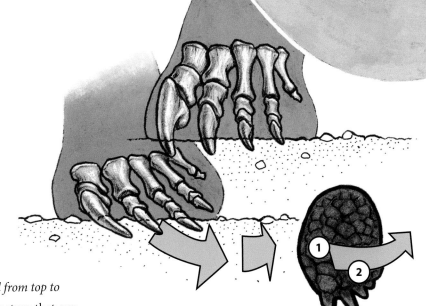

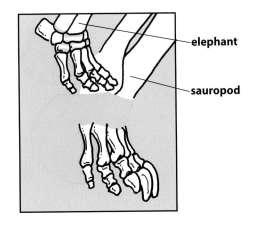

elephant

sauropod

gravitational force on a weight *pushes downward from top to bottom* more or less vertically throughout a structure that supports the weight from below, like a pillar holding up a building or an animal's leg holding up its body. In *tensional force*, gravitational force on a weight *pulls in an outward direction at an angle* throughout a structure that's located away from the weight and anchors onto a foundation big and strong enough to resist the pull, like a steel cable holding up a bridge or the ligament supporting an animal's head from its body. Coming down on four legs instead of two supports weight more evenly, and this pushed the prosauropods and ancestral sauropods, like *Aardonyx* and others, into becoming quadrupedal, while still retaining their bipedal capability.

Four-footed animals that evolve large size also need stronger limbs. When an animal reaches a certain size, the typically tube-like cross section of their upper limb bones, especially the femur, often changes to become *eccentric* or compressed, somewhat like an industrial I-beam. The strongest muscles of the upper limbs are located anterior and posterior to these bones, and while these can correct for stress and bending in these directions, more strength is derived from the

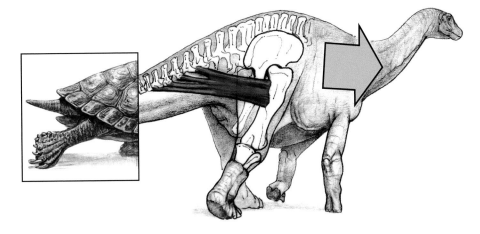

By attaching lights to the foot of a moving elephant and recording these as time-motion sequences, biomechanics expert **John R. Hutchinson** was able to infer the probable limits of a sauropod's limb movements and possible speeds.

cross section of the bone being transversely wide, offsetting the tendency for unstable, lateral bending. Since a vertical support holding up lots of weight is weaker the farther it bends, big, heavy animals typically move their lower limbs less at the elbow and knee and more at the shoulder and hip. Less bending at these joints helps to better resist gravity, while still maintaining long legs for stride length. Because the muscles for flexing and extending the elbow and knee are now less important, their size and areas of attachment on the lower leg bones (radius/ulna, tibia/fibula) become smaller, while the muscles powering the shoulder and hip grow bigger and need more space for attachment on these bones. Compared to prosauropods like *Massospondylus*, ancestral sauropods' pelvises have greatly expanded areas anterior, posterior and dorsal to the femoral attachment for the bigger muscles needed for moving the leg. Of these, the *posterior iliac process* (or postacetabulum) played a role in improving the strength of the rear thigh muscles by extending much farther ventrally and providing a more effective source of attachment. This was the origin for large muscles that attached to the femur and pulled the leg back. Although these contributed to give a sauropod's hind leg great thrust in pushing the animal forward, the most powerful and important femoral retractor muscles were the paired *caudofemoralis*, which arose from the anterior caudal vertebrae and inserted onto the upper rear femora. In some sauropods, especially the later titanosaurs, the anterior edges of the ilia expanded to form large, transverse crests. These were anchoring points for tendonous sheets that helped to support these sauropods' tremendously wide, bulky abdomens.

Finally, another change that happens when a big animal becomes a big quadruped is that the length and number of bones in each of the toes decrease. A longer foot capable of extension and flexion gives way to a short, compact one that bears weight more efficiently, like the base of a column. All of these adaptations for weight bearing, called a *graviportal* limb condition, begin to appear in the skeletons of the true sauropods, which by the Early Jurassic were much bigger than their prosauropod-like ancestors.

WALKING WITH GIANTS

At the same time that sauropods' limbs became adapted to supporting their growing frames, their feet also evolved in a distinctive way to make it possible for a giant move easily and tirelessly. With the need both to support their huge bodies and to move over sometimes great distances to find food, sauropods, like elephants, developed short, compact fore- and hindfeet to efficiently spread their weight as they walked. Tough, elastic cushions of fibrous cartilage tissue at the bottoms of both fore- and hindfeet provided support and made steps springy and resilient. Here, however, the resemblance to elephants ends.

While an elephant's hindfoot (*pes*) *metatarsal* and *phalangeal* bones are almost vertical (digitigrade), these bones in a sauropod's foot became somewhat horizontal, or *plantigrade*, like those of bears, extinct ground sloths and humans, which allowed for even greater weight to be supported but still retained some ability to flex and extend. Unlike an elephant's flat toenails, big claws curved out anteriorly and laterally from the pes. Researchers such as Matt Carrano (National Museum of Natural History, Smithsonian Institution) have discovered that sauropod hindfeet worked in a special way. Most hindfeet (like ours) work by extending down and back, pushing against the ground and forcing the whole animal forward. This is also how a sauropod's foot worked, but at the same time the hindfoot *rotated or twisted outward* as well as down. The big, laterally slanted pedal claws helped by digging into the ground, and these provided traction on uneven terrain but could be "wrapped" or pulled in around the foot if the going was more stable. With their extrapowerful muscles and lateral foot rotation, the pelvis, hindlimb and pes were the sauropod's powerhouses, like the large wheels on a steam locomotive that create the main forward thrust.

The forefoot, or *manus*, became even more specialized than the hindfoot. Here the *metacarpals* grew thicker and came to form an arch that became more tube-like, close-fitting and thus stronger. Except for a big claw on the first digit that stuck out to the side, the toes became nubbins that in some taxa disappeared altogether. The base of the manus, unlike that of an elephant, had, instead of one big pad, *two smaller ones* on each side, giving eusauropods and most neosauropods a characteristic, horseshoe-shaped footprint. In later titanosaurs, however, the space in between the two smaller pads disappeared and gave way to a single, more elephant-like pad. On many of these the sometimes large, first digit "thumb claw" doesn't make an impression, suggesting that it was kept cocked well upward to keep it off the ground and relatively sharp. For extra traction, some sauropods had dozens of tiny, tough, pointed spikes on the undersurface of the manus pads (and probably the pes), as shown by some trackway evidence. The *carpal* (wrist) and *tarsal* (ankle) bones of elephants and other giant mammals are constructed like a series of close-fitting construction blocks, with thin coverings of smooth (*hyaline*) cartilage that allowed these joints to glide over each other. As sauropods evolved, however, their carpal and tarsal bones grew smaller and their support function became replaced by much thicker, more elastic *fibroid,* as well as thinner hyaline, cartilage that allowed the ankles and wrists to absorb, release and push back weight with every step. The knees, elbows and hips, like the feet, had similar, oversized cartilage cushions, firmly anchored into roughened (*rugose*), pitted joint surfaces for better attachment. This means that sauropod joints could bear proportionately greater weight than those of giant mammals. This probably gave the huge animals, as with elephants, a smooth, gliding walk, or, when more speed was required, a faster walk or amble. In his experiments with living elephants, John R. Hutchinson (Royal Veterinary College) inferred that sauropods, because of their similar overall limb proportions, could not gallop or run, but in an amble they might have been able to reach 25 mph.

In quadrupedal mammals, the *scapula*, or shoulder blade, plays a role in walking, swinging forward along with the forelimb and foot when a step is taken, contributing its length to increase the forward stride; the scapula and forelimb act as a unit. This is possible because the scapula has no attachment to other bones, and only muscles hold it in place. In sauropods and other archosaurians, however, the scapula is rigidly attached to another bone, the shield-shaped *coracoid*, and this, in turn, articulates with one of the paired *sternal plates* that attach to the rib cage. If the scapula was to contribute to forward stride, this potentially non-flexible attachment could have been a problem. Gregory S. Paul and Phillip R. Platt (Brontobuilder), both independent workers who have tackled sauropod scapular motion, have suggested that the solution was for the articulation between the coracoids and sternal plates to become curved and smooth. This allowed the thin, convex ventral rims of the coracoids to glide backward and forward to a limited degree along the

In addition to the femoral head itself, the acetabulum, or hip socket, of a sauropod (*left*) had an enormous space to accommodate the mass of fibrous cartilage that, like the ends of the limb joints (*blue areas*, *below*), absorbed stress from the animal's weight as it walked. The shoulder blades, as in some ungulate mammals, had a cartilaginous extension (*black arrow*), the suprascapular cartilage, that provided both additional area for muscle attachment and leverage to increase the stride length of the forelimb. Fore and aft mobility (*orange arrows*) was made possible by the rounded margin of the coracoids, which slid past the corresponding edge of the sternal plates (*blue areas*, *cartilage*).

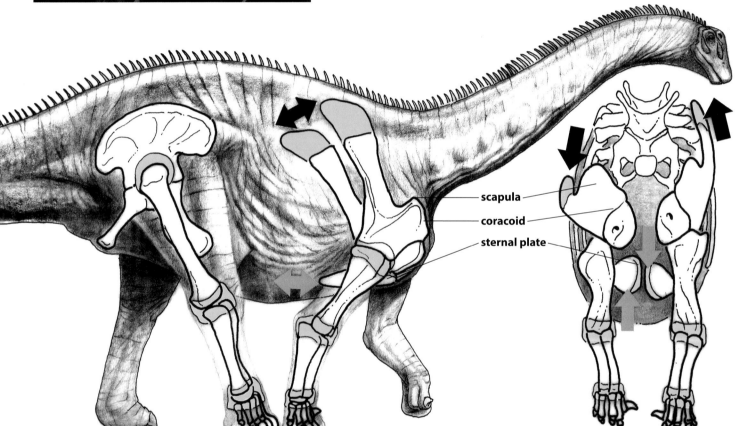

scapula
coracoid
sternal plate

Land Striders

In quadrupedal mammals, the scapula plays a role in walking, swinging forward along with the forelimb and foot when a step is taken, contributing its length to increase forward stride—the scapula and forelimb act as a unit. This is possible because the scapula has no bony attachment to the ribs, and only muscles hold it in place. In sauropods and other archosaurians, however, the scapula connects to another bone, the shield-shaped coracoid, which in turn articulates with one of the paired sternal plates that attach to the rib cage. If this had been a rigid, inflexible connection, it would have created a problem for some sauropods, because the shorter forelimb would have had to have moved faster than the hindlimb to keep up. Although they couldn't swing their forelimbs as freely as mammals, there was enough motion between the coracoids and sternal plates (*orange arrows*) to allow limited fore-and-aft movement of the shoulder as in modern crocs and varanid lizards. This gave the entire forelimb a wider range of motion (*black arrows*). Since sauropods moved more slowly and had a more limited arc of motion than other dinosaurs, this was enough to equalize their limb stride while walking.

grooved concave surface of the sternal plates, while the scapula tilted up and down. For sauropods this would have helped to slightly lengthen forelimb stride length during walking and produced a somewhat mammalian type of gait. The scapulae also bore cartilaginous extensions from their distal ends, which provided extra length for upper muscle attachment and still more stride.

KEEPING IT LIGHT

Eating takes up most of the time for an herbivore, and for the majority of plant eaters this involves both finding food and efficiently harvesting it. A remarkable thing about the skeleton of a sauropod is that it was engineered not only to hold up the animal's great size while it was stationary but also to allow both support and movement while the animal was actively obtaining food. Harvesting arboreal browse by reaching and growing ever larger to support their huge food-processing systems was the primary adaptation and main thrust of sauropods during their evolution. While greater size enabled early sauropods to harvest and digest food in bigger and bigger quantities, it was also necessary to evolve longer necks to reach it: like the sauropods themselves, their food plants, the conifers, were increasing in diversity and in height, and animals with longer necks had an advantage. The earliest true sauropods started out with the respectable number of 12 cervicals, but as time went on some ended up with as many as 19, as in mamenchisaurids, all to increase the length of the neck. Neck length increase was accomplished in four ways: (1) the transformation of the first few dorsals into cervicals, (2) the multiplication of cervical vertebrae, (3) a combination of the first two and finally (4) by some vertebrae becoming lengthened, so that in certain cases a vertebra was 6 times as long as its diameter. As with larger size, an animal's skeleton when faced with evolutionary pressure to achieve height and reach must change, but it has to do so within the bounds of gravity, stress and other physical limitations. A longer neck allowed greater reach, but it also imposed problems in creating an efficient support system while not becoming too heavy and unwieldy. The cervical vertebrae had to simultaneously be both strong and yet light in weight, as well as the rest of the spine. The answer partly lay in the gradual elimination of any bone tissue that didn't contribute to the vertebra's strength, while reinforcing areas that needed it. A sauropod's spine is an amazing system of hollow spaces, masts and struts that in their construction share many of the same design principles as modern human bridges and industrial cranes. The other part of the answer lay in developing ***pneumaticity*** (hollowness) inside the bones themselves. Hollow spaces in tetrapod bones are common, but here we need to distinguish between bones that are hollow and filled with marrow tissue (almost all tetrapods) and those that are hollow and contain only air, the second condition being uncommon. In addition to the cranial sinuses of mammals and archosaurs, however, postcranial pneumaticity only occurs in *ornithodiran,* or "bird-line," archosaurs, specifically in pterosaurs and saurischian dinosaurs. The sauropods inherited a very simple version of this from their ancestors, and the spaces were lined with membranes and connected with a system of multiple, extended passageways called ***diverticula*** (singular, *diverticulum*) that lodged against vertebrae, creating fossae, or depressions. These fossae, in turn, later evolved into holes, or ***foramina***, that invaded the outside of the vertebra and came to occupy the interior. Some actually came to occupy a space above the spinal cord. Although the diverticula themselves played

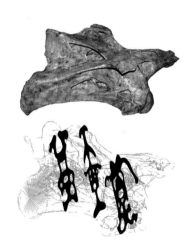

A cervical vertebra of *Giraffatitan* shown in a normal photo (*top*) and a digital model with CT cross sections (*bottom*). The three black CT slices show the cross section of the vertebra, and the green lines show the distribution of stresses computed by finite element analysis (FEA—see page 167).

Sauropod Skeletons

In supporting their ever-increasing size and weight, the skeletons of sauropods such as *Camarasaurus* (*below*) evolved a remarkable balance between dense, reinforced bone where needed and an alternating system of struts and spaces where it wasn't. While massive limb bones distributed gravitational stress to short, blocky pedal bones for support, lightness and strength were simultaneously achieved in a series of highly flexible vertebrae whose excavated, air-filled spaces made the entire animal's frame up to 50% lighter than it would otherwise have been.

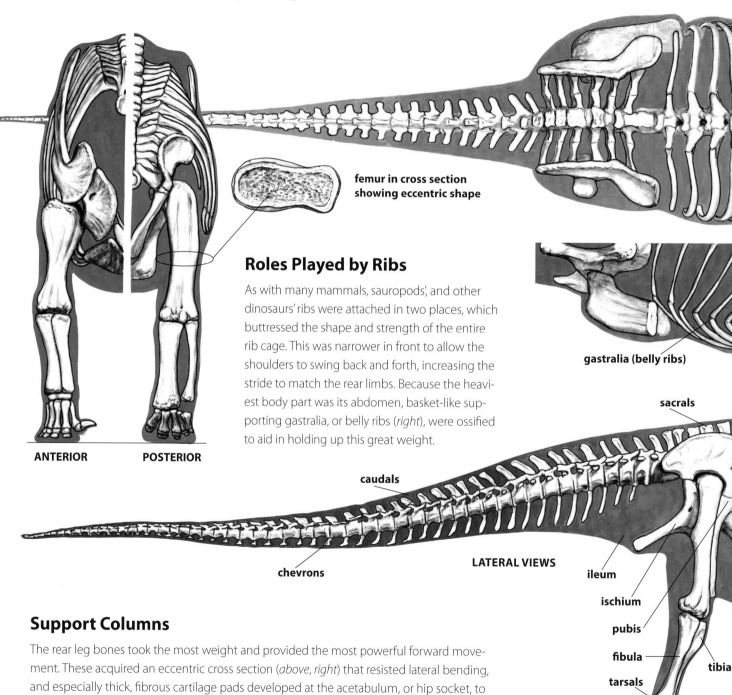

femur in cross section showing eccentric shape

Roles Played by Ribs

As with many mammals, sauropods', and other dinosaurs' ribs were attached in two places, which buttressed the shape and strength of the entire rib cage. This was narrower in front to allow the shoulders to swing back and forth, increasing the stride to match the rear limbs. Because the heaviest body part was its abdomen, basket-like supporting gastralia, or belly ribs (*right*), were ossified to aid in holding up this great weight.

ANTERIOR   **POSTERIOR**

gastralia (belly ribs)

sacrals

caudals

chevrons

LATERAL VIEWS

ileum

ischium

pubis

fibula

tibia

tarsals

metatarsals

Support Columns

The rear leg bones took the most weight and provided the most powerful forward movement. These acquired an eccentric cross section (*above, right*) that resisted lateral bending, and especially thick, fibrous cartilage pads developed at the acetabulum, or hip socket, to cushion stress at this place. Broad, bladelike pelvic bones, shoulder arms and paired sternal plates allowed for the attachment of extensive muscles needed to power the limbs.

A Flexible Mast

In *Brontosaurus* and most other sauropods (*photo, right*), the neck was habitually carried above shoulder level most of the time in a gentle sigmoid (S-shaped) curve. Long to short cervical ribs, depending on the species' feeding adaptations, gave stability while the neck was in motion by helping to control torque, or twisting. Intervertebral cartilages allowed for considerable vertical and lateral flexibility, most of which was at the base and the upper end of the neck.

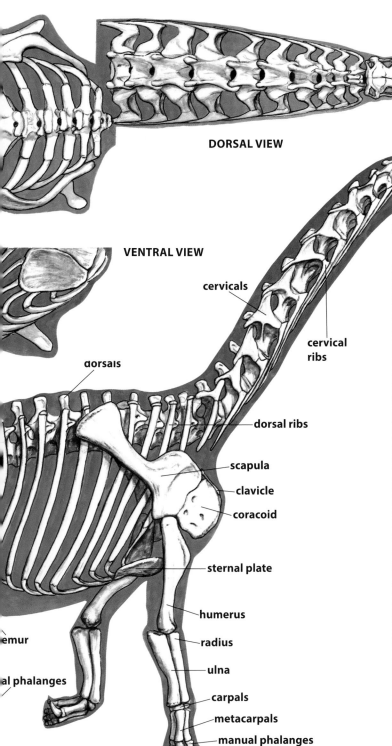

DORSAL VIEW

VENTRAL VIEW

cranium

hyoid bone

mandible

(For individual bones of skull see p. 62)

cervicals

cervical ribs

dorsals

dorsal ribs

scapula

clavicle

coracoid

sternal plate

humerus

femur

radius

ulna

al phalanges

carpals

metacarpals

manual phalanges

Modified from an original reconstruction by Gregory S. Paul

Strength with Lightness

A mid-cervical vertebra from the neck of *Apatosaurus*, showing the extreme economy of bone structure. Most of the loading occurred through the drum-like centrum, but strut-like laminae and additional articulation points, the pre- and postzygapophyses, redirected and evened out the stress, while cervical ribs helped stabilize against sideways torque. The elaborate, weight-saving hollow spaces, or pleurocoels, featured fossae and foramina, whose pneumatic sacs acted to rid the body of unwanted interior heat.

(1) neural spine, **(2)** prezygapophysis, **(3)** lamina, **(4)** condyle, **(5)** diapophysis, **(6)** cervical rib, **(7)** parapophysis, **(8)** foramen, **(9)** pleurocoel, **(10)** cotyle, **(11)** centrum, **(12)** fossa, **(13)** neural canal, **(14)** postzygapophysis

One way to find an *approximate* amount of volume for a dinosaur's body is to immerse a scaled-down model in water—the more detailed and accurate the model is, the more likely the accuracy of the results.

(1) Fill a graduated container (with measuring marks) with enough water to completely immerse the model and big enough to allow for water displacement.

(2) Before immersing the model, note the level of water at a measuring mark.

(3) Immerse model completely, below the surface of the water, noting the new water level caused by the model's displacement.

(4) Subtract the volume measurement of the water level noted earlier from the new volume measurement of the water with the model in it. This should produce a very rough idea of the sauropod's or other dinosaur's volume.

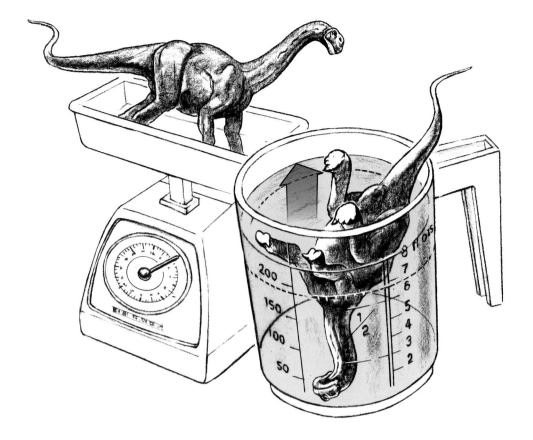

no part because of being dead air spaces, they connected to **air sacs** that acted like bellows to aerate the lungs. In sauropods a progressively more and more elaborate system of small and tiny passageways came to be connected with these respiratory air sac extensions. As solid as the axial skeleton looks from the outside, it's actually honeycombed with many separate and sometimes interconnected spaces beneath the bone's outer surface, structured in such a way that during life the vertebrae and some other bones remained strong but were incredibly light in weight.

The most basal sauropods had large fossae on each side of the vertebral centrum accommodating diverticula that partially occupied the interiors. More advanced forms like *Camarasaurus* had simple, paired, hollow chambers inside the vertebrae called **camerae** (singular *camera*, "chamber"), which later developed smaller accessory areas, as in *Diplodocus*. In the most advanced sauropods, such as the titanosaur *Saltasaurus*, the camarae became a complex, interconnected series of tiny chambers called **camellae** (singular *camellus*, "small chamber") in which some diverticula actually filled in some of the space above the spinal cord. This is actually already present in some early eusauropods, such as *Barapasaurus*. In addition to the vertebral interiors, the diverticular sacs also lay in the outer **pleurocoels**, the spaces around and in between the lamina. Different lineages of sauropods developed this system on their own as their size and need for lightness increased. The vertebrae were not the only places where pneumatization took place, however, and the ribs of some sauropods, especially *Brachiosaurus*, *Euhelopus* and some titanosaurs, have foramina that lead to internal chambers. For giant sauropods pneumatization was a great structural innovation, since it meant that their bodies were much less heavy, and less energetically costly to move, than they would have been. In *Diplodocus*, the well-known skeletons of which have allowed coauthor Mathew Wedel and others to map the extensive presence of the diverticula, pneumaticity may have made the entire dinosaur's body as much as 10% lighter than it would have been otherwise, while the neck was about 50% lighter.

HOW MUCH WAS AIR?

The fact that advanced sauropods' axial skeletons were so highly pneumatic is crucial in determining how much *mass*, or weight, the animals had. Why is this so important? Body mass has a direct effect on such questions as how fast an adult sauropod could move, how much oxygen it required, how many eggs it could produce, how fast baby sauropods grew, how much it needed to eat and other considerations. Answers to these questions could, in turn, help us to understand what demands these animals made on their environments and how many sauropods these could have supported.

There are two ways of estimating mass in dinosaurs and other large animals. One is by a method called **allometry** (*allo*, "different"; *metry*, "measurement"), in which changes in places like a large animal's limb bone thickness (typically, a cross section of the femur at midpoint) are compared with the corresponding bone thickness in a smaller, close relative. Large animals' bones are not simply larger versions of smaller ones. When a large sampling of closely related species of different sizes is measured and compared, we find that the **proportional thickness of bones increases along with an increase in the animal's mass.** Once the average relationship between mass and limb bone proportions is known, this can be used to estimate the mass of the animal being studied. This method has limitations. You have to have the same parts of the animal's skeleton preserved to compare with its relatives, and you have to have a big enough sampling of these relatives. Another limitation is that different groups of animals have different relationships between limb bone thickness and mass, such that an equation that works for mammals doesn't apply to birds. This, of course, is a real problem in determining mass for sauropods, for which we have no close living relatives with which to compare. Another is that allometry can sometimes lack precision, since some living animals with the same limb bone proportions can vary in mass by a factor of two or more: it doesn't help us much to know that the mass of *Apatosaurus* could be anywhere from 15 to 30 tons, a variation so wide as to be useless.

Another method of estimating an extinct animal's mass is by **measuring volume**, or internal space. This is sometimes done by first taking as accurate a scale model of the animal as possible and immersing it in water or a fine material like sand, and then measuring the amount of water or material that the model displaces (this can also be measured mathematically by sectioning the actual or virtual model and measuring each small section or slice, and then adding these together). **The amount of displaced material is equal to the volume of the model.** Next, the model's volume is multiplied by the *scale factor*, or amount of enlargement, of the actual animal. For example, if *Brachiosaurus* is estimated at 5.6 m (560 cm) tall at the shoulder in life and the scale model is 14 cm at the shoulder, then the scale factor is 560 divided by 14, equaling 40—a live *Brachiosaurus* would then be 40 times longer, and wider, than the model, giving us its actual *linear* dimensions. We're looking for *volume*, however, so this means taking the measurement and mul-

Sauropods like *Giraffatitan* (*below*) eliminated weight in their long necks, making them more controllable, by developing an extensive system of **air sacs** (*blue areas, above*) within the fossae and interiors of the cervical and other vertebrae. These also helped to channel and eliminate excess heat from the body's interior.

Many industrial cranes use a double system of cables (*left*) to equalize weight of the boom, or lifting arm, and to help control its torque, or tendency to twist when in motion. Here the cables run up through a mast and extend to the end of the boom to pull it up. Although some sauropods that browsed quadrupedally generally needed only a single ligament to support their necks, those that reared bipedally to feed had, like a crane, a system of double supraspinal ligaments running from the dorsal tall neural spines (the equivalent of a mast) and attached to bifid neural spines of the neck (the equivalent of a boom), which distributed tensional loading and, like a crane's cables, also equalized support of the neck.

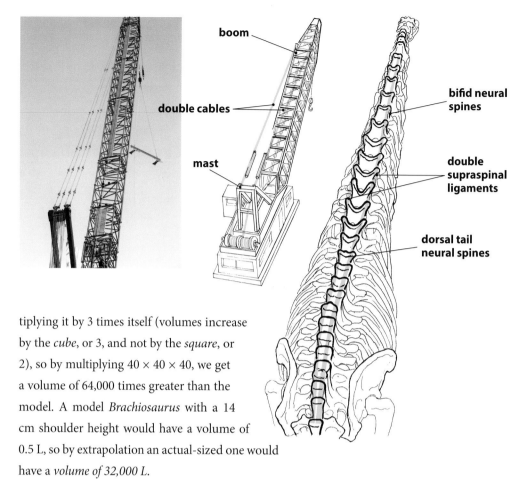

tiplying it by 3 times itself (volumes increase by the *cube*, or 3, and not by the *square*, or 2), so by multiplying 40 × 40 × 40, we get a volume of 64,000 times greater than the model. A model *Brachiosaurus* with a 14 cm shoulder height would have a volume of 0.5 L, so by extrapolation an actual-sized one would have a *volume of 32,000 L*.

Finally, the estimated *volume* of *Brachiosaurus* must be multiplied by *density* to obtain its **mass**, our end goal. Because the estimated density (expressed as weight) of living animal tissue is conveniently very close to that of water (approximately 1 kg = 1 L), according to this reasoning *Brachiosaurus* would have then ended up having a mass (weight) of 32,000 kg, or 32 metric tons.

There are, however, other factors that affect this estimate. Obviously, the air sacs, diverticula and lungs won't be anywhere this dense, so these areas have to be taken into account. Here we come to a crucial question: how can we know the actual mass of *Brachiosaurus* and other taxa unless we have a realistic understanding of how much of their bodies was air? This is why the study of pneumaticity in sauropods, and mapping their known and probable areas in the body, is so important in coming up with realistic mass estimates and ultimately in understanding their biology. Researchers R. MacNeill Alexander (Leeds University) in 1989 and Donald Henderson (University of Bristol) in 1999 separately compensated for the decrease in mass by methods such as creating lung-sized holes in the 3D model and virtual ones, along with air spaces and diverticula, in the digital model. Paul in 1987 used calculations of pneumatization for the cervical vertebrae, which he combined with those for the abdominal air spaces, to arrive at density values of 0.9 kg/L for the trunk or torso and 0.6 kg/L for the neck. Coauthor Wedel has pointed out the need to recognize four distinct areas in a sauropod's body if air volume and, as a result, density are to be correctly calculated: (1) the trachea or windpipe, (2) the "core" respiratory system of lungs and air sacs, (3) the diverticula outside the skeleton (between the body organs, muscles and under the skin) and (4) the skeleton and its pneumatization, especially the vertebrae. For determining skeletal pneumatization, he used the technique of *computerized tomography* (see Chapter 7), in which a series of bone sections are first traced from a two-dimensional (2D) image, and after transferring this image to a computer, the bone and empty spaces are digitally "filled in" with separate-colored backgrounds. The number of pixel units is then counted for each section and added together, rendering the *air space proportion* (ASP) for each bone. By doing this systematically, the overall vertebral pneumatization of several diplodocoid sauropods and some others is now known, showing a gradual increase in these forms that reached its greatest development among the diplodocids like *Apatosaurus*, *Barosaurus* and *Diplodocus*. Although these dinosaurs were not the largest known sauropods, they were among the lightest in weight for their sizes, and their necks were especially light.

BRIDGES, MASTS AND CRANES

If we think of a sauropod's neck as essentially a structure designed by the evolutionary process to rise high enough for the head to harvest arboreal conifer and other vegetation, how was such a structure engineered? A neck that could reach and exceed 10 m (33′) and weighing hundreds of pounds in some species, such as *Mamenchisaurus* and *Omeisaurus*, had to be strong but also light. It also needed extra support when movement subjected it to even more stress. Any big, long object attached at only one end must be designed for these two needs, and the technology of building bridges, cranes and other giant architectural structures can help us to understand how sauropods dealt with this. Gravity is a constant factor that imposes itself on everything, and in pulling down on any structure, both living and non-living, it creates a *load* (or force) that requires support. This is accomplished in two ways. First, using our anatomical terms *dorsal* and *ventral* for "top" and "bottom," the tensional load on a *cantilevered* (sticking out from a stationary place) structure can be countered by having **dorsal bracing**. Often taking the form of a cable, this starts from the fixed part of the stationary form and attaches at the tip, and sometimes at several places along the way. The cable must be strong enough to at least equal, and for safety surpass, its share of the load to avoid breaking, and the point of origin where it starts out also must be high enough to provide the cable with an efficient angle from its origin to where it attaches to the point bearing the load. Giant suspension bridges (like the Golden Gate, San Francisco) use this system, with long cables on each side coming down from vertical *masts*, or towers, to (with the aid of smaller vertical cables) support the load of the horizontal *beam* (in this case the roadway) underneath. A giant sauropod's spine was also supported like this, with the neck featuring long *ligaments*, the equivalent of a bridge's cables. Ligaments are strong, relatively elastic tissues. In mammals, the *nuchal* or main dorsal ligament has the ability to stretch yet snap back like a broad rubber band. Although sauropods didn't have a single large, nuchal ligament, it's likely that other ligaments did the same thing and played a primary role in stabilizing the neck's mass while permitting up-and-down, as well as sideways, movements of the head. These ligaments probably ran both in between the dorsal sagittal surfaces of the cervical vertebrae, where they would be known as an *interspinous ligament* (*inter*, "between"), and in the second connecting the tips of the spines, where they would be called a *supraspinous ligament* (*supra*, "above"). Both ligaments

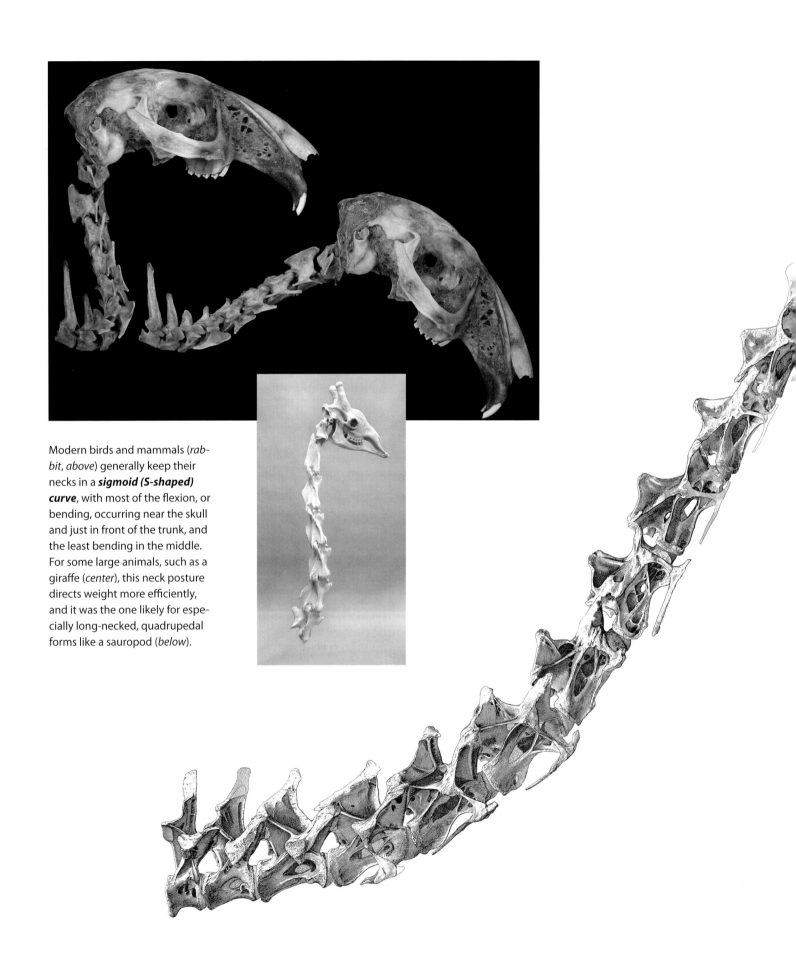

Modern birds and mammals (*rabbit, above*) generally keep their necks in a ***sigmoid (S-shaped) curve***, with most of the flexion, or bending, occurring near the skull and just in front of the trunk, and the least bending in the middle. For some large animals, such as a giraffe (*center*), this neck posture directs weight more efficiently, and it was the one likely for especially long-necked, quadrupedal forms like a sauropod (*below*).

probably formed a dorsal bracing system in sauropod necks, along with the powerful dorsal neck muscles.

There's also a second solution, **ventral bracing**. Here the load is supported underneath by rigid structures, attached to the beam making up the load. On a suspension bridge, these run along the sides, and underneath, the roadway and collectively provide stability. Sauropods, unlike a suspension bridge, didn't use ventral bracing to hold up their necks. They did, however, have a similar system of slender bones, attached in pairs in front of and underneath each cervical vertebra, called *cervical ribs*. Each pair of cervical ribs was attached to its vertebra, and to other cervical ribs before and behind it, by strong yet elastic ligaments that permitted them to slide past one another. Cervical ribs, although they didn't contribute to holding up a sauropod's neck, did help to control another serious problem that long cantilevered structures can have: *torque*. This is the tendency of a long, heavy object to twist and become unbalanced both under its own weight and when it's in motion. Controlling torque is especially important when a large structure has to move, and for this reason many industrial cranes or derricks, designed to lift loads up, down and sideways, have a system of diagonal and right-angled struts that collectively make the beam (or boom) rigid and resistant to torque. Sauropod necks could not be rigid, however, since their function was to reach up, down and around, not to lift weights. In sauropod species with long cervical ribs, the ribs made sliding contact along their ventral surfaces. With the help of smaller ligaments and muscles, the cervical ribs prevented twisting and uncontrollable sideways motion when the animal moved its neck laterally and/or ventrally. These reached extreme lengths in omeisaurids and mamenchisaurids such as *M. sinocanadorum*, in which a single cervical rib can extend under and along *six* subsequent vertebrae.

The tension created by both the load of the crane boom itself (the mechanical equivalent of a neck) and the cargo is often handled in part by a *double cable system* that originates from an A-frame *gantry* (or *mast*) near the crane's base and runs to the tip of the boom. Having two cables gives more control, like having two handlebars on a bike to steer instead of just one stick. In sauropods, the dorsal and sacral vertebral spines were the equivalent of a crane's gantry. While some sauropods had single, low cervical (and sometimes dorsal) spines that provided attachment for the above ligaments and dorsal muscles that held the neck up, others to varying degrees had *double* cervical spines, especially among bipedal/tripodal rearing specialists like diplodocids. In these, the double supraspinous ligament and a double set of dorsal muscles would have permitted the bilateral support and movement needed in a tall, flexible living structure often held vertically or subvertically. This, as with torque control, is a case of a naturally evolved structure and human technology obeying the same mechanical principles, but the marvel here is that the sauropod neck allowed both stability *and* flexibility.

As we remember, the earliest sauropod ancestors combined their already bipedal posture with their long necks to reach higher than other dinosaurs. Even when sauropods grew larger and heavier, there was still a great advantage in retaining the ability to rear up to reach this food, and for this reason the skeletons of many forms became specialized to allow them to continue

Brachiosaurids, mamenchisaurids and other sauropods (*left*) that fed from a quadrupedal position had only moderately tall dorsal vertebrae and neural spines over the dorsals and sacrum (*blue areas*). In upright-feeding diplodocids (*right*) these were very tall to provide the ligaments and muscles of the spine with extra leverage for tilting up and back, far more than elephants that occasionally do this (*middle*), and indicating that this was a habitual posture for these sauropods.

to browse bipedally, if they needed to, as well as quadrupedally. While industrial cranes rely on a winch and winding cables to raise and lower the boom, sauropods had a system of dorsal, ventral and lateral neck muscles for this. Because there were many cervical vertebrae (up to 19 in some species), these collectively allowed for more gradual, and more controlled, movement than found in even the most sophisticated industrial crane, and the shapes and pneumaticity of each bone made the entire neck very lightweight for its size. All sauropod types didn't feed from the same heights and in the same way, however, and as a result there are differences in both the height and shape of the vertebral spines. Types that habitually tended to feed quadrupedally like *Brachiosaurus* or *Omeisaurus* often have necks with very short vertebral spines for dorsal bracing and very long cervical ribs for ventral torque control, while specialists for rearing like *Diplodocus* and *Brontosaurus* had tall dorsal and sacral spines, shorter cervical ribs and moderately to strongly bifurcated cervical and dorsal vertebrae. *Diplodocus* and its close cousins *Apatosaurus* and *Brontosaurus*, among the best-known sauropods, were members of the family Diplodocidae, which, in addition to their spinal adaptations, had relatively short forelimbs, extra robust pelvises and extremely long, heavy tails. As pointed out in the 1970s by Bakker, these features, and the expanded chevron bones that reinforced the underside of the caudals from the 20th to the 27th vertebrae, were all adaptations to rear bipedality (see Chapter 6), and the reinforced tail actually helped prop up the animals into a tripodal feeding position. A typical sauropod's center of body mass was located just in front of the pelvis. The effective arrangement of strong muscles (like the paired *caudofemoralis,* running from the anterior tail to the femur and others) made it much easier for sauropods to rear than elephants, which in the wild occasionally do this to reach high tree foliage. Since stability was especially important for a huge animal browsing from an almost vertical position, a sauropod's big, outwardly slanting pedal claws probably played the same role as the horizontal extensions on a backhoe excavator, digging sideways into the ground to keep the tall browser steady while it moved its neck. Another specialization that evolved to contribute to stability when rearing (and secondarily in walking) was the downward relocation on the anterior proximal femoral surface of a *process* (or attachment origin) called the *lesser trochanter*. Originally a place for the insertion of the *iliofemoralis* muscle originating from the ilium to pull the leg forward, this process moved to a position on the femur's proximal and lateral surface, becoming a ridge or bump. In this new spot, it made the muscle effective in swinging the upper leg *back and away from the body* rather than forward, which helped the animal brace its hind legs outward to broaden its base of support and create more stability.

THE BEST FIT?

Not all sauropod specialists agree with this theory. Some researchers, such as F. Martin (Leicester City Museum), Michael J. Parrish (Northern Illinois University) and Kent Stevens (University of Oregon), feel that the above anatomical features indicate that most sauropods (with exceptions like *Brachiosaurus* and some other "high-shouldered" forms) were *medium-height to low browsers*. As a result, they would have had a generally much more horizontal, "head-down" feeding orientation for browsing on shorter plants, like ferns. In support of this, Stevens and Parrish argue that the articular surfaces of the zygapophyses, the upper connecting points for each adjoining vertebrae, should naturally overlap with each other as much as possible when an animal is in a resting or **osteologically neutral pose** (or ONP, known informally as "best fit"), and never less than 50% when active. Starting with this concept and using approaches such as manipulating dissected cervicals from turkeys and other present-day birds, as well as taking illustrations and photos of various sauropod cervical series and digitally orienting them, these researchers formulated sets of ONP configurations for the cervical vertebrae of *Diplodocus*, *Apatosaurus* and *Brachiosaurus*, as well as some other forms. Based on these programmed images, they then established what they considered limitations on the amount of movement of which they felt each of these sauropods' necks were capable,

Quadrupedally feeding sauropods like *Omeisaurus* (*right*) tended to keep the middle of their necks relatively straight (*dark orange bars*), which was partially reinforced and supported by the long middle cervical vertebrae and ribs, while feeding. In these most neck movement was anterior-posterior, with flexion occurring primarily at the base and near the skull (*yellow, light orange bars*). In upright-feeding diplodocids like *Barosaurus* (*left*) and some others, however, the cervicals tended to be shorter, with much shorter cervical ribs. This allowed the neck to have a lot of lateral as well as fore-and-aft flexion.

This drawing shows the posture most sauropods would have most of the time if they were subjected to an osteologically neutral position, or ONP. Not only would their eyes have difficulty seeing forward because of the downtilted head, but their out-stretched, habitually horizontal necks would make a tempting target for a theropod ambush. While small, short-necked types like *Nigersaurus* and *Brachytrachelopan* were exceptions, long-necked, arboreally browsing sauropods probably adopted this posture when only low browse was available.

using the terms "dorsiflect" (move dorsally) and "ventriflect" (move ventrally), as well as "lateral flexion." According to this, a digital *Apatosaurus* could "dorsiflect," or raise its head, about 20 degrees above shoulder level, while *Diplodocus* was capable of 15 degrees. In "ventriflection," or downward motion, both species had a range of about 30 degrees; they also had a digitized lateral flexion range of 30 degrees. The two researchers also used their theory of ONP to test the up-and-down limitations of *Apatosaurus* and *Diplodocus* skulls. Based on these findings, Stevens and Parrish concluded that the potential feeding envelope of these sauropods, including *Brachiosaurus*, was much more limited than if these species had more flexible necks. Regarding the possibility of bipedal or tripodal feeding, both point to potential problems of balance and neuromuscular coordination that they feel would make this unlikely. In general, the Stephens and Parrish theory of ONP postulates that sauropods habitually carried their necks in a much more horizontal, and in some cases slightly downward-curving, posture and not in an elevated, upwardly curved position. They concluded that the "best fit" concept would keep diplodocids from being able to reach very far above shoulder level, and that these dinosaurs always fed from a quadrupedal stance close to the ground. Like diplodocids, brachiosaurids' necks came straight out from the shoulders and in ONP usually were held, except for feeding, at only about 20 degrees.

Is this correct? In their research into the cervical biomechanics of *Brachiosaurus*, Andreas Christian and Wolf-Dieter Heinrich (Museum für Naturkunde) analyzed and then calculated the compressive forces that would have acted on this species' intervertebral discs when the neck was oriented in three positions, (1) horizontal, (2) at 30 degrees and (3) vertically, all in the configuration of a gentle *sigmoid* (S-shaped) curve. They found that the overall stress placed on the cervical series decreased considerably as the neck became more elevated, with the centra bearing the loads most efficiently as the pose became more vertical. This vertical neck position for *Brachiosaurus* is very similar to the one postulated as early as the 1970s by Bakker and Paul. These findings, as well as Christian and Heinrich's studies, actually suggest an upright, highly elevated neck position for many long-necked, quadrupedal forms.

WHAT REAL NECKS SAY

Coauthor Wedel and his colleagues Darren Naish and Michael Taylor (Palaeobiology Research Group, University of Portsmouth) have demonstrated that in a wide variety of present-day animals (mammals, birds, etc.) at rest, the zygapophyseal connections of tetrapods' cervical vertebrae actually do not need to overlap very much to stay in safe alignment. For almost all of these forms, a neck in ONP typically "dorsiflexes" (or more correctly stated, extends) *upward* at its base where it arises from the dorsals, straightens out slightly and then "ventriflexes" (or just *flexes*) *downward*. This results in a *sigmoid* (S-shaped) *curve*, partly made possible by the flexible cartilage in between the cervicals. Unless sauropods' necks followed completely different biomechanical principles than modern forms and had completely different bony articulations, then it's reasonable to think that their necks would also have had a sigmoid pose in life. This would mean that these dinosaurs had an elevated neck pose, and not a ventrally curved or horizontal one. Observations have shown that when live giraffes, camels and other long-necked large mammals bend their necks laterally there is sometimes almost no zygapophyseal overlap, but the strong, resilient ligaments and connections between the centra prevent dislocation. Here the intervertebral cartilage in a live animal is also an important factor to consider, since this permits greater flexibility than would be possible if the cervicals were directly in contact. Different techniques of manipulating the vertebral series of present-day and extinct animals can also yield different results. Even though Stevens and Parrish's digital cervicals of *Diplodocus carnegii* specimen CM 84 indicated a downward-curved and relatively inflexible neck, the actual manipulation by hand of the same specimen's fossil cervicals by researcher Ken Carpenter (Denver Museum of Natural History) resulted in a posture that could be extended farther both dorsally and laterally. Again, if we remember that sauropods like present-day forms also had intervertebral cartilage, then their cervicals would have been capable of a much greater range of motion. This is also suggested by a series of dissections by Wedel on the necks of rheas and some other birds, archosaurian cousins as a group to the sauropods.

ON THE LEVEL

Both living and extinct forms, including sauropods, have spaces within their skulls that are, or were, occupied by *semicircular canals* (SCCs), vertical and horizontal liquid-filled tubes that let an animal know whether its head is level or tilted. Some researchers who have studied these claim that living animals keep their horizontal semicircular canals (HSCCs) level in ONP, and this would also be true of extinct forms. In reality, however, there is a wide range of variation in the way living animals habitually hold or tilt their heads, regardless of HSCC position, with rabbits at about 16 degrees, cats at 20 degrees and humans at 22 degrees. These figures vary much more widely than the 5–10 degrees of possible variation one researcher has claimed. Stevens and Parrish's theory of ONP requires that the skulls of *Diplodocus*, *Apatosaurus* and *Amargasaurus*, because of the orientation of the *foramen magnum* and *occipital condyles*, must be oriented almost straight down, at an acute angle to the long axis of the first cervical (*atlas* bone) to which they're connected. Do the theories of level HSCCs and ONP corroborate, or support, each other? For the moment, let's consider the skulls of *Diplodocus* and the macronarian *Camarasaurus* with their HSCCs actually horizontal. If we stay with this limitation, the occipital condyle and atlas connection would have allowed both sauropods to tilt their heads *at least* 30 degrees upward and 20 degrees downward. Interestingly, these were also the habitual orientations found for many

A versatile feeder, *Diplodocus carnegii* (*overleaf*) and other diplodocids carried their necks in a U-shaped or sigmoid curve, and were capable of feeding both quadrupedally and bipedally as the need arose. Studies by **Hallett**, **Wedel** and **Taylor** (*above*) now cast doubt on the ONP hypothesis that many sauropods walked holding their necks horizontally and curved slightly downward.

birds by the ornithologist M. Duijm (University of Amsterdam) and agree with the range found in many modern mammals. In doing this and keeping their HSCCs level, both sauropod skulls would actually attach to necks *angled upward* at about 25–30 degrees, a posture consistent with the flexed neck seen in modern amniotes. ONP, on the other hand, requires the skulls on a horizontal, downward-curved neck to be attached at such acute angles that the two sauropods' heads would actually be tilted ventrally and posteriorly, which would have made it hard for them to see in front. The first described posture is more typical of what we can observe from living tetrapods than it would be from a more or less horizontal or downward-curving neck, and it also allows the probable line of forward (and downward) vision to be less obstructed by the long anterior and dorsal portions of the sauropods' skulls. Taken as a whole, these and the neck postures of living animals strongly contradict the ONP hypothesis.

STANDING TALL, A SAUROPOD SPECIALTY

Regarding the question of whether sauropods could habitually rear upright, can we learn anything by comparing sauropods to living animals? No modern form closely resembled these dinosaurs, but the bulky torso and column-like limbs are in some ways similar to those of elephants. Since African elephants in the wild can and do occasionally rear to reach food with their trunks, and Asian circus elephants are trained to do this regularly, observations of living elephants can provide valuable insights into the biomechanics of sauropods, as noted by Paul. Unlike forward-heavy elephants that have to carefully keep their centers of gravity above their

strongly flexed (and stressed) hind legs as close to the ground as possible when rearing to avoid tipping over, sauropods have significant tails. These provided a counterbalance for the rest of the body, which was already very light for its size because of pneumaticity. Sauropods have proportionately larger, stronger dorsals, which progress in size and height posteriorly toward the hips, a key bipedal adaptation that allowed them to support the entire forward weight of the body. The dorsal and sacral spines evolved to an almost fantastic extreme among the diplodocids, like *Apatosaurus*. The proportionately larger pelvises, with their massive cushions of fibrocartilage and oversized hind legs, bore much more of the body's mass than an elephant's. Sauropods and other dinosaurs were ***hindlimb dominant*** compared to the more robust forelimbs of elephants and other mammals; the short forelimbs of rearing specialists like *Apatosaurus* and other diplodocids were most likely an adaptation to reduce more forward weight in rearing. Even long-armed, high-shouldered forms like brachiosaurids may possibly have been capable of this, although as Heinrich Mallison (Museum für Naturkunde) noted, a brachiosaur's higher center of mass and relative instability would have made this position much less secure than for a diplodocid.

In conclusion, sauropods took the possibilities of bony support literally to heights never before or since seen in the animal world. With certain exceptions, vertical reach and upright feeding dominated sauropod evolution and explain their amazing skeletal structure.

Although the possible heart size of sauropods is currently being debated, those of the most enormous may have rivaled that of the blue whale, seen here as a life-sized model.

CHAPTER FIVE

BRONTOSAUR BIOLOGY

TO IMMENSITY AND BEYOND

For almost any multicellular organism's body to perform efficiently, many intricate systems must work simultaneously. This is greatly compounded when an animal becomes huge, as large as a whale or a dinosaur. Many of these organ systems evolved with and to permit increasing size over time, and recent studies of modern animals are giving us insight into the paleobiological questions raised by sauropods.

I T'S ALL ABOUT THE CELL. Gigantic creatures like sauropod dinosaurs, whales and elephants are so awesome in their dimensions that we sometimes forget that they were, and are, subject to many of the same biological constraints as the tiniest, most fragile insect. It's a marvel of nature that such creatures got to be so big by simply keeping to the laws of physics and biology, and that they succeeded in achieving such size and stature. In Chapter 4, we explored the skeletal/muscular anatomy of sauropods and came to some conclusions about how these dinosaurs' giant bodies operated on a macroscopic level, but what about the other end of the scale? We can understand and appreciate this by considering some basic biological rules a sauropod or a gnat must live by, and this starts with the cell. Cells, those little bags of protoplasm that make up the bodies of complex, multicellular organisms from dragonflies to dinosaurs, all

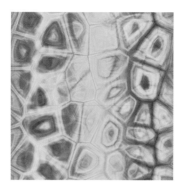

When cells aggregate in large enough numbers, like residents in a growing community, they require an ever more complex network of blood vessels to supply nutrients and remove wastes. For giant animals, this situation is compounded many times because of their massive internal volumes.

have similar, fundamental requirements, and these are basically to obtain food and oxygen and to excrete wastes. We sometimes don't realize how much the anatomy of living things is shaped by the needs of their cells, and this was as true for a giant sauropod as for any creature. Single-celled organisms deal directly with the outside environment, by **diffusion**, or transfer across their cell membranes. That works for the smallest multicellular organisms, too—mosses need no special tissues for transporting water and sugars among their cells, and earthworms and some salamanders have no lungs, getting all the oxygen they need through their moist skins. But at some point as organisms got larger and more complex, they needed specialized structures to collect and distribute the materials that their cells required (oxygen and food). They also had to eliminate wastes (carbon dioxide) and the by-products of digestion. For this reason, multicellular animals and plants have sophisticated systems that provide these functions. It was a problem nature solved millions of years before the dinosaurs or other vertebrates evolved.

In considering a huge animal like a sauropod, let's begin with the fact that the problems of supplying raw materials to an animal's cells and getting rid of its wastes become more complicated as it gets bigger. As a 3D object increases in size, its **volume increases three times, not twice, in proportion to its linear dimensions**. If we had a 2″ long beetle, for example, its volume would be 8 cubic inches ($2 \times 2 \times 2 = 8$). By increasing the beetle's length to 4″, the linear dimensions double, but not the volume, which is now *64* cubic inches. Now our beetle has *8* times as many "insides" (cells) to service, even though its linear dimensions have multiplied only *twice*—for far larger creatures, the overall internal volume becomes almost astronomical. Almost all of the processes necessary for life occur across membranes. The surface area, which increases or decreases with linear dimensions, limits this because larger organisms have much more internal volume relative to their outside envelopes than small ones, and these present a challenge for these proportionately smaller surfaces to deal with.

Trees avoid this mismatch between volume and surface area by "cheating." Most of the volume of a tree is wood—dead cells—with only a thin layer of living cells around the outside, under the bark. This layer of living cells grows proportionally to the surface area of the trunk, not its volume. Trees also multiply their surface area by making more and more leaves up top, as well as more and more roots underground. Most animals, however, don't have the option to simply grow more surface area outside to compensate for when their volumes increase. As a result, most multicellular animals have specialized tissues that *internally* multiply surface areas. Here the extra surface area is hidden inside the body, in systems of complex chambers, folds, pockets, and tubes. If a tiny, simple village grew into a huge city, you could get more people living in the same geographic space by having tall, interconnected apartments or condos instead of single houses. Of course, now that you've got a city, thousands of complex alleys and roads are needed to reach deep into the now vast community to supply its basic needs. In larger animals, the tissues are organized into complex organs such as the lungs to absorb oxygen for the rest of the body while eliminating carbon dioxide, intestines to take in dissolved food and kidneys to eliminate cell wastes—all these systems must keep up with the demands of their cells. While most worms have a digestive tract that consists of a single tube running more or less straight from mouth to anus, the intestines of large backboned animals are coiled up to fit more length and absorptive area inside (your own intestines are 3–4 times as long as you are tall). They're lined with internal ridges, which are themselves covered in tiny finger-like extensions called *villi*, which in turn are covered in even *tinier* hair-like tubes called *microvilli*. As a result, an average adult human has an intestinal surface area the size of a football field. Many of our other organs are similarly complex, all with the purpose of serving the ultimate microscopic customer: the cell. This principle even governs the shape of our brains, which have lots of folds to fit more cells (volume) inside the enveloping skull (surface area). Most animals use the same tricks to fit large surface areas into small volumes. For example, birds evolved complex brains independently of mammals, but both groups hit on the same solution, that of folding the surface of the brain. Birds and mammals took very different paths in evolving complex respiratory systems. Considering that as archosaurs both birds and sauropods are related to each other, this is one key to understanding how sauropods got so big. Everything that's needed to keep cells alive becomes a challenge as animals get bigger. First, let's start with one of the most basic issues—food, the one that started the sauropods on their path to immensity.

FEEDING THE BEAST

It's a question that naturally comes up when we think of sauropods: how did they obtain enough food to maintain fuel for their enormous bodies? To explore this, we have to confront two misconceptions that frequently come up regarding these dinosaurs: head size and jaw strength. The small heads of most sauropods are often pointed to as a severe limitation in how much food they could harvest in sustaining their massive bodies. This assumption is considered, in turn, to have automatically excluded sauropods from being animals with consistently high mammalian or bird-like levels of activity, since a small head leads to the idea that sauropods could not have satisfied such caloric needs with the limited amount of food they could have ingested per day.

If we look at a typical mammalian herbivore's head strictly from the standpoint of being a food-gathering device, we see that a lot of the mouth and head space is devoted to the teeth used in chewing or grinding, the premolars and molars, with very little taken up by the incisors, the teeth that actually bite off or collect the food. If the animal didn't need to chew, its head could be much smaller. This is precisely the case with sauropods, which maintained the saurischian system of basically cropping food but not chewing, and then sending it down to the stomach, sluiced down by plentiful saliva and a muscular gullet. While the heads of sauropods *were* small, the space occupied by food-cropping incisor teeth was far greater in proportion to their skulls than those

While an increase in surface dimensions (here a beetle) is easily understood, what's not as apparent is the way **volume** changes. Taking the insect and doubling the surface dimensions (length and width) makes it only *twice* as large, but the volume or interior increases cubically, making it 8 times bigger inside, and this means 8 times the number of cells to be serviced. The body of a sauropod (represented here by just a foot) had many thousands of times this volume of cells, and its body evolved complex and amazing organ systems to deal with their needs and keep the animal alive.

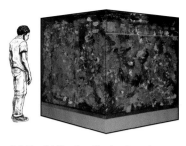

A 3.5 m³ (4′ x 4′ x 4′) plastic cube from the traveling exhibit "The Largest Dinosaurs" contains a facsimile of what might have been the daily amount of browse for one of the larger species of sauropods.

of a mammalian herbivore. Compared with a modern large horse whose skull was roughly similar in size, *Diplodocus* had a mouth that was 2.5 times broader, with 4 times the number of cropping incisors for harvesting food. Next to the giant extinct rhino *Paraceratherium* (*Indricotherium*), the largest known terrestrial mammal, a *Diplodocus* with a torso mass equal in size again still had the broader mouth, as well as more cropping dentition.

Could such heads and teeth bring in enough fodder to fuel a high metabolism? Paul has estimated that a hypothetical, tachymetabolic (*tachy*, "swift"; *meta*, "to change") *Brachiosaurus* weighing 35–45.5 tons might have required about half a ton of fresh browse per day. If a dinosaur with a mouth capacity of 1.5 bushels averaged 16 hours of feeding per day and took from one to six bites per minute (a typical feeding rate for large herbivores like giraffes), then each bite would have equaled from 2/3 to at least 1/10 of a kg (0.22 lb). Feeding at this rate, a 40-ton sauropod with a 10 m (32′) long neck and a 150 kg (330 lb) head could have cropped enough bites to bring in enough food to equal 1.5% of its body mass per day. This is comparable to the dietary needs of an elephant. *Brachiosaurus* might well have taken larger bites than this, thus shortening its feeding time. A second misconception regarding sauropods as food gatherers has been that the jaws were weak, probably owing to the long-held assumption that they were eating soft, aquatic plant material. Weakness, of course, in this case is a subjective term: sauropods certainly had weaker jaws in terms of absolute strength compared to that of most theropods with similarly sized skulls, but what we are concerned with here is how effective the jaws were in cropping. In the case of *Diplodocus*, the mandible, based on methods of bite-force calculation and computer modeling, could exert a force of about 85 lb/sq. ft. This compares favorably with the bite force of 88.5 lb/sq. ft. displayed by a large horse with a similar-sized head.

The skulls of sauropods all show that they are strongly built to take the stresses of either chopping or pulling off foliage, and the tooth-bearing areas (*alveoli*) of both upper and lower jaws are reinforced by extra deep bone margins that cover the *labial*, or outer, surfaces where the shearing force applied to the teeth would be greatest. The mandible, or lower jaw, is deepest and most robust at the area of the *symphysis*, or joining between the two jaws, further strengthening the tooth-bearing areas of the jaws against the demands of harvesting foliage. Sauropod skulls also indicate overall jaw action and type of bite. Most macronarian skulls are high and relatively short, with jaw closure (adductor) muscles that attached at a steep angle to the mandible. They have jaw joints that are relatively short from front to back. This would have permitted the lower jaw to swing up and down but allowed little or no front-to-back motion, and it was well adapted for cutting and chopping through clumps of foliage powered by muscles that provided a basically upward pull. The long, low skulls of *diplodocoids*, on the other hand, had jaw-closing muscles that were oriented at a shallower angle to the mandible, with longer jaw joints that allowed a forward-and-back, as well as up-and-down, motion to the mandible. This kind of structure is better suited for grasping and pulling backward rather than chopping or shearing, and as we'll see in Chapter 6, these two basic jaw movements, in combination with sauropod tooth adaptations, allowed these dinosaurs to efficiently harvest a variety of conifer foods. This hypothesis is supported not only by anatomical and biomechanical studies of the postcranial skeleton but also by recent analyses of the jaws, teeth and dental microwear of known forms.

Diplodocid skulls like those of *Diplodocus* and *Apatosaurus* show a highly unique feature that set them apart from all known sauropods. Two thin *laminae*, or sheets of bone, one set from the maxillae of the cranium and the other from the dentary bones of the mandible, form an enclosed space at the sides of the jaws when the mouth is shut. As pointed out by Barrett and Upchurch, diplodocids probably had a wide gape to crop a large mass of foliage, and the laminae may have prevented the mass of cropped leaflets within the mouth from falling out after a bite was taken and these had been raked off. If so, it may have been roughly the functional equivalent of a mammalian herbivore's fleshy cheeks; by eliminating lost food after a bite was taken, these would have maximized the amount of food ingested, thereby increasing the animals' feeding efficiency. This characteristic evolved independently, and was developed even further, by a titanosaur, *Bonitasaura salgadoi*, from Late Cretaceous Patagonia, Argentina. Although the maxilla isn't known, the dorsal border of the mandible behind the teeth is thin, ridge-like and filled with abundant vascular tissue. This suggested to the worker who described the sauropod, Sebastian Apesteguia (Museo Argentino de Ciencia Naturales), that there may have actually been keratinous, horny blades on the upper and lower jaws that sheared past one another to crop vegetation, as well as holding the food mass in place.

Once a mouthful of a particular plant was cropped, it had to be swallowed. With no ability to chew, sauropods of all kinds probably needed both strong, well-developed tongues and large salivary glands to help sluice the bulky food, aided by contractions of a muscular, capacious esophagus, on its long way down to the stomach. Although no direct evidence for it exists, some diplodocids like *Diplodocus* and *Dicraeosaurus* may have taken this further. In addition to being an area where replacement lower incisors were formed, the presence of a pointed bump or "chin-like" process underneath the mandible *may* have been an attachment for a tongue that might have had prehensile capabilities, like a giraffe's. Such a tongue, if it actually existed, might have been as useful in helping to pull conifer branches into the mouth for a diplodocoid as would a giraffe's in harvesting thorn acacia foliage. After being swallowed, the food mass (known as a ***bolus***) was pushed down the long esophagus by peristaltic action into the stomach by the combined action of longitudinal and circular muscles that work the same way as ours do. Probable sauropod foods (conifers, ferns, horsetails and ginkgoaleans), like all plants, contain a compound known as ***cellulose***, the ingredient in plant cell walls that helps give them shape and support and makes up most of the fiber in what herbivores eat. Cellulose contains valuable carbohydrates, and in modern *ruminants* like sheep and cattle up to 80% of digestible energy is from cellulose that's been broken down by bacteria and microbes in the animals' gut. Cellulose, however, is extremely tough and hard to break down, so vertebrate herbivores deal with this by having long, capacious digestive tracts with specialized fermentation regions to house the microbes. There are two basic systems, ***foregut*** and ***hindgut*** fermentation. With *foregut fermenters* (ruminant hoofed animals like cattle, sheep and deer) the chewed food is sent to a complex stomach, where it's shunted between specialized, microbe-filled chambers (cattle, for example, have four of these) until it yields its nutritional payoff. *Hindgut fermenters* (horses, rhinos, elephants, gorillas, some birds and others) do it differently. Here the food is only *partially digested* in a simple stomach. It then travels to a specialized part of the intestine called a ***caecum***, where

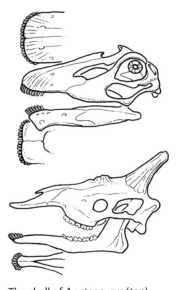

The skull of *Apatosaurus* (*top*), drawn to the same scale with that of a giraffe (*bottom*). In spite of being of similar size, the head of this sauropod has about 7 times the amount of cropping dentition (*blue*) as the mammal. This, and the fact that sauropods didn't need to chew, meant that they could crop and swallow proportionately a far greater quantity of browse per bite and per hour than a mammal, an adequate amount to sustain their huge bodies. (Modified from original art by Gregory S. Paul.)

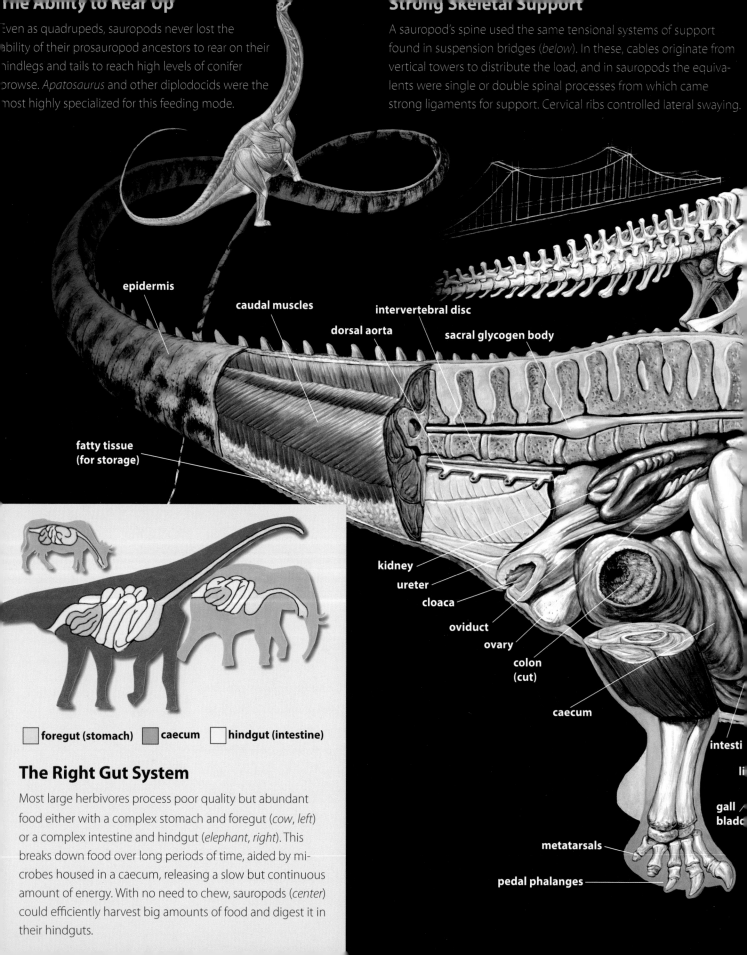

The Ability to Rear Up

Even as quadrupeds, sauropods never lost the ability of their prosauropod ancestors to rear on their hindlegs and tails to reach high levels of conifer browse. *Apatosaurus* and other diplodocids were the most highly specialized for this feeding mode.

Strong Skeletal Support

A sauropod's spine used the same tensional systems of support found in suspension bridges (*below*). In these, cables originate from vertical towers to distribute the load, and in sauropods the equivalents were single or double spinal processes from which came strong ligaments for support. Cervical ribs controlled lateral swaying.

epidermis

caudal muscles

intervertebral disc

dorsal aorta

sacral glycogen body

fatty tissue
(for storage)

kidney

ureter

cloaca

oviduct

ovary

colon
(cut)

caecum

intesti

li

gall /
blad

metatarsals

pedal phalanges

☐ foregut (stomach) ☐ caecum ☐ hindgut (intestine)

The Right Gut System

Most large herbivores process poor quality but abundant food either with a complex stomach and foregut (*cow*, *left*) or a complex intestine and hindgut (*elephant*, *right*). This breaks down food over long periods of time, aided by microbes housed in a caecum, releasing a slow but continuous amount of energy. With no need to chew, sauropods (*center*) could efficiently harvest big amounts of food and digest it in their hindguts.

A Flexible Crane

Cranes have rigid beams formed by diagonal struts, which are controlled by cables that originate from a mast. A sauropod's neck, though similarly constructed, was made flexible by many vertebrae connected by elastic ligaments, and moved by contractions from a complex muscular system.

Big Hearts

The heart of a sauropod was proportionately larger than an elephant's, with a particularly huge, thick-walled pair of ventricles to pump blood up into the long neck. The arterial system may have had special muscular swellings to help pump blood, and in the brain a rete mirabile vascular net to prevent hemorrhaging in the brain by absorbing blood pressure.

dorsal vertebrae

spinal cord

stomach (cut)

lung

supraspinal ligament

cervical muscles (superficial)

cervical muscles (deep)

trachea

esophagus

carotid artery

jugular vein

right atrium of heart

heart

brachial muscles (upper forelimb)

right ventricle

metacarpals

manual phalanges

In the Belly of a Brontosaur

Although no single species is typical of all the others, the body plan of *Apatosaurus excelsus* shows the main features of a sauropod's main organ systems, which were probably in some ways similar to those of modern archosaurians like crocs. As did their saurischian cousins, sauropods inherited a unidirectional lung system like that of birds (see next spread) and evolved massive, hindgut-dominated digestive tracts similar to those of large mammals like elephants to slowly process and release nutrients. A four-chambered heart kept oxygenated and non-oxygenated blood separate and helped to fuel a metabolism that was actually or virtually endothermic, like that of a bird's or mammal's.

The Right Size Brain

Tiny in absolute size, the brain of a sauropod was actually as large as it had to be to direct body movement and behaviors. It was well protected from changes in blood pressure from large, distendable vascular sinuses.

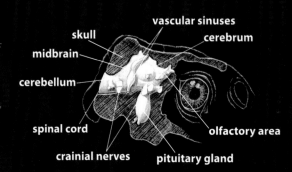

vascular sinuses

skull

cerebrum

midbrain

cerebellum

spinal cord

crainial nerves

pituitary gland

olfactory area

a concentrated community of microbes that can break down cellulose and other tough plant structures goes to work. These dissolve the partly processed food into a broth that the animal's own digestive enzymes can now deal with, and which now allows for the extraction of valuable fatty acids and other foods in other parts of the intestine. For the microbes to accomplish this, the plant material needs to stay in the caecum for enough time to break down, like a compost bin that releases the products of its decomposition to create rich soil.

The amount of time, as well as the amount of food, that the partly digested plant food spends in the caecum is critical if this process is to work efficiently. In their recent experiments to find out the nutritional value of certain types of plants that could have served as sauropod foods, Jürgen Hummel and Marcus Clauss (Steinmann-Institut für Geologie, Mineralogie und Pala-ontologie, Rheinische Friedrich-Wilhelms-Universitat Bonn) discovered that some plants only gave up their nutrients fully after they had been digested for a minimum amount of time. For one conifer genus, *Araucaria*, an unusual quality of this foliage was that when ground up under laboratory conditions and made to chemically ferment, ***more energy was produced when it was retained for very long amounts of time*** (up to 30 hours). This is similar to the length of time it would probably spend in a hindgut fermenter's digestive tract. Another interesting result from the same experiment is that as the gut retention time of the food is increased, ***the larger is the size of the food particle that can be digested.*** In other words, you can digest a bigger bite if you keep it in your gut longer. This explains how sauropods were able to get away with not chewing their food: with big, extensive hindgut systems they could process and extract high-quality nutrition just by keeping it inside, and in big enough quantities, for this to work. This has led James Farlow (Indiana-Purdue University), who has extensively considered the possibilities of dinosaur diets and digestive physiology, to hypothesize that most dinosaurian herbivores were hindgut specialists. For sauropods already adapted to harvesting huge amounts of low-quality but abundant food, a bigger digestive tract was a better one, but it also meant having to evolve a bigger, stronger frame to carry it around.

AIR AND OXYGEN: KEEPING IT MOVING

For most living things, breathing or assimilating air into the lungs or other organs is usually automatic. Technically called *ventilation*, this is only the first stage in breathing. The next stage, *gas exchange*, happens inside, across the delicate membranes of our lungs, where ever-narrowing, finely branching tubules deliver the air to blind-ended, bubble-shaped air sacs called *alveoli*. Each of these is ringed by capillaries, in which the blood pumped from the body surrenders its heavy, concentrated freight of acidic carbon dioxide (for animals, a waste product) when the animal breathes out, or exhales. The carbon dioxide diffuses into each alveolus for its return journey as exhaled air. This process is followed by breathing in, or inhalation. Here oxygen is captured by the red blood cells, which contain hemoglobin, an iron oxide–based protein with an affinity for absorbing oxygen. The newly invigorated blood is now pumped from the lungs to the body's waiting tissues, to which the red blood cells deliver their oxygen through diffusion across cell membranes. Take a deep breath and you may feel a tiny burst of energy and clarity as the incoming oxygen floods your bloodstream.

Small animals, especially those with slow metabolic rates, require comparatively little surface area for this process of gas exchange. In lungfish and amphibians the lungs are simple sacs with little or no internal structure, like balloons. One obvious way to increase the surface area of such a lung without increasing its size is to pack the balloon with membranes—to fill the big sac with lots of smaller sacs. And that's just what most animals with more sophisticated lungs have done. Mammals have taken this to an extreme—in a healthy adult human, each lung contains 200–300 million alveoli, which give the lungs a combined surface area of about 70 m² (750 sq. ft.). That's a lot of area for gas exchange, but we are limited in other ways. The most obvious limitation is that the air in our alveoli just sits there, in the brief span between the end of inhalation and the beginning of exhalation. So we rely on simple diffusion of oxygen and carbon dioxide between our blood, which is moving through the lungs, and the inhaled air in our lungs, which remains motionless with respect to the blood. Faster movement of air in and out of the alveoli by breathing more quickly—hyperventilation—enables more gas exchange but doesn't speed up the process.

It doesn't have to be this way. In the gills of fish, for example, there is **countercurrent** exchange of oxygen and carbon dioxide. The gills have small ridges or plates called *lamellae* that are lined

In contrast to a daytime photo (*opposite*), a thermal photo of an African elephant (*above*) reveals which areas of its body contain the greatest amounts of heat (cooler areas blue-green, warmer areas yellow-red). As elephants do with their ears, sauropods used body areas like their long, air-filled necks to eliminate heat from the body's interior (*below*).

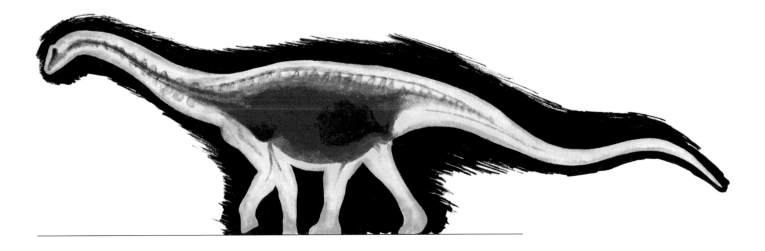

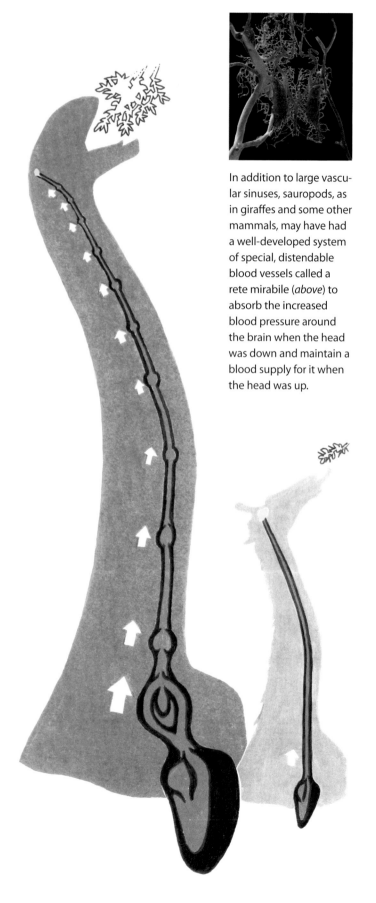

In addition to large vascular sinuses, sauropods, as in giraffes and some other mammals, may have had a well-developed system of special, distendable blood vessels called a rete mirabile (*above*) to absorb the increased blood pressure around the brain when the head was down and maintain a blood supply for it when the head was up.

up parallel to the flow of water. As water flows over the gills from front to back, the blood in the lamellae flows from back to front. So the moving blood is not exchanging gas with a static pool of water but with a constant stream, and the farther forward within a lamella the blood travels, the fresher the water it encounters. This has the effect of supercharging the blood with available oxygen. Alas, the key word is *available*—even the most oxygen-rich water only holds about 1/30 as much oxygen as air. Fish that can breathe air, like tropical lungfish, enjoy a huge advantage—as long as they can keep their respiratory membranes from drying out, they have access to far more oxygen in air than in water, and this was probably the main reason for the evolution of lungs in the first place.

So with their gills, fish have a very effective gas exchange mechanism—countercurrent exchange—but they are limited by the low oxygen content of water. We mammals breathe air, but we do so using a relatively inefficient form of gas exchange—simple diffusion with a static pool of air. The *ultimate* respiratory system would be one that combined the mechanism of (1) countercurrent exchange with (2) the vastly greater oxygen concentration of air. As far as we know, such a system has only evolved once among vertebrates, in the archosaurs: the crocs, birds, dinosaurs and some others. Crocodilians use a simple form of this flow-through breathing, but it reaches its fullest development in birds, and it's probably the same mechanism that powered the metabolism of sauropods. Let's take a closer look.

TO BREATHE LIKE A BIRD

Imagine blowing up a balloon and letting the air out, over and over. Blowing up a balloon means forcing air in using *positive* pressure, whereas our lungs draw air in using *negative* pressure. It's like pulling the sides of the balloon apart to cause air to rush in through the neck, as with a bellows or an aspirator bulb (sy-

A giraffe (*left*) has an especially strong heart and thick neck arteries to deal with the extreme gravitational pull on its blood while pumping it upward. Although there is no evidence for it, some physiologists think that sauropods may have had highly contractile arterial enlargements (*far left*), which, along with an especially massive heart, may have acted like "auxiliary hearts" to pump in relays (*arrows*) to counter the extreme gravitational pressure in the blood column of their necks. (Modified from a drawing by Choy and Altman.)

ringe) used to clean the nostrils of infants. If you can imagine a simple sac being inflated and deflated, you've got a pretty good model of mammalian breathing—just multiply the number of sacs to about 500 million. It's conceptually and mechanically pretty simple. In contrast, the breathing of birds is so different that at first glance it hardly looks like the same activity (see picture essay, "Breathing like Birds," p. 100). To understand it, let's follow the path of air through the respiratory system of a bird. Like us, birds have nostrils, nasal cavities, a ***pharynx*** (the space where the nasal cavities and the oral cavity come together), a ***syrinx*** (similar functionally to our *larynx*, or voice box, although birds also have this in a less developed form) and finally a ***trachea***, or windpipe, that leads to a pair of lungs. Also as in mammals, movement of the ribs and sternum causes the rib cage to expand during inhalation. This creates negative pressure inside the thorax and causes air to be drawn through the nostrils, nasal cavities, pharynx, larynx and trachea and into the lungs. After this, things get weird, at least to our mammalian sensibilities.

Whereas our lungs are composed of tiny blind-ended sacs—the alveoli—*the lungs of birds are made up of tiny tubes*. These tubes come in two sizes. The larger ***parabronchi*** are between 0.5 and 2 mm in diameter, and each parabronchus is surrounded by a spaghetti-like tangle of smaller ***air capillaries***, which are at most 1/10 as large. And stranger still, most of the air drawn in when a bird inhales doesn't go into the lungs. Although a small amount of freshly inhaled air actually does go into the lungs, *most of it bypasses the lungs* and flows into large ***air sacs*** below and behind the lungs. These air sacs have few blood vessels in their walls, and not much gas exchange takes place inside them. Instead, the air sacs serve as bellows to blow air *back through* the lungs. Our imaginary bird's packet of air has now been inhaled, through the nostrils and windpipe, past the lungs, and into the posterior air sacs. When the bird exhales, the air in these large posterior sacs is blown into the lungs—and not just into the lungs, but again *through* them. The air stream is broken up by an array of branching tubes to reach the thousands of parabronchi and the millions of air capillaries. The dense network of air capillaries is tangled up with an

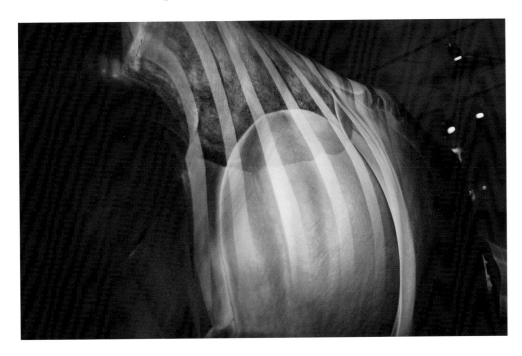

The transparent, illuminated abdomen of a life-size model of *Mamenchisaurus* by Hall Train Studios (*left*) for the traveling exhibit "The Largest Dinosaurs" shows how the air sacs and other internal organs were protected by long, wide ribs.

Sauropods like *Brachiosaurus* would not have been able to easily drink without bending their heads downward. They would have had a problem with their nasal openings being below water level if the nose openings had not retracted upward toward the top of the head. Giraffes solve this problem by spreading their forelegs apart while drinking, but this would have offered insecure support for a massive sauropod.

equally complex set of tiny blood capillaries, and this is where gas exchange takes place. Carbon dioxide passes out of the blood and into the air of the air capillaries, and oxygen passes from the air into the blood. Some of the blood capillaries run at right angles to the moving air, allowing *cross-current exchange*, and some blood flows in the opposite direction of the air, to allow *countercurrent exchange*. The cross- and countercurrent exchange allows birds to extract much more oxygen from the air than we can—up to *2.5 times* as much as any mammal.

So much for the blood leaving the lungs—what about the air? In our bird we've followed air from outside the body into the posterior air sacs on inhalation, and from the air sacs into the lungs on exhalation. The air in the lungs is now "spent," with much of the oxygen extracted and replaced by carbon dioxide. It will take another cycle for this air to leave the body. On the second inhalation, the spent air is drawn out of the lungs and into another set of air sacs, this time anterior to the lungs. On the second exhalation the spent air is blown from the anterior air sacs up the trachea, through the upper airways, and out into the environment. Our and other mammals' sac-like lungs have to perform both parts of respiration—moving air around and exchanging respiratory gases—whereas birds have *separated* these functions. The air sacs are responsible for blowing air through the lungs but perform almost no gas exchange. The lungs are free from the demands of ventilation and can be specialized for extracting as much oxygen as possible per breath. This separation of functions gives birds two great advantages over us when it comes to breathing. First, they get fresh air blown through their lungs on both inhalation *and* exhalation; we only get fresh air when we inhale. Second, the air is blown *through* their lungs, through those millions of microscopic tubes, and doesn't just sit in blind-ended sacs as it does in our lungs. Consequently, birds can perform aerobic—and aerobatic—feats unmatched by mammals, like flying for hours or days on end when necessary, and reaching altitudes of more than 29,000′ without passing out or requiring bottled oxygen.

Bird-like breathing offers at least three other, less obvious advantages as well. The first of these is that the air sacs of birds are much larger in terms of collective volume than the lungs of similarly sized mammals, so they can more easily overcome **tracheal dead space**. Tracheal dead space is the volume of spent air left in the trachea at the end of exhalation. Imagine breathing in and out through a length of garden hose, or a snorkel. A short tube would be no problem, but if the tube were much over a meter (3.5′) in length, you'd eventually suffocate. The reason is that the volume of the tube would be equal to *the tidal volume* (the amount of air equal to the total of what goes in and out) of your lungs. When you inhaled, you would only be able to draw in the spent air in the tube, and not receive any fresh air from outside (underwater snorkelers can avoid this by exhaling the spent air through their nostrils instead of blowing it back out the tube). Tracheal dead space works the same way—whenever we breathe in, the first air that reaches our lungs is the spent air that was left in our upper airways at the end of the last exhalation. It's not a trivial amount of air—when we breathe during resting and we take shallow breaths, the re-inhaled spent air can account for up to a *third* of the air we take in. Tracheal dead space is much less of a problem for birds, because the volume of spent air left in their airways is simply countered by the immense volume of their air sacs. In a typical bird, the air sacs are collectively *4 times* as voluminous as the trachea and lungs combined. Almost all birds take advantage of this by having a trachea that is relatively large in diameter. The larger the diameter of the trachea, the less the resistance of the airway to moving air, and the windpipes of birds are on average 3 or 4 times as wide as those of similarly sized mammals. Imagine breathing through a straw—to birds, this is how mammals go through life. Like birds, sauropods probably had large-diameter tracheas to supply huge volumes of air with little resistance. Even more importantly, skeletal evidence from a nigersaurine diplodocoid, *Tataouinea hannibalis* from Tunisia, shows that this and other sauropods possessed abdominal air sacs—and a unidirectional breathing system—like that of birds.

The second advantage of bird-like breathing is more efficient control of body temperature, called *thermoregulation*, and we're already pretty familiar with the ways our own bodies do this. When we're cold, we shiver, our muscles contracting and relaxing to rapidly build up body heat, whereas when we're hot, we flush; here, countless tiny blood vessels in the deep layers of our skin fill with blood, releasing excess body heat through the surface of the skin. When we're really hot, we perspire, or sweat—the fluid on our skin surface extruded from sweat glands evaporates, carrying away the unwanted heat into the cooler air outside. Shivering and flushing are common in other animals, but sweating isn't, since, unlike humans, most mammals can't produce enough sweat to cool off (non-mammals can't sweat at all). *Panting* is common in non-mammals, and we know, of course, that mammals like dogs can pant too. The problem here is to avoid *hyperventilation*, where too rapid an oxygen/carbon dioxide exchange can cause a blackout or unconsciousness. Animals that pant solve this by only taking rapid, shallow breaths that move air over the moisture (and heat-laden) tissues of the mouth and throat, sucking as little air into the lungs as possible. It's here that birds have a real advantage. Since birds have separated the processes of ventilation and gas exchange, they can blow air in and out of their air sacs while bypassing the lungs—they can ventilate independently of gas exchange. As a result, a bird's panting is much more efficient than a mammal's, and while getting rid of the unwanted heat, they sacrifice only half as much valuable body moisture as a mammal the same size. Pretty cool, literally and figuratively, and we'll see how sauropods may have used this saurischian–avian adaptation to deal with thermoregulation in the next section.

Breathing like Birds

Mammals have a system of bidirectional breathing, in which the lungs perform both ventilation, or pumping air in and out, and also gas exchange, the actual trade-off of oxygen with carbon dioxide when the air is inhaled. Saurischian dinosaurs, including sauropodomorphs and birds, evolved a different, and much more efficient, system called unidirectional breathing, where oxygen is absorbed not only when the animal inhales but also when it exhales. Here the work of ventilation is done exclusively by a system of air sacs. During inhalation, these expand (*black arrows*), moving air one way (*blue arrows*) through a capillary-rich system of tubes called parabronchi (here simplified) in the rigid lung. Channeled by valves, these capture some oxygen and surrender carbon dioxide (*inset boxes*). When the sacs contract during exhalation, the same (less rich, but still oxygenated) air goes back through the parabronchi, which again capture most of the remaining oxygen before exiting back in a different direction (*blue arrows*) through the trachea. The advantage of unidirectional breathing over bidirectional breathing is that gas exchange not only takes place twice, once during inhalation and a second time during inhalation, but also does so in much greater proportions, permitting birds to sustain much greater aerobic activity than mammals.

Bird—Unidirectional Lungs

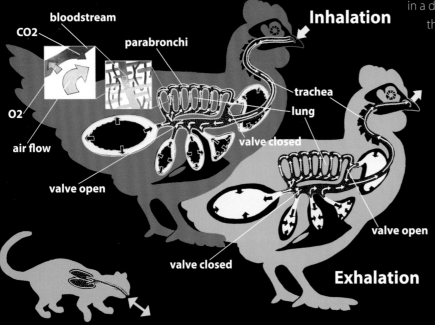

bloodstream

CO2

parabronchi

O2

air flow

valve open

Inhalation

trachea

lung

valve closed

valve open

valve closed

Exhalation

Cat—Bidirectional Lungs

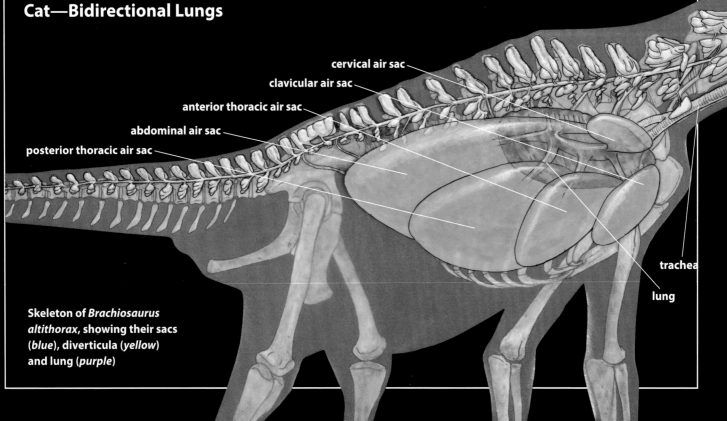

cervical air sac

clavicular air sac

anterior thoracic air sac

abdominal air sac

posterior thoracic air sac

trachea

lung

Skeleton of *Brachiosaurus altithorax*, showing their sacs (*blue*), diverticula (*yellow*) and lung (*purple*)

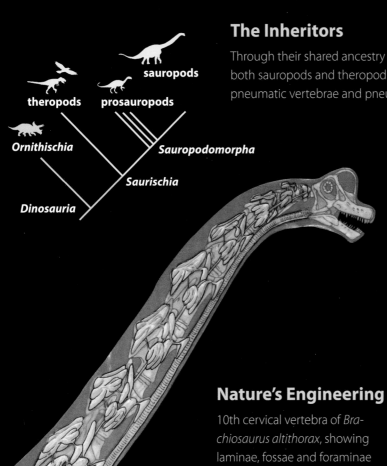

The Inheritors

Through their shared ancestry as saurischian dinosaurs, both sauropods and theropods retained a system of pneumatic vertebrae and pneumatic air sacs.

A System of Air Cells

A sauropod's system of diverticula, or air sacs, was incredibly elaborate and extensive, beginning with the cervical vertebrae and extending through to the anterior caudals and, in the case of basal titanosauriforms like Brachiosaurus (shown here), the ribs and pelvis. Unlike the air sacs, diverticula could not be expanded or compressed, but because they connected with the air sacs, they contributed to keeping the sauropod from dangerously overheating from within.

Nature's Engineering

10th cervical vertebra of *Brachiosaurus altithorax*, showing laminae, fossae and foraminae (*right*) and with diverticular sac system (*below*) in yellow. The interconnected diverticulae joined with the air sacs and were a sauropod's air conditioning, relieving the interior of unwanted heat.

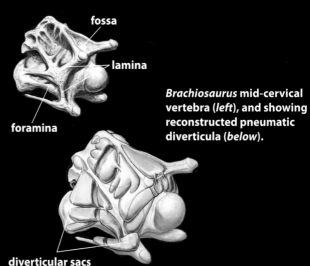

fossa

lamina

foramina

Brachiosaurus mid-cervical vertebra (*left*), and showing reconstructed pneumatic diverticula (*below*).

diverticular sacs

Lightness and Strength

All sauropod vertebrae, in addition to some other areas of the skeleton, developed pneumatic cavities and diverticula that combined lightness with strength by keeping bone only in those areas where support was needed. This was most extensive in the cervical vertebrae, which in earlier sauropods like *Haplocanthosaurus* first existed as simple, paired chambers called camarae, then became more and more finely honey-combed to become camellae, as in *Sauroposeidon*. The diverticula entered a vertebra or other bone through special depressions on the outside of each vertebra (pneumatic fossae) and openings (pneumatic foramina).

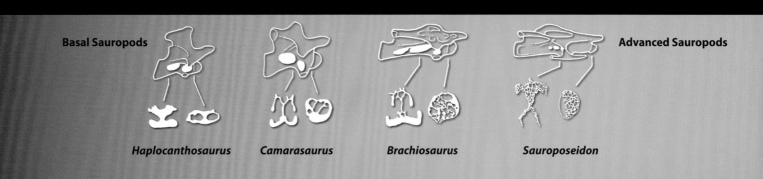

Basal Sauropods

Advanced Sauropods

Haplocanthosaurus   *Camarasaurus*   *Brachiosaurus*   *Sauroposeidon*

Getting a Drink

The moderately to highly retracted bony nares of most advanced sauropods may be an adaptation for such a tall, long-necked animal for drinking. As sauropodomorphs evolved away from smaller species like *Adeopappasaurus* to become huge, long-legged and longer-necked ones (remaining species, from left to right), it became necessary to incline the head at a steep angle to reach water while standing (*below and opposite*), deeply immersing the jaws to make it possible to suck or pump water upward with the tongue. Because at these angles submerged anterior nares would have made breathing impossible, increasingly retracted nares evolved so that breathing and swallowing water could alternate without withdrawing the head (skulls to same scale; orange dot represents probable nare location).

THE SAUROPOD CONNECTION

Along with being better able to cool off when overheated, there's a *third* advantage to bird-like over mammalian respiration. In birds (and in their extinct archosaurian relatives, the saurischians) ventilation doesn't just stop at the air sacs: these are connected to **diverticula**, or ever-branching, membranous (like the air sacs themselves) passageways. The diverticula actually snake up into **pneumatic**, or hollow, spaces within certain bones like some vertebrae (especially the *cervical*, or neck, vertebrae) and other places. Mammals and other vertebrates also have air-filled compartments like the sinuses within the skull, so in this sense we're all airheads. But pneumatic spaces, or *pneumaticity*, in vertebrates' postcranial skeletons (all of the skeleton following the skull) is much less common, and like our sinuses, there has to be a route from the outside to the inside of the bone. How did this evolve? When a membranous diverticulum comes into contact with a bone's surface, it can cause the bone tissue to break down and become reabsorbed by the body, after which the diverticulum can expand to fill the newly created space. The diverticula still maintain their connection to the air sacs and lungs—to visualize this, we can imagine an inflated rubber glove: the part that's the palm of the hand is like the air sac, while the fingers are like the diverticula. In theropod saurischian dinosaurs, the ancestors of birds, the pneumatic spaces (and diverticula) are on their way to becoming well developed, suggesting that these early ancestors had already acquired a basal form of the sophisticated avian respiration. In birds themselves, bony pneumaticity is developed to the greatest extent of all, extending to sometimes the remotest places within the skeleton. Why? For birds, this considerably lightened the skeleton without sacrificing strength, very important for fliers. And this is where we get to sauropods.

THE INHERITORS

As pointed out in Chapter 2, the sauropodomorph dinosaurs were, as saurischians, inheritors of the same highly efficient breathing system found in living birds. We can see this when we look at the vertebrae of most sauropods, in which the outer and inner pneumatic spaces become ever more elaborate and complicated as we go from known basal (for example, *Barapasaurus*) to highly derived forms (*Sauroposeidon*). In these the first vertebrae to become pneumaticized are those of the cervical bones, followed by the dorsals (trunk) and sacrals (hips)—this is the same front-to-back developmental pattern that paleontologists have identified more recently in the theropod ancestors of birds, as well as the birds' evolutionary "uncles," the sauropods. At this

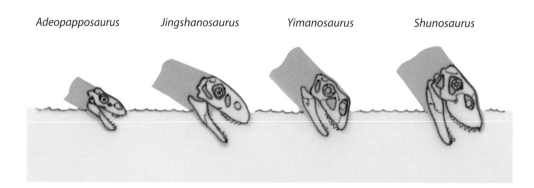

Adeopapposaurus   *Jingshanosaurus*   *Yimanosaurus*   *Shunosaurus*

time we still don't understand how or why pneumaticity first developed in saurischians and later sauropods and birds, but the patterns seen in these suggest that there was a selective advantage in making this system become ever more sophisticated.

While the greater efficiency of an early avian-style breathing system was certainly an advantage to the earliest sauropods and their ancestors for all the reasons described above, it conferred another, tremendously important side benefit to these increasingly larger and heavier animals: *it lightened their skeletons*. Any animal reaching sizes as gigantic as those of sauropods must pay a greater and greater toll as gravity exerts its pull on their increasing mass, and whatever reduces this without sacrificing strength is going to result in greater stability and mean less calories consumed in moving its body around. This was especially important for a dinosaur in which the main specialization involved the evolution of a long and potentially ponderous neck. It explains why, in addition to the dorsal vertebrae that supported the massive trunk, sauropods at an early stage first developed pneumaticity in their cervical bones. This made their necks far lighter than they would have been otherwise, and in some advanced species like *Diplodocus* the neck of the living animal is estimated to have been 50% lighter than it would have been without pneumaticity. In the vertebrae any bony material that didn't contribute to the overall strength of the vertebrae disappeared and was replaced by a honeycomb of interconnecting, small to tiny pneumatic spaces. Although we don't know exactly how far the diverticula extended in life, they probably went far back into the spaces, where they terminated as blind sacs. At this small size they probably didn't contribute much (if at all) to ventilation, but they still would have performed the vital function of lightening the skeleton and the overall mass of the animal. This is one situation in which we can reasonably construct the living biology of a long-gone animal species. Although we're probably never going to find fossilized air sacs and diverticula in sauropod remains, we can be sure that these animals possessed them because of the fact that they had generally the *same kind of pneumatic cavities* that their archosaurian cousins, birds, have. In short, the extent of these highly distinctive bony structures in sauropods and their similarity to those of birds mean that they also had the same kind of respiratory system. In addition to more efficient respiration, the sauropods' pneumatic spines and some other skeletal areas gave them a vital adaptation to growing huge—reducing their mass. As we can see from the increasing complexity of the pneumatic recesses, this remarkable adaptation served sauropods well throughout their evolution and was a key to their ability to grow to the sizes they did.

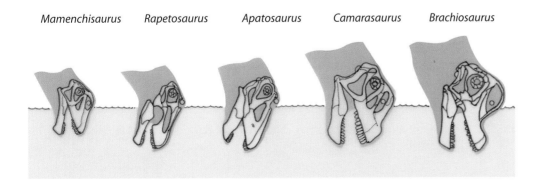

Mamenchisaurus   Rapetosaurus   Apatosaurus   Camarasaurus   Brachiosaurus

GETTIN' HOT IN HERE

All metabolic processes generate heat, and managing this was as much a concern for sauropods as for any animals. The challenge for really big animals, from elephant size on, is keeping cool. Remembering that as an animal's size increases its volume grows much faster than its surface area, this means that there are more metabolically active cells producing heat on the inside than there are on the outside getting rid of it. If an animal is going to become really big, it has to evolve some tricks to deal with this or end up cooking inside its own skin. Elephants can teach us a lot about eliminating heat. They have huge ears for their body size, and these often make up one-fourth of their entire skin surface. The ears are well supplied with blood vessels, and when it starts getting too warm for an elephant, it starts flapping its ears slowly but continuously. Hot blood from the body core gets pumped into the ears and is cooled by the outer air flowing over them, transferring heat (a process called *convection*); the ears are like giant refrigeration units tacked onto the sides of the head. But the ears aren't enough. Recent studies have shown that when the weather is hot, elephants simply can't dump enough heat this way to keep their internal temperatures from rising. Instead, they *tolerate very hot temperatures* during the day, experiencing what would be a critical, fever temperature for a human, and then *release it during the cool evening hours*. Ironically, bigger elephants have an advantage because their larger bodies have more **thermal inertia.** Their bigger volumes in proportion to their skin surfaces let them heat up more slowly, and although it also takes them longer to cool off, they have more capacity to store heat. This is one reason why African elephants living in deserts and open savannas are larger than their forest relatives and have bigger ears, since this helps them deal with the daily high heat loads. Among living elephants, the largest max out at 6–7 tons, but the biggest sauropods were *5–10 times* more massive, each mature adult the equivalent in weight of a *small herd* of elephants, with individuals of some species perhaps 75–80 tons. Their cooling requirements were incredible, and although we probably may never know all their secrets of temperature maintenance, or **thermal physiology**, we can make some educated guesses.

While sauropods didn't have big ears to reduce their temperatures, they *did* have structures with large surface areas, their necks and tails. To investigate the possibility of how sauropods may have used these as radiators, Donald M. Henderson (Royal Tyrrell Museum) built 3D digital computer models of many sauropod species to compare their surface areas with their volumes. He reasoned that if sauropods were actually using their necks and tails in this way, the neck and tail surface areas would show **positive allometry**, meaning that these would be proportionately larger in larger animals. When he evaluated the results, however, he found that in part this wasn't true: the tails of bigger forms were no larger in proportion to their sizes than those of smaller ones. The necks, though, were another story. These *did* scale with positive allometry, strongly indicating that necks in sauropods, like the ears in elephants, had value in reducing heat loads. Of course, there are certainly other valid reasons for bigger sauropod necks—to reach food, for example—but a body part can have more than one function. Sauropod tails could have functioned as well to eliminate heat, but since large sauropods didn't get any special benefits from longer tails than smaller ones did, it doesn't seem as if these were specifically contributing to heat reduction.

Which Nose?

The retracted position of the bony nares of many sauropods, especially diplodocoids, has traditionally led paleontologists to assume that the fleshy external nares were also located here as well (*bottom right*), with air (*blue*) entering through these before passing through the skull and downward into the trachea (*bottom left*). Recent digital scanning and research have led Larry Witmer and his laboratory associates, however, to hypothesize that diplodocids and most other types may have actually had fleshy nostrils near the anterior muzzle as in most living vertebrates, and that paired tubes may have led from these to the bony nares (*top left and top right*). Only one mammal, *Macrauchenia* (*photo, below*), is known to have had similar retracted bony nares. The reader is invited to make her or his own decision as to which is correct.

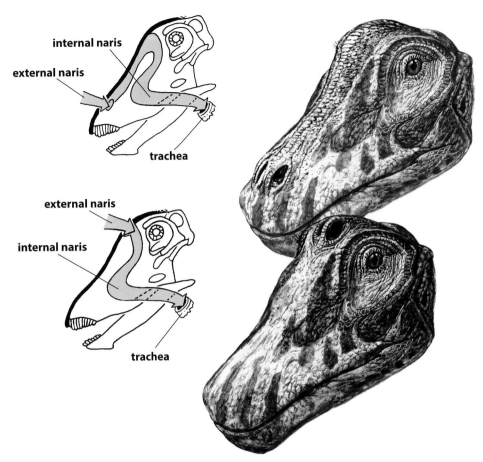

internal naris

external naris

trachea

external naris

internal naris

trachea

The long necks may have functioned as heat exchangers on the inside, too. In studies by Steve Perry (University of Bonn) and his colleagues, the researchers found that a sauropod's trachea could have efficiently eliminated heat through the processes of moisture and humidity exchange. For the gas exchange to take place in an animal's lungs the air has to be moist—completely at 100% humidity, and totally saturated with water vapor. The outside air, however, is usually drier and has to be humidified by the upper respiratory tract (nasal airways, pharynx and trachea) before reaching the lungs. In mammals this mostly takes place in the inner linings of the nose, but in birds the trachea is just as, if not more, important. Glands in the windpipe's lining secrete mucus, whose moisture is picked up by the incoming air through evaporation. Evaporation is the key because water has a high specific heat, meaning that while it takes a lot of energy to heat up a certain amount of water, when that water evaporates it also takes a lot of the heat with it through convection, a process called *evaporative cooling*. That's why we have to first sweat to cool off, and why misting fans are so effective on hot, dry days. For sauropods, evaporative cooling in the trachea might have worked like this: as a sauropod inhaled, the relatively dry air coming in picked up moisture from the tracheal walls. By the time the air reached the lungs, it had been warmed to the core body temperature and moistened to the optimal 100% humidity. When the sauropod exhaled, the hot water vapor in the outgoing breath carried unwanted body heat to the outside. This is the way evaporative cooling works for most animals, and it seems to

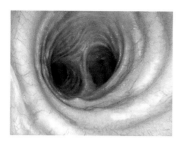

The necks of sauropods, like those of rheas, swans and other long-necked birds, may have had large diameter **tracheas**, or windpipes. Wide tracheas would have supplied large volumes of air with less resistance, and probably served as heat and moisture exchangers as well.

really solve the problem of overheating in larger animals—but not always. In desert ostriches, exhaled air isn't at 100% humidity, the reason being that in a desert ostriches need to hold onto valuable water *more* than they need to cool themselves. So on the return trip from the lungs to the nostrils, air goes through the reverse process that it did coming in: water vapor *recondenses* on the walls of the trachea, and the heat the animal lost through evaporation stays in the body. That's the dilemma: you can exhale moist air and thereby lose heat, but you do so at the cost (if you need to conserve water) of losing moisture. On the other hand, you can exhale drier air and keep that precious water, but then you'll be stuck with the unwanted heat. What would a sauropod have done? Probably whatever worked best in a particular situation. When they were hot and water was plentiful, they may have used their long tracheas for evaporative cooling (ostriches and other long-necked birds do this, as well as some mammals, by panting) and exhaled moist air. When they were in an arid environment, however, where the importance of conserving water outweighed cooling their bodies, sauropods may have recaptured the water and waited until nightfall to cool, as elephants do. Camels make this kind of trade-off as well, since they can rarely afford to waste water by losing heat, and instead tolerate even greater temperature buildups than elephants, from 34°C (93°F) in the morning to 40°C (104°F) just before sunset.

GOING WITH THE FLOW

Besides food and oxygen, the cells of all living things require water. How did sauropods obtain it, and once it had become a waste product, how did they get rid of it? To start with, there are a lot more ways of getting water than just by drinking it. Some of it comes into the cell along with its dissolved food, and still more is produced in the cell itself as a result of the way energy-rich compounds are metabolized. In the process of **cellular respiration**, cells "burn" glucose and oxygen to create energy, and these, in turn, make carbon dioxide and water as by-products. Water created by the metabolism of glucose is called **metabolic water**, and some animals are very dependent on this. Small desert dwellers like collared lizards and kangaroo rats have very few opportunities to drink and gain little moisture from their food. For these animals metabolic water is their primary source of this, and they actually get almost all they need in this way.

As handy as metabolic water is, however, for large animals it's not enough. This is dramatized by documentary scenes of parched animals gathering around a water hole, reflecting the need for big species to drink on a regular basis. This was probably also true of sauropods, which may have obtained some of their water from their browse but also needed it in liquid form, perhaps for smaller forms obtained by dipping their heads up and down like birds or in huge ones by pumping the water into the mouth using their tongues, like tortoises. We can imagine the awkwardness and vulnerability of a sauropod at the watering hole by seeing a giraffe's difficulty in doing this, and these huge creatures probably evolved ways to maximize their efficiency (and reduce the time required) in getting a drink. Because of the problems in blood pressure in continually raising and lowering their heads, it's unlikely that larger sauropods practiced "head dipping" as in birds, instead preferring to keep their heads down in the water while a tortoise-style "tongue pump" action sucked it upward into the mouth. The peristaltic action of a muscular esophagus would then send it upward—and then down—on its long journey, by the quart or gallon, to the stomach until the animal had had enough. Here we get into some unproven but reasonable speculation, postulated by coauthor Hallett. Sucking or pumping water into the mouth doesn't work as well when the mouth (especially a large one) is partially open and above water level, however, and this means keeping the mouth well down in the water, and preferably below the surface, to counter the tendency to pull in unwanted air. For the increasingly tall, quadrupedal sauropods this meant keeping the head at an acute angle to water level when they drank (remember that they couldn't safely spread their forelimbs apart like those of a lanky giraffe), so it's possible that *high, retracted nostrils*, far enough above the water when the anterior muzzle was dunked, appeared early on in all sauropods. This allowed them to alternately breathe and then take in water without raising their heads. As with tortoises, strong, muscular tongues would have helped, and these are anchored to **hyoids**, rod-shaped bones just below the upper throat. In relatively complete sauropod skeletons the hyoid bones are well developed, suggesting that in these the tongue was an important organ. All this might have helped sauropods to get a long drink.

TO PEE, OR NOT TO PEE?

Once the body is through with it, excess water is excreted through the cell walls and is taken up by the bloodstream, now along with **urates**, ammonia and other dissolved cell wastes, to become distilled by the kidneys as **urine**. The TV documentary series *Walking with Dinosaurs* in one episode showed an early Triassic archosaur, *Postosuchus*, marking its territory with this, but it's extremely likely that urine elimination was retained in later archosaurs, including sauropods. Crocs and birds, the closest living relatives of dinosaurs, usually do not urinate, but shed their metabolic wastes in the more solid form of uric acid. In most birds the urates in the uric acid within urine are highly distilled into a solid—the white pasty stuff that, along with the brown-colored feces, or "poop," is characteristic of most archosaur waste. This doesn't mean that some birds can't make urine as well, since ostriches are capable of storing this in their cloacas and releasing it when water is plentiful. The ancestors of archosaurs, the reptiles, evolved in the supercontinent of Pangea's enormous, largely arid interior, and in addition to eggs that could develop out of water and scaled, moisture-retaining skins, they also developed kidneys that efficiently extracted water, as well as a bladder that could store it as urine. As with modern desert tortoises, the water in this may have been recycled if needed. Since kidneys and bladders unfortunately don't fossilize, it's hard to reconstruct the evolution of these organs. Although crocodilians and birds simply store urine in their cloacas and have no bladder (in the first case because they're animals that live in an aquatic environment and have lost the need to conserve water, and in the second because they, as flyers, would have been hindered by the weight of stored liquid), bladders are found in primitive living sphenodontids (the two species of Tuatara, *Sphenodon punctatus* and *S. guntheri*), turtles and some lizards. So bladders appear to be basal for saurians, or possibly are easy to re-evolve. In addition to liquid wastes, of course, sauropods also had to process and eliminate solid wastes in the form of **feces**, and it's time to step into this.

After extracting all the available energy from the long retention in its intestines, a sauropod finally defecated the unusable residue. During its lifetime, a sauropod produced several thousand tons of feces, and the remaining plant matter supported not only dung beetles and other arthropods but also untold numbers of microorganisms.

DINOPOO

However long the retention time of digesting food in a sauropod's gut was, the intestinal wastes, or *excreta*, had to eventually come out, and when it did, it would have been prodigious. The largest known species like the titanosaurs *Argentinosaurus* and *Puertasaurus* at 75–80 tons were about 10–11 times larger than a 6.5- to 7-ton African elephant male, and by extrapolation we can guess that individuals like this were eliminating over 1.5 tons of dung each day. If we further assume that some of these formed pods of, say, 10 subadult to adult individuals, we're looking at about 15 tons of dung deposited on the landscape every day. In spite of this abundance, however, very few coprolites, or fossil feces, can be assigned to sauropods themselves, but there are exceptions that may cast light on sauropod food sources. One example is the "Type A" coprolites found in the Late Cretaceous central Indian Lameta Formation by Dahnanjay M. Mohabey (Geological Institute of India). If these belong to the titanosaur *Isisaurus colberti*, whose remains also come from this geochron, it shows that in addition to conifers, these later sauropods had adapted to eating some graminiform (grass family) angiosperms, since phytoliths from this family are present. Another is a large coprolite from the Aguja Formation of the Late Cretaceous Campanian stage in Texas. A coprolite from the Two Medicine Formation described by Karen Chin (University of Colorado, Boulder) and Bruce D. Gill (Agriculture Canada), although not eliminated by a sauropod because of its age and location, like the Type A coprolites was made up of conifer fragments and shows these plants to be a viable food source for some large dinosaurian herbivores. It also contained dung beetle burrows. By the Cretaceous and probably earlier, dinosaur dung was the focus of activity by specialized and extinct types of *coprophagus* (feces-eating) cockroaches and scarab beetles, which must have acted much in the same way as modern scarabs by removing dung balls and burying them to act as food and incubation chambers for their grubs. George and Roberta Poinar Jr. (Oregon State University) have estimated that the fecal output for a pod of 10–15 titanosaurs could have supported a population of 200–250 million beetles of one species alone. Sauropod poop probably acted as a system of transport and nourishment for some types of plants as well, whose digestion-resistant seeds were distributed far and wide by the wandering pods to new locations, extending their range (see Chapter 10).

The unimaginably massive amounts of dung produced by sauropods and other dinosaurs supported a complex community of insects like dung beetles, which were able to make use of the undigested plant matter both for themselves and for their larvae.

STOUT HEARTS AND HIGH BLOOD PRESSURE

Another problem facing sauropods was blood circulation. Once you solve the problems for providing a giant with the food, water, waste removal, temperature regulation and oxygen it needs, then you have to come up with a delivery (cardiovascular) system that can do all this. This means a heart that can pump blood upward, against great gravitational pull, to the level of the animal's head when (and if) it reaches upward (see Chapter 6) as much as 12 m (40′). Any tetrapod's heart must generate enough blood pressure within its circulatory system to push blood, with its load of oxygen and nutrients, to the tiniest recesses of the body, while the blood must simultaneously remove carbon dioxide and other dissolved wastes. When blood is pumped up a tall, vertical column (as is done in a long-necked animal like a living giraffe or a sauropod), it has to overcome more and more gravitational pull the higher the neck is held. This wouldn't be too much of a problem if sauropods had habitually held their necks horizontally, but what would happen when they raised them in a vertical position, such as when they reared up high to feed in trees or when a male mounted a female during mating? Roger Seymour (University of Adelaide), Harvey Lillywhite (University of Florida) and some other researchers concluded that a rearing sauropod's heart would fail in the attempt to pump the blood to the required height (anywhere from 12 to 18.5 m [40′ to 60′]) when the head and neck were raised. In lowering the neck from this position, fatal hemorrhaging would occur within the capillaries from the sudden and enormous increase in blood pressure in these areas, especially around the brain. These same problems affect the modern giraffe, although on a smaller scale: a large adult may have 17 pints of blood within the neck, while its blood pressure may be 5 times the amount of an average human's 85–90 mmHg. When the head and neck are raised, this pressure drops to about 150 mmHg, while a giraffe that lowers its head to drink must deal with a raised pressure of about 300 mmHg.

In sauropods, these factors would be multiplied many times. The necks of many were anywhere from 2.5 to 3 times as long as the average adult giraffe's, and in hyperlong-necked, vertically rearing forms the heart would have been pushing an arterial blood column 40′ or more upward. This would have resulted in the need for a staggering blood pressure of perhaps 700 mmHg, acting

Karen Chin (University of Colorado, Denver) is a worldwide authority on both the taphonomy, or post-life history, of dinosaurs and other ancient organisms and also *coprolites*, or fossilized feces. These trace fossils are particularly important in yielding clues about dinosaur diets, and in the case of sauropods they may one day yield decisive information as to what plant species a given sauropod ate.

against a gravitational force of 1 g. Based on the assumption that sauropods had a mammal's or bird's four-chambered heart and could pump blood with the same oxygen-carrying abilities, Seymour and Lillywhite used various calculations to arrive at the conclusion that the left ventricle (the chamber of the heart that pumps blood into the body) alone of the diplodocid *Barosaurus* would have occupied 5% of the dinosaur's estimated body mass. To pump the same amount of blood per beat as a finback whale (*Balaenoptera physalus*) of equal weight, they estimated that the ventricle walls would have to be 5 times thicker and therefore 15 times heavier. The massive hearts of the largest sauropods would need to have been spectacular in their pumping capacities and would have accounted for huge portions of the animals' resting metabolic rates, let alone the rates that resulted from their active behaviors like foraging, migration and defense.

While the above workers conclude that the energy costs and other physiological restraints would make such hearts improbable and therefore make it impossible for sauropods to raise their heads and necks significantly above a horizontal plane, others disagree. Some physiologists have suggested that sauropods may have had an upscaled version of the modern avian cardiovascular system. Birds, both flying and flightless, have larger hearts, higher blood pressures and higher cardiac outputs than most mammals of equivalent size. Among many mammals (with the exception of giraffes) the heart is 0.5% of body mass, while those of many birds are typically 1.0%. *Systolic* (the contractive heart phase) blood pressure in mammals, which is usually around 110–120 mmHg, is often in birds 150–175 mmHg and in domestic turkeys this can actually be up to 400 mmHg.

If we choose a lower (and more conservative) weight estimate of 30 tons for *Barosaurus*, the heart at 1.0% of its body mass would weigh 300 kg—not shocking if we consider that the heart of the finback's cousin, the larger blue whale (*Balaenoptera musculus*), is about as big as a Volkswagen Beetle. Granted, the heart-to-head distance in whales is smaller than those of sauropods, but it implies that such huge hearts could theoretically work quite well. In mammals and birds the heart/body mass ratio does not change with increasing size, and if that of a sauropod was similar to a bird's, it probably could maintain an adequate flow of blood to the brain, even with the head erect. Some computer models suggest that weights for sauropods like *Barosaurus* may have even been less than the above, which would make the corresponding size of the heart even more capable of pumping blood up the long neck. In 1992 the research team of Daniel S. Choy (Lenox Hill Hospital and Cardiac Catheterization Laboratories) and P. Altman (Ventritex Inc.) independently suggested that a sauropod's heart may have been assisted by auxiliary swellings in the arteries of the neck, which could contract or pulse in rhythm to each heartbeat, helping the heart to send blood upward. Because of the gravitational force acting on the blood column, valves, like those present in many tetrapod veins, would be necessary in these places to prevent backflow. While such arterial structures are unknown in living animals, they aren't impossible. Arterial walls already possess a muscular layer that assists the heart in pumping, and it would be a relatively small evolutionary step to develop larger, more powerful sac-like structures that could pulse in rhythm with the heart, possibly aided in their contractile ability by fibers of the protein *elastin*, which would aid the musculature in its contractive phase. Valves could also then evolve at these points to prevent arterial backflow, a condition already present in the cerebral arteries of some mammals, such as rats. Recent 3D digital analyses obtained by Ryan C. Ridgely and Larry Witmer (both Ohio University College of Osteopathic Medicine) of sauropods like *Ampelosaurus* show that the neurovascular sinuses in these sauropods were relatively large. Such sinuses were probably capable of expansion in receiving sudden and large amounts of blood sent to the cranial cavity when the head was lowered, as well as keeping blood in reserve to ensure an adequate supply of oxygen to the brain if blood pressure dropped when the head was held up. In addition, sauropods could have possessed a structure that could have controlled blood pressure near the brain called the **rete mirabile** ("wonderful net"). This is a network of capillaries with highly elastic, extendable walls, which are also capable of expanding to receive a sudden, high amount of blood from the arteries supplying the brain. This kind of pressure increase would normally lead to fatal hemorrhaging within the brain in the case of typical capillaries, but the structure's ability to absorb the increase prevents this. In living giraffes and whales, the rete mirabile deals with the problems of sudden pressure change that occur in the first when the head is lowered and in the second during dives and in coming to the surface. Giraffes have also evolved other ways of dealing with blood pressure. In addition to having huge, thick-walled hearts and arterial systems to push the blood column up, their bodies tolerate a much greater degree of *hypertension*, or adaptation to high blood pressure, than shorter-necked ungulates, which contributes to the efficiency of their blood circulation. In most mammals the lower limbs are encased in tough, sheet-like tendons called *fascia*. These are the equivalent to human orthopedic support hose, and the tight, non-elastic "stockings" prevent gravity from allowing the body's lymph fluid to accumulate in the lower extremities, keeping blood pressure up high where it's most needed. Lower limb fascial sheets are especially strong and well developed in large, long-legged ungulates like giraffes. Most mammalian ungulates also possess highly vascularized, supportive pads or cushions underneath the foot bones, and when the act of bringing the foot down on a hard surface compresses these, the blood within them is squeezed and is assisted in its upward return to the heart. Sauropod limbs were likewise probably enclosed in *fascia*, as well as possessing highly vascularized foot pads to help keep blood pressure where it was most needed.

BRAINS AND NERVES

Sauropods, in addition to other things, are well known for their very small heads, which of course contained equally small brains. This, along with the question of how they could obtain enough food to nourish themselves, raises the question about the way these creatures might have controlled their huge bodies with such (from a mammalian viewpoint) limited brain and neural equipment. But from the aspect of function, was it really limited? As humans, brain development is one of our most distinctive attributes, and we tend to judge other animals' potential behaviors from this. To properly understand and assess the capability and quality of a sauropod's neural system, including the brain, we have to first remember that the original, basic purpose of this apparatus in all animals is to relay information (a *stimulus*) obtained at the outside of the body and to coordinate an appropriate response (*a reaction*) to whatever may affect the organism. In vertebrates, this means that, in addition to the nerves that automatically supply biochemical impulses to maintain inner body functions, the ones that receive a stimulus (known as

The fastest speed that a neuron or nerve can transmit impulses (*above*) is about 400' per second. This speed is maintained by fatty myelin sheaths that enclose the axon and keep the electrons from "leaking" out and weakening the impulse. Because of their hyperlong extremities, sauropods probably had a sophisticated system of peripheral nerves to the head, neck and tail to sense and respond to predators.

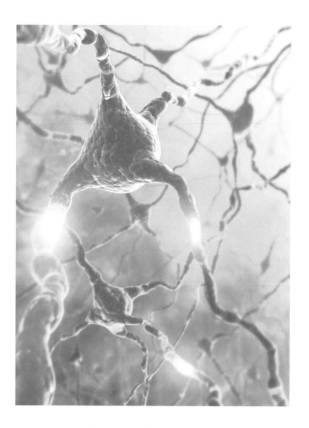

afferent nerves) take it to the spinal cord, where short, associative nerves, called *interneurons*, transmit the impulse to *efferent nerves*. These send an impulse to the *voluntary* or *skeletal muscles* that move a limb or other body part. In neurophysiology this is called a *reflex arc* and is what normally takes place in situations such as when a mature, non-infant animal walks: it's a "no-brainer" because the balance, timing of steps and messages to the leg muscles are all under unconscious neural control, traveling from the sensation to the motor response, via the nerves and spinal cord. Under some circumstances, however, the brain overrides this automatic function, for example, if the same animal has to make judgements about stepping over uneven ground to avoid a fall. Here the most primitive part of the vertebrate brain, the *amygdala,* and a more advanced region, the *cerebellum*, can intervene. The amygdala operates on a very basic, instinctive level. It's the region that takes over the rest of the nervous system when the animal's life is at stake, making split-second decisions in whether to escape, attack or initiate other behaviors. As vertebrates became more anatomically complex and advanced, neural structures like the cerebellum evolved to augment the coordination of the voluntary muscles in a more finely tuned or sophisticated way, as well as the *cerebrum*, whose main function, when needed, is to initiate and control the motor function of the voluntary muscles. In animals with larger and more complexly folded cerebra, this motor control not only is highly sophisticated but also has areas capable of *cognitive function*, which includes the ability to process external stimuli, store memory (learn) and initiate behavioral responses. In all animals many behaviors are based on neural "hard-wiring" and instinctive response, but in the most advanced, highly derived mammals and birds behaviors can be greatly determined by cerebral cognitive function.

In archosaurs, which in many ways are (and were) the most biologically advanced saurians, the cerebrum is usually only moderately developed compared with that of more derived mammals, but in pterosaurs and *avetheropods* (the dinosaurian ancestors of today's living birds) the cerebrum is relatively large and contains rudimentary folds. Since many living birds (like crows, jays and other corvids) are capable of very sophisticated cognitive function (loosely translated, learning, remembering and "thinking") on the level of some advanced mammals, we can safely assume that dinosaurs that possessed similar features were also capable of cognitive function at this level. Although cognition is at its highest level in something like a crow or raven, all saurians can definitely learn, and both crocodilian and avian archosaurs are capable of some very sophisticated behaviors relating to caring for young and predation.

Unlike mammals, the crania of reptiles and more advanced saurians don't usually reflect a precise mold of the actual brain, which is actually smaller than the inner shape indicates when you factor in the membranes and fluids that enclose and protect the brain itself. Virtual images or cranial endocasts from well-preserved sauropod skulls indicate that in these dinosaurs the general structure and proportions of the brain were very similar to those of crocodilians. When we compare the cranial endocasts of known sauropods with crocs, we see that in addition to their auditory (hearing) regions sauropods had well-developed **olfactory** (smelling) and **optical** (visual) brain lobes, a fairly typical condition for most reptiles and the more derived archosaurs. As with other dinosaurs, the **pituitary body**, which in vertebrates secretes hormones that stimulate and control body size, is comparatively huge. This isn't unexpected in creatures as enormous as sauropods and many other dinosaurs, and it probably reflects the important role the pituitary played in governing their amazing initial growth rates and stabilizing this growth as they reached optimal body sizes. What about actual sauropod cognitive abilities, what we'd call "intelligence"? Although we all usually think we know what "intelligence" is, it's a very subjective concept. Sauropod cerebra are only modestly developed, but it doesn't mean that they were, from a typically human standpoint, merely stupid automatons, capable of only the most fundamental body functions and behaviors. Sauropods, although their brains (and those of other dinosaurs) were *absolutely smaller* in proportion to their overall body sizes than those of most mammals, actually had brains that *were the right size* for archosaurians. What does this actually mean?

THE RIGHT BRAIN FOR THE RIGHT BODY

First, large animals have relatively smaller brains than related small animals, and this remains extremely consistent throughout the major clades, such as all saurians and all mammals. As you go from small to large (in saurians, tiny lizards to Komodo dragons, or in mammals, mice to elephants), brain size increases, but not as fast as body size. Brains actually grow only about two-thirds as fast as bodies, and as a result, large animals have low ratios of brain to body weight. Since an elephant is certainly no less smart than a mouse, or a crocodile than a skink, it seems reasonable to conclude that brain size isn't all about intelligence. Large animals just require less brain to function equally as well as small animals. This is especially true when we remember that most of the cerebrum's purpose is that of governing motor control, not cognition. Second, the relationship between brain and body size isn't identical in all vertebrate groups. Although all large animals share the same *rate* of brain size decrease with body size, small mammals have much larger brains than similarly sized small reptiles with the same body weight. When you put these two facts together, it's not unreasonable to expect that a huge saurian would have a proportionately smaller brain than an equally sized, gigantic mammal. When neurophysiologist Harry Jerison (formerly Antioch College) in a 1973 study plotted the endocranial cast measurements of 10 different dinosaur taxa from six major clades against a curve with projected brain dimensions for projected large body masses, he found that each one fell very close to the extrapolated theoretical sizes for giant saurians. Several years later, James A. Hopson (University of Chicago) refined this study by actually assigning a value, an *encephalization quotient*, or EQ (not to be confused with "IQ," or "intelligence quotient"), to the volume of each dinosaur endocast. If a given dinosaur fell at a particular brain/weight point on the curve projected for its estimated mass, it received a standard value of 1.0; those that were above got a higher value and those that were under, a lower one. As we've seen in Chapter 4, body mass can be difficult to determine accurately, but using Hopson's parameters, these, like Jerison's, came in close to what was expected. Sauropods, with their enormous discrepancy between brains and body masses, received in this study values of EQ from 0.20 to 0.35. Again, this doesn't mean that sauropods were stupid, simply that their brains were a size sufficient for controlling their body movements and behaviors.

A MATTER OF CONTROL

An old joke among paleontologists goes like this: *how did the sauropod know when to wake up the next morning?* It first bit itself at the tip of its long tail, and then went to sleep. Later it awoke, on time, after suddenly feeling the pain that had finally reached its brain eight hours later! Maybe not the best of a paleontology teacher's possible beginning-of-semester openers, but it does prompt the question of how fast the neural sensitivity and response were in such long, massive animals. Given that they had small brains for their sizes, could the largest sauropods efficiently coordinate their movements and response times, especially if attacked by a giant theropod? Like other vertebrates, sauropods needed to have individual sensory (afferent) nerves that ran from extremities like the tips of their tails and soles of their feet, as well as their skin surfaces, to their brain stems. In the longest sauropods, these nerve cells, or *neurons*, were probably about 44 m (150′) long, perhaps the longest known cells in the history of life. These, of course, will never be found as fossils, but since they were amniotes, we know that sauropods must have had them. The *fastest* nerve impulses recorded in modern vertebrates travel at about 50–70 m (400′) per second, or 270 miles per hour. A signal from the very tip of a 150′ sauropod's tail would therefore take about half a second to reach the brain, probably fast enough if you were nipped there to do something about it.

Is it possible that sauropods had accelerated nerve conduction velocities, to bring in those distant signals faster? To the brain, probably not—the only way to speed up a nerve impulse is to increase the diameter of a nerve fiber itself. Although some invertebrates and a few vertebrates have evolved "giant neurons" to accomplish this when an almost instantaneous reaction is needed, this solution is rare. That's because, even in a huge animal, it would ultimately result in nerves, which are collectively made up of many individual fibers, so thick that they couldn't fit within the body. To achieve faster conduction velocities, most vertebrates wrap their individual nerve fibers in a special, fatty tissue called *myelin*. This is like the plastic insulation around the metal wire in an electronic cable that improves transmission speed by preventing the flowing electrons from "leaking" to the outside; without such sheathing, the conduction would be slowed. The thicker the myelin sheath, the faster the nerve impulse will travel. But as with thicker nerve fibers themselves, there are limits to how thick the myelin sheath can be, since many of these would, like individual giant nerve fibers themselves, result in nerves that would become impossibly big in diameter. We know that the spinal cord openings in sauropod vertebrae were surprisingly small, as discovered by John Bell Hatcher when he studied them more

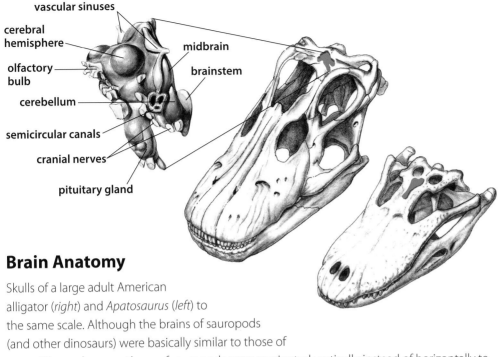

vascular sinuses

cerebral
hemisphere

olfactory
bulb

cerebellum

semicircular canals

cranial nerves

pituitary gland

midbrain

brainstem

Brain Anatomy

Skulls of a large adult American
alligator (*right*) and *Apatosaurus* (*left*) to
the same scale. Although the brains of sauropods
(and other dinosaurs) were basically similar to those of
crocodilian archosaurs, those of sauropods were reoriented vertically instead of horizontally to
accommodate the tilting of the back of the head and retraction of the nostrils. They were well
protected from sudden changes in blood pressure by large, distendable vascular sinuses and
possibly a rete mirabile network as in some mammals.

than a century ago. So there's a trade-off—sauropods could have had a relatively limited number
of very fast, fat nerve fibers, resulting in fairly poor sensory ability at their extremities, *or* more
nerve fibers and better coverage, but at the expense of slower response time. Could predators
have taken advantage of this comparatively long conduction time in preying on sauropods?

Probably not. As we've seen, simple reflex arcs are governed by the spinal cord instead of the
brain, and this would have meant a much shorter trip for the nerve impulse. A stimulus/reac-
tion would have taken less time here than it would for the "tail-to-brain" pathway, certainly less
than half a second, and this wouldn't give a theropod much time to formulate any meaningful,
on-the-spot attack behavior. For the sauropod, any nerves outside the spinal cord could, up to
a certain point, have been more heavily myelinated (for example, each of the twin sciatic nerves
running down the backs of your thighs is much larger in diameter than your spinal cord). This
would give them correspondingly faster reaction-response times. If these outside, or *peripheral*,
nerves were engineered for fast conduction, they might have been bigger—and faster—than
anything around today. A 150′ sauropod like *Supersaurus* would have weighed between 50 and
100 tons, and with its massive tail it would have been capable of inflicting incredible damage to
even the largest theropods, which probably maxed out at 8–10 tons. The predator would also be
subject to the same neurophysiological constraints as its prey. Even though theropods were certainly likely to be much faster and more maneuverable than sauropods, they were also affected
by the same nerve conduction delays, limiting whatever advantage they might have had—you

Help from Behind

Tiny as it was in absolute size compared with that of an elephant, a sauropod's brain was as large as it needed to be to direct body movements and behaviors. Like an elephant's and other vertebrates', it was aided by brachial sacral and other nerve plexi, which helped to channel incoming responses and coordinate the limbs and other body areas without going directly to the brain. An especially important one for sauropods and other dinosaurs may have been the sacral body (*inset*), located in a cavity within the sacral vertebrae above the hips. Large nerve plexi like these may have been the reason for a sauropod's surprisingly small spinal cord diameter (*bottom right*) compared with the elephant's (*top right*), since the auxilliary centers may have made it unnecessary to have such a thick spinal cord to handle so many transmissions.

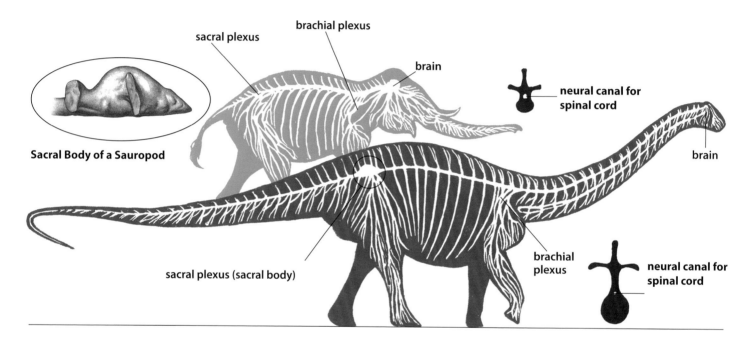

Sacral Body of a Sauropod

sacral plexus

brachial plexus

brain

neural canal for spinal cord

brain

sacral plexus (sacral body)

brachial plexus

neural canal for spinal cord

can't be both massive *and* agile. Finally, it's probable, as Heather L. More (Simon Fraser University) and her colleagues have hypothesized, that large to gigantic animals then as now may have evolved a process to overcome slower nerve conduction called ***sensorimotor prediction***. If correct (and although at present poorly understood), it's the possible ability of the central nervous system to compensate for neural delays by using an instinctual, internal awareness of the body's reaction-response time, taking into account delayed and incomplete sensory information and "predicting" or "programming" the best future motor response. For dinosaurs, this would mean the ability of a sauropod to instinctually "anticipate" an oncoming theropod's bite and start swinging its tail, while for the theropod it would be the split-second sensory warning to swerve away before getting hit.

A SECOND BRAIN?

In many sauropods, stegosaurs and some other archosaurians, the neural openings that run through the vertebrae when they reach the sacrum become enlarged into a chamber, the ***lumbo-***

sacral expansion of the spinal cord. The discovery of this long ago led to the popular misconception that these dinosaurs had a sort of "second brain," which, because of the assumed limitations of the tiny real brain, evolved to control the movements of the tail and rest of the hindquarters. Humans and other vertebrates actually have this also, but it isn't another "brain"—it's there to accommodate the increased need for space to house the nerve *plexi* (major spinal cord concentrations and connections) and, as explained earlier, help your brain in taking over the job of operating your limbs. There's also one at the base of a vertebrate's neck (the ***brachial expansion***) to control the forelimbs. The lumbosacral expansion in sauropods also helped in running the hind end of the body, but it also may have had another purpose as well. In living avian archosaurs (birds), it stores an energy-rich carbohydrate compound called ***glycogen***, and for this reason it is sometimes called the *glycogen body*. Although poorly understood and studied, it's occasionally hypothesized that it helps with balance and is also a store of dissolved nutrients that support the growth and maintainance of the nervous system. At present we just don't know, but as in most animals this might have a dual or multiple purpose.

THE SAUROPOD GAME: TO IMMENSITY AND BEYOND

So far, we've explored the possibilities of how sauropods, the largest known land animals that lumbered the earth, could have supplied oxygen and nutrients to the tissues of their giant bodies while removing wastes through a hyperefficient breathing apparatus and powerful heart and circulatory system. A nervous system that adequately coordinated all this, as well as a subtly engineered skeletal and muscular system, was in place to serve the needs of these enormous creatures. The fact that sauropods did exist on the physical scale that they did confirms that all of the problems and anatomical solutions known in living, much smaller vertebrates had been solved to permit huge size. The question remains: *why so big?* Sheer size isn't an advantage in itself unless it somehow serves the need of the organism to successfully live long enough to reproduce its own species. The answer to this lies in what kind of foods a sauropod ate, how it gathered them and, finally, how such a large group of gigantic herbivores could have shared an ecosystem's limited resources.

In a Rhenland forest, a *Giraffatitan brancai* feeds from *Sequoiadendron* foliage in the company of flying, basal avians. Quick and alert, these benefited from insects stirred up by the giant herbivore and probably returned the favor by screeching a warning of nearby theropods.

CHAPTER SIX
CONIFER CUISINE

Any animal, living or extinct, is or was the product of its attempts to adapt to its environment, and a large part of this adaptation arises from its efforts to obtain food. Why were sauropods often so enormous, and how can we explain their spectacular body structures? Why were they so successful in terms of their taxonomic diversity and geologic longevity, which spanned from the Early Jurassic until the very end of the Late Cretaceous?

SUCCESS OF THE CONIFERS. The backdrop of sauropod evolution was the development and radiation of the *conifers*, best known to us as redwoods, pines, firs, junipers and other living forms. These shrubs and trees today cover huge areas of Earth's surface. While in our time they now share the world with the more diverse *angiosperms*, or flowering plants, they should still be regarded as one of the most successful kinds of vegetation, not only in terms of diversity and geological longevity but also in terms of adaptation to a wide array of environments. These range from the driest deserts to wet, humid rainforests. Although some of the known conifer families are now extinct, others have survived essentially unchanged since the Mesozoic era. Among these are the monkey puzzles and closely related forms (*Araucaria* and others), the swamp cypresses (*Taxodium*), the redwoods and sequoias (*Sequoia)* and the podocarps (Podocarpaceae). Conifers are a major subdivision of the *gymnosperms,* or "naked seeds," which also include cycads, the extinct bennettitaleans and the ginkgos, or maidenhair trees. Living forms have characteristic woody stems (in trees and shrubs) and leaflets or needles that are generally narrow and have a thick outer layer or cuticle to resist moisture loss. These features make conifers capable of adapting to climatic extremes ranging from dry conditions to moist

The spread of the conifers (*right*) and their caloric potential offered sauropodomorphs a diverse and abundant food source, as well as the horsetails (*opposite, top*) and to a lesser extent ferns, such as the Royal and Australian Giant Fern (*opposite, center*). Likewise, ginkgos (*opposite, bottom*) offered a reliable food source. Some conifer families, such as the monkey puzzle of the *Araucaria* family, however, evolved spines, great height and probably toxic phytochemicals that forced sauropods to adapt to and coevolve with their food plants.

environments. While these features must certainly have contributed to their success, a revolutionary development early in conifer evolution, described below, gave them an overwhelming advantage over the other existing plants in their reproductive capability. This led to the global rise of conifers, ultimately making possible the success of the sauropodomorph dinosaurs and especially the sauropods themselves.

Like most plants, individual conifers usually bear both male and female reproductive organs. Also, the sperm-bearing **pollen** of the male conifer is deposited by wind or animal pollinators on the egg-bearing **ovule** of the female. In more primitive plants a tube is produced into which sperm are released, and these swim and come into contact with the ovule, fertilizing the *ovum*, or egg. Conifers are different: instead of releasing swimming sperm, the pollen transfers the sperm's nuclei directly into the egg. Rather than wasting large amounts of stored food in ovules that may never get fertilized, the female conifer ovule only begins receiving food from the parent plant *after* pollination, a tremendous savings in the plant's reproductive energy budget. This does not occur in other gymnosperms, and it would have given primitive conifers a huge competitive edge in producing a maximum number of viable seeds, with a minimum amount of resources invested. Furthermore, the ultralightweight conifer pollen, equipped with balloon-like air sacs, can be dispersed for hundreds of miles by the wind. As Earth's single gigantic landmass Pangea began its breakup, the daughter supercontinents of Laurasia and Gondwana carried their evolving floras with them into new climatic zones, and the conifers swiftly spread. By the Mid-Jurassic they occupied most of the globe and had diversified into several families, and in terms of size and height they became the dominant types of trees in most areas of the world. By this time, conifers shared their environments with an equally diverse array of their gymnosperm cousins, the ginkgos, cycads and bennettitaleans, some other tree-like plants and many types of ferns and smaller forms. The *angiosperms*, those flowering plants that now dominate so much of Earth's surface, would not appear for millions of years.

A DIFFICULT DIET

For the herbivorous dinosaurs that were coevolving with this burgeoning new Mesozoic salad, the new conifers presented opportunities but also challenges. As a potential food plant, most modern conifer foliage, as well as conifers' reproductive bodies and resins, is high enough in caloric value to be a significant food source; it also responds to pruning by regenerating quickly. If extinct species were similar to their closest living relatives, both the relative density and the amount of conifer foliage available would make the exploitation of this resource by an herbivore worthwhile. The challenge lay partly in accessibility: by the middle of the Jurassic many, if not most, conifers had evolved into tree-like forms. This maximized height and space for food-producing foliage, gave them a competitive edge for gathering sunlight over smaller, under-story plants and also created a higher launch platform for the release of pollen to the wind. For animals, however, it also made the foliage hard to reach. To get at it, a dinosaurian coniferous herbivore would therefore have to either climb or grow tall.

A second problem is that the foliage of many modern conifer species contains a high amount of indigestible, sometimes toxic *phytochemicals* in most leaves and needles, making them relatively unappetizing to most animals. Such phytochemicals do not always attack the animal's digestive system, but they can interfere with or resist attempts to break down the plant matter. As a result, few modern vertebrates exploit this food source, although in the past some did, such as the extinct American mastodon (*Mammut americanum*), as well as living species such as Holarctic moose (*Alces alces*), snowshoe hare (*Lepus americanus*) and white-tailed ptarmigan (*Lagopus lagopus*). Like all plants, conifers also contain a compound known as *cellulose*, the main ingredient in the cell walls of plants that helps give them shape and support, and which forms most of the fiber in plant food. Cellulose contains valuable carbohydrates, and in modern ruminants such as ante-lopes and cattle up to 80% of the digestible energy from plant matter is derived from cellulose that has been broken down by bacteria in the animals' guts. Cellulose, however, is extremely tough and relatively resistant to digestion, and vertebrate *folivores*, or leaf eaters, must deal with the problem by having long, specialized digestive tracts with fermentation chambers for the bacteria. As we learned in Chapter 5, there are two basic kinds of fermentation systems: foregut fermenta-tion, where the food is broken down in a specialized group of stomachs, and hindgut fermenta-tion, in which the plant material stays in a special branch of the intestine known as a caecum. Hindgut fermentation is used by horses, elephants, gorillas and some birds as well. Therefore, other archosaurs could have done this too, and this is the system most likely used by sauropods.

As if this weren't enough, a dinosaur conifer eater would have to deal with a third and final problem. Most of the known Jurassic conifer leaflets actually were small to medium-sized, blade-shaped structures that were sometimes closely adjoined to the woody twigs to which they were attached, more like the foliage of the living monkey puzzle tree than the long, slender needles of pine trees. Small, dainty feeders like the white-tailed ptarmigan can use their small beaks or mouths to pluck needles precisely and selectively, but a large herbivore must resign itself to taking in some quantity of wood with each bite. They are necessarily bulk feeders, too big to be precise and too hungry to be picky. Wood contains *lignin*, even more difficult to digest than cellulose; no known living vertebrate can digest this at all. Termites are one of the few known

Modern herbivores like the sooty ptarmigan (*above*) and others rely on conifer buds as a calorie source, but their digestive systems have to deal with the phytochemicals, lignins and other relatively indigestible materials in these plants. A spiky araucarian (*below*) had defenses such as spines to deal with as well.

multicellular organisms that can digest lignin, owing to protozoa in their intestines that secrete enzymes that break it down. As an overall food source, however, tree-like conifers still had possibilities: they were reasonably abundant, and if a way could be found to deal with the toxins and lignin present in their foliage, they could offer a sustainable food source to any animal capable of reaching it and somehow extracting enough nutrition once it was ingested. Reaching that food required a long neck, and digesting it required an effective digestive system.

James Farlow (Indiana Universtity–Purdue University, Fort Wayne), who has extensively considered the possibilities of dinosaur diets and digestive physiology, thinks that most known dinosaur herbivores, including sauropods, were likely to have been *hindgut specialists*. When ingested food reached the area of the intestines, it was retained for long periods of time in organs like a caecum, where *microflora*, or a community of microorganisms that specialized in breaking down resistant plant structures, dissolved these materials into a broth that the animal's own enzymes could begin digesting. These fermenting chambers yielded nutrients at a slow but steady rate, allowing the animal to extract valuable fatty acids and other usable food energy from tough, poor-quality plant products. To make this process pay off in terms of energy yield, the animal had to take in and retain very large quantities of vegetation that were continually in the process of being broken down, much like a compost bin that releases its products of decomposition to produce soil. This would have meant having an equally large abdomen and a big body to carry it around.

Besides having enough material for hindgut digestion to continually release energy, both the caloric potential and *period of time the food is retained* are also important. In their recent survey to determine the possible food potential of known Mesozoic plants and their living relatives, Jürgen Hummel and Marcus Clauss (Steinmann-Institut für Geologie, Mineralogie und Palaontologie, Rheinische Friedrich-Wilhelms-Universitat Bonn) researched the nutritional value of a wide spectrum of plant species. Some conifers were discovered to give up their nutrients fully only after they have been digested for a minimum amount of time. For one genus, *Araucaria*, an unusual quality of the foliage was that when ground up under laboratory conditions and induced to chemically ferment, a greater yield of energy was produced when it was retained for very long periods of time (up to 30 hours). This is similar to the length of time it would probably spend in a hindgut fermenter's digestive tract. Following araucarians in terms of caloric value and digestibility were members of other conifer families, the Cupressaceae (including the giant coast redwoods *Sequoia sempervirens*, junipers such as *Juniperus communis* and others), Podocarpaceae (*Podocarpus imbricatus*) and the Pinaceae (the Norway spruce, *Pinus abies* and others). In addition to conifers, the researchers also found that the genus *Equisetum*, the horsetails or scouring rushes, were also high in caloric value, as well as ginkophytes, now represented by the sole surviving member of this once extensive family, *Ginkgo biloba*. In an independent study, Carol T. Gee (also University of Bonn) ranked the araucarians, ginkgos and horsetails as the most likely sauropod foods.

EXAPTATIONS: THE RIGHT STUFF

We've seen that the largest dinosaur herbivores were the *sauropodomorphs*, a group that by the Carnian stage of Late Triassic times included the prosauropods and (perhaps) the very earliest true sauropods. Early prosauropods had diversified considerably from their carnivorous saurischian cousins, and some may have been omnivorous, but the mainstream of the group became herbivores. The long necks and bipedal capability of prosauropods allowed them to exploit any vegetation from the ground up to about 3.7 m (12′), and they were the only known dinosaurs at that time that could reach into and crop at the lower tree levels. Besides having the means to stand and reach upward, the digestive systems in some groups, such as plateosaurids, were already becoming extensive, as evidenced by their large, barrel-like rib cages. Large bellies mean big digestive systems, adaptations for processing an abundant but relatively low quality nutritional source. If a given type of plant resource isn't very nutritious but there's a lot of it out there, and it's there most of the time, it may be worthwhile to develop a long, specialized digestive tract to process large quantities of the stuff to get the amount of energy you need. Many types of mammals (ungulate grass eaters and primate foliage specialists) have followed this strategy, evolving wide, capacious rib cages and supporting pelvic structures to hold up their massive gut systems. The earliest sauropods were already **exapted** (*ex*, "from"; *apted*, "fitted for," or having body structures already set up to take advantage of this trend) for such a lifestyle, and they would take it to an almost unbelievable level.

In some parts of Laurasia, the ancient plant communities dominated by forms like the seed-fern *Dicroidium* were, by the Late Triassic, undergoing replacement by a new floral assemblage composed primarily of cycadophytes, ferns and conifers. This may have coincided with, and possibly contributed to, the extinction of most of the low-browsing reptile herbivores like rhynchosaurs, anomodonts and transversodonts, which probably fed on the old Paleozoic flora. The new dinosaur herbivores began to radiate, and it was at this time that prosauropods, with their high reach and increasingly high-volume digestive systems, rose to prominence, notably in Europe, South Africa and South America. By the Early Norian stage of the Late Triassic, prosauropods like *Melanorosaurus readi* and *Plateosaurus longiceps* had surpassed their earlier ancestors in size and girth, while still retaining their bipedal capability and evolving even longer necks. This suggests that they were utilizing an abundant food source that offered enough calories to make such size and continued specialization worthwhile. Importantly, some of these forms secondarily developed a quadrupedal body posture, developing longer, more robust forelimbs and modifications to their forefeet that contributed to supporting a larger, more massive abdomen (see Chapters 2 and 4). As the sauropodomorphs evolved, other dinosaurian herbivores, the early *ornithischians*, were also adapting to the new plants. Small to medium *fabrosaurs* and later *heterodontosaurs* were also bipedal, but their smaller stature would have restricted them to lower browsing levels, and their narrow muzzles gave them a ptarmigan-like ability to select only the choicest foliage. Although the evolutionary radiation of ornithischians would produce giant, broad-snouted bulk feeders like the hadrosaurs or duck-billed dinosaurs later in the Mesozoic, in the Triassic ornithischians and sauropodomorphs had already taken different paths: ornithischians were generally staying small and feeding selectively, while sauropodomorphs were tending to grow big, standing tall and eating everything within their reach. This, of course, had to include plants other than trees for at least juvenile sauropods, for which this food was obviously out of reach until they were old (and tall) enough.

Although some fossil sauropod sites indicate the presence of possible gastroliths, it is now believed that they did not use these to digest their food, as do birds such as turkeys, whose crop here is full of stones.

SIDE SALADS

What could these have been? Gee's recent evaluations of modern representatives of the major plant families present during the Late Jurassic, when sauropod diversity was at its maximum, have indicated that certain kinds, in addition to conifers, have high caloric value. Because young sauropods grew fast (see Chapter 8), they would have depended on short plants that not only offered high-quality nutrition but also, like conifers, were abundant and could regenerate quickly after being cropped. Although abundant, cycads and their close relatives the bennettitaleans would be a bad choice, not only because of their low nutritional value but also because they were slow growing and tough. Some ancient fern types, related to the Australian giant fern (*Angiopteris evecta*) and European royal fern (*Osmunda regalis*), were not much more nutritious but at least were abundant and quick to regenerate. An advantage for vulnerable baby and juvenile sauropods in eating these would have been ferns' tendency to occur under dense, closed forests (with the moist and humid conditions preferred by most types) that would have provided protective cover. A third plant group, the "horsetails" (or "scouring brushes," *Equisetum*), not only has a high energy content but is incredibly abundant, grows fast and is quick to regenerate. Deep rooted and growing wherever there is enough moisture, *Equisetum* shoots and growing tips would have provided an ideal food source for the fast-growing young. It's significant that for some species of modern geese, this plant is the exclusive food source for juveniles where it occurs; it also probably formed an important item in the diets of some adult sauropods. *Equisetum* and the lesser-quality fern species, if they occurred in enough abundance, might not only have supplemented the possible nutritional needs of adults but also may have meant the difference between life and death if the preferred conifer browse was unavailable. Fungi may also have contributed to the diets of juvenile sauropods. Commonly found in the form of both various mushrooms and other species in moist environments, in modern forms these are not only a source of dietary fiber but also a rich source of many vitamins and folic acid, essential in most animals for the metabolism of carbohydrates, lipids and amino acids.

BIG GUTS AND GIZZARD STONES?

Although so far it seems that early sauropodomorphs were definitely capable of accessing a relatively elevated food source like conifer foliage, what about their ability to gather, consume and digest this plant matter? Compared to the more sophisticated chewing dentitions of the emerging ornithischian dinosaur groups, prosauropods retained relatively primitive jaws and dentition. Two of the best-known forms, *Massospondylus* and *Plateosaurus*, have long, deep skulls with leaf-shaped teeth that bear large denticles, and the teeth are oval to laterally compressed in cross section. Grinding or chewing didn't take place as in fabrosaur or heterodontosaur dentition—the teeth were adapted simply to crop. Modern birds, which are toothless, depend on a special section of the foregut called a *proventriculus*, or gizzard, equipped with thick, muscular walls and sometimes an interior lined with horny projections, to crush and break down food. Like some living and extinct birds and their distant crocodilian cousins, the proventriculus might have been aided by "gizzard stones," or *gastroliths*, small pebbles or stones that the animal may have selectively collected and swallowed. Firmly nestled in the walls of the proventriculus, these would have taken the place of teeth in grinding and crushing food, a process called a "gastric mill." Could sauropodomorphs have possessed this also?

Alleged gastrolith material, claimed to be in association with some sauropod skeletal remains such as the ones described near the anterior thoracic area of *Diplodocus* (*Seismosaurus*) *halli*, is usually formed of hard, asymmetrically rounded types of rock and bears a high polish. If these actually are gastroliths and were a part of the regular strategy sauropods used in dealing with ingested food in the manner of birds, we should expect that they would have occurred in association at some point with the well-preserved, known in situ skeletons. To date, however, very few gastrolith associations with sauropod skeletons are known, and the few that are known are problematic. In a 2011 study by Oliver Wings and P. Martin Sander (both Steinmann-Institut für Geologie, Mineralogie und Palaontologie, Rheinische Friedrich-Wilhelms-Universitat Bonn), the workers conducted detailed field tests on domestic ostriches to explore the question of how gastroliths are processed, as well as resulting implications for how these would have benefited sauropods. This makes sense because since ostriches are archosaurs and therefore sauropod cousins, they possess a gastric mill and their fibrous plant diet approximates the diet of sauropods. This was done by providing the ostriches with three basic types of rocks in three levels of hardness: quartz (the hardest), granite (second hardest) and limestone (third hardest). These were pre-sized when ingested and the rates of abrasion were measured, and then they were collected from the stomach contents at particular intervals. As a result of the tests, it was found that, depending on their hardness, the ingested rock units lost much of their mass through acidic dissolution in the ostriches' stomachs. Also, none of the ostrich gastroliths, or those documented from other herbivorous living birds, had the asymmetrical shapes and high polish reported in the case of claimed sauropod gastroliths. Extrapolated from the volume of gastroliths averaged in each ostrich stomach, the workers determined that the estimated stomach volume of a sauropod with the body mass of *D. halli* (about 50,000 kg [1,100 lb]) would have contained a very large amount of gastrolith

Until recently, some sauropod workers had hypothesized that these dinosaurs might have used gastroliths (*above*), to assist in the breaking down of food, either in the stomach or within a specialized gizzard, as in herbivorous birds. Based on recent experiments with ostriches, however, it now seems more likely that sauropods simply swallowed their food, which was then broken down over long periods of time without any external aids.

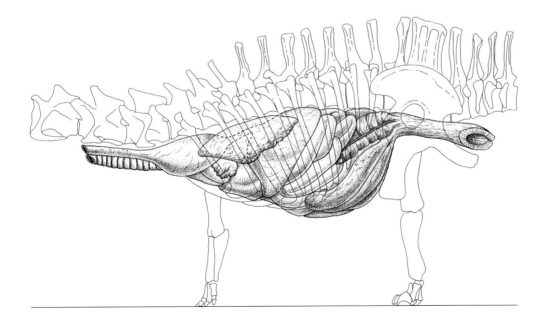

Possibly aided by a collecting sac or crop, a sauropod's stomach and massive gut system (*colored area, left*) were safely enclosed by a strong rib cage and supported by bony *gastralia*, or belly ribs.

Although some workers hypothesize that diplodocids were low browsers that swept their necks sideways to maximize their range of low forage on the ground (*bottom*) as they moved forward, others have demonstrated that these sauropods' skeletal specializations would have made them far more effective as arboreal browsers (*top*), whose flexible necks in this situation gave them a much greater feeding envelope, in either remaining stationary or walking forward, than on the ground.

material, about 500 kg (110 lb). This is much greater than the extremely small amounts of material associated with *Diplodocus* and the other two sauropods, *Dinheirosaurus* and *Cedarosaurus*, to which gastrolith associations have been attributed. In light of this, it appears that sauropods, unlike some ornithischians and non-avian theropods, did not process their food with gastroliths.

In the apparent absence of a gastric mill, many sauropod specialists have come to the conclusion that these dinosaurs simply swallowed their food and allowed the hindgut tract to finish the stomach's job of initially breaking it down. The plant material then underwent fermentation, through the help of microflora, in the caecum and hindgut as described earlier. This process of digestion may be slow, but if enough material is accumulated, the process could provide a steady and continuous source of rich, metabolically ready sugars and fats via the bloodstream to the tissues. Like a compost bin, it would enable a steady supply of nutrients available for energy and tissue building if there was always enough material in the hindgut composter to be broken down. This meant taking in a lot of food, but it would pay off if enough of it could be harvested.

Was there any relationship between the evolutionary spread of conifers and the sauropodomorphs growing bigger guts? During the Early Jurassic, the time when conifer families were evolving into tall forms and becoming the dominant types of trees throughout Laurasia and Gondwana, the earliest true sauropods were the *vulcanodontids*, represented by *Vulcanodon*, *Tazoudasaurus*, *Isanosaurus* and *Aardonyx*. These show some skeletal specializations for the support of greater body weight, and with even bigger gut capacities, than their prosauropod-like ancestors, but they were nevertheless still potentially capable of bipedal rearing to reach into lower tree levels. Here it is important to remember that throughout their entire existence, sauropodomorphs and other dinosaurs remained what is termed **hindlimb dominant**: the skeleton and musculature of the pelvis and rear legs were almost always more robust compared to the scapulae and forelimbs, contributing more thrust and leverage to the animal's movements than the forelegs. This was especially true of the diplodocoid sauropods. Even among some of the later advanced sauropods that developed long forelimbs for primarily quadrupedal feeding, strong hindquarters were retained and gave them the potential for rearing. Additionally, these earliest true sauropods had longer necks, and here we come to a critical question: *what specific factors played a role in sauropods evolving long necks?* What seems like an obvious answer, to reach upward to feed, has been challenged by some paleontologists, and we need to examine some other possible alternatives that have been proposed.

LONG NECKS AND THE NEED TO REACH

How did earlier paleontologists view the use of the great necks? When the anatomy of sauropods first became known in the late 19th century, Cope in 1877 proposed that long necks may have been employed not only to reach high vegetation, like a modern giraffe,

but also to reach downward to pull up submerged aquatic freshwater plants, a technique used by some species of swans. Other paleontologists also considered bipedal rearing as at least a possibility, and Henry F. Osborn commissioned the artist C. R. Knight in 1907 to portray a *Diplodocus* in the act of high browsing while standing on its hindlegs and tail. With these occasional exceptions, however, the question of sauropod diets and feeding postures was rarely touched upon until the 1970s, when Bakker began questioning older, entrenched ideas about dinosaur paleobiology in scientific papers, popular articles and books. Bakker argued against many long-held assumptions about dinosaur paleobiology, including the lives of sauropods, and by the early 1980s, the subject was being freshly examined by a new generation of paleontologists.

In the late 1980s and 1990s, researchers J. Martin, V. Martin-Rolland, E. Frey (Museum of Leicester, Bristol University) and some others made their case for what can be called the "vacuum-cleaner" hypothesis for some macronarians and diplodocoids in the use of the long neck. In this, the animal would have employed a sweeping, side-to-side (and occasionally up-and-down) cone of motions limited to a more or less 45-degree angle from the base of the neck while walking slowly forward, gathering vegetation of many types. The nearest living example would be the way geese forage their way across an open space, nibbling here and there on what takes their fancy. Geese, of course, have much more flexible necks, have short legs that set them closer to the ground and mainly consume grass, but they still provide a reasonable model for this idea. The logic here is that a sauropod's long neck would optimize its range of foraging, while the animal would expend a minimum amount of energy in moving toward the food source. In support of this idea, Martin and others point to the skeletal morphology of the neck in the "cetiosaurids" and similar forms, claiming that the system of buttressing and joint articulations were mainly adaptations for holding the neck out sideways, like an industrial beam (see Chapter 4). According to this, using the neck in a vertical high reach mode is ruled out for most sauropods, except for possibly brachiosaurids, camarasaurs and euhelopodids, which had high shoulders and long forelimbs.

In the early 20th century a variation of this idea was proposed by O. Hay, W. J. Holland and G. Haas (Yale Peabody Museum) and was restudied much later by Kent Stephens (Oregon State University) and Michael J. Parrish (Northern Illinois University). These workers variously suggested that the pencil-toothed, generally long-necked diplodocoid sauropods may have specialized in feeding on soft, aquatic plants (and possibly bivalve mollusks, according to Holland) from a basically horizontal plane, utilizing a similar side-to-side series of sweeping motions as they foraged along the shore of a lake or actually from the water itself. In this scenario, the elongate, cylindrical teeth were used to seize and swallow the aquatic vegetation, while the presumably elevated position of the nostrils allowed the animal to breathe while the muzzle was under shallow water. Again, the neck is considered in human terms to have been engineered more like a horizontal, high-beam construction crane than a subvertical crane or derrick, for close-to-the-ground or at the most understory feeding rather than for high reach.

There are problems with the idea that a large animal would gain in feeding efficiency by evolving a long neck to extend horizontal reach, based on the assumption that this would reduce its energy expenditure in foraging. Moving toward a relatively close, standing food source consumes only a trivial percentage of an animal's total energy budget. This is proportionally even less for large animals; although their absolute caloric expenditure in walking is greater than that of smaller animals, they use *fewer calories relative to their size* than a small animal. This book's authors also agree that an animal needs long reach only to obtain access to foods to which it cannot otherwise get. Long-legged, herbivorous forms that graze or browse from the ground or lower levels (elephants and ungulates, for example) have well-developed trunks or moderately long necks to get their food-gathering apparatus down to what they are eating, while others (such as giraffes) evolved their necks to reach an otherwise inaccessible food source, such as the thorn tree foliage of subtropical Africa. Simply stated, a large animal doesn't have to evolve a horizontal structure to reach a nearby food source on its own level: it can simply walk up to it. Connected to this is the fact that big animals, although they need absolutely larger quantities of food, actually need *less* per kilogram of body mass than a small species. This is why large species (elephants) can wait for their food to release its energy, while a small bird or mammal (a shrew) might starve to death in a single night, being unable to lower their metabolic rates during this time.

The idea of a sauropod using its neck for aquatic browsing also runs into problems. In envisioning an aquatic feeding scenario for diplodocids, Holland and more recent workers based their hypothesis on the then poorly known stratigraphic and *lacustrine*, or lake, deposits that were assumed to have existed in the Morrison Formation of North America during the Late Jurassic. Although some large seasonal lakes and shallow, braided streams occurred during this time, these are unlikely to have supported the amount and variety of aquatic vegetation that could sustain the several types of diplodocids that lived during that time and in that area. These limitations would also have applied to bivalve mollusk populations, which Holland considered as a possible food source. Had there been geographically extensive, deep enough bodies of water with habitats that could have supported varied amounts of plant and animal food, it might be argued that the differences in necks and length among Morrison diplodocid sauropods were adaptations to avoid competition by harvesting differing parts of an aquatic food source, a condition known as *resource*

Apatosaurus and other diplodocids, owing to their intervertebral cartilage, short cervical vertebrae and cervical ribs, could achieve considerable extension and flexion in their necks and may have even tilted the head backward. Like camels, giraffes and other browsers, these vertebrae were kept in alignment during such extreme flexion and extension with tough, elastic ligaments. (Muscle figure based on an original restoration by Gregory S. Paul.)

partitioning. Recent studies, however, indicate that the Morrison Formation's several ecosystems probably had a strongly seasonal rainfall, with only a few large temporary lakes that, in addition, were likely to have been alkaline and therefore unsuitable for supporting more than algae. We must search further for ways to explain sauropod feeding strategies and how these cosmopolitan dinosaurs could survive in such a wide set of conditions throughout the Mesozoic.

ENGINEERED FOR HEIGHT

In Chapter 4, we took an in-depth look at the sauropod dinosaurs' remarkable skeletal engineering. Among these attributes were an axial and appendicular bony framework that simultaneously allowed strength and lightness, as well as great size and stability. Although these features are also characteristic of many of the megamammals like mammoths and giant rhinos that appeared long after the dinosaurs' extinction, sauropods are unique in the degree to which the bones of the neck were adapted in bearing weight and stress, while also permitting movement. We also saw how extensive is the *pneumaticity* of the cervical vertebrae, and the way in which these vertebrae are braced, as well as the fact that there are distinctive modifications to the spine and pelvis for each group of sauropods.

Taking into account these unique skeletal modifications, and the premise that most sauropodomorph physical features were probable adaptations to eating conifers as a main food item, this book's authors and some other workers think that the existing evidence supports the idea that most Late Jurassic sauropods (macronarians and diplodocoids) **were basically medium- to high-reach specialists,** whose heads, generally long necks and other postcranial features were adaptations for browsing from low- to high-level conifer tree foliage, as well as conifer fruiting organs and possibly resins. In this strategy, some families maintained a primarily quadrupedal stance while reaching into the lower- and middle-story foliage but were capable of tilting backward on their hindlimbs, if necessary, to reach up even further. Others may have habitually assumed a *bipedal* or, if the tail helped in support, *tripodal* stance to feed. Here the closest functional analogy in human engineering terms would be that of a cherry picker, with most of the skeletal morphology adapted to the stresses and strains of maintaining the neck in a vertical or subvertical feeding posture. Although in this hypothesis conifers of differing sorts were the main dietary item for most sauropods, other tree-level foliage was probably important in the diets of other species, depending on its seasonal availability and regional abundance. Some forms actually became low browsers of ground-level vegetation.

At this point the obvious mammalian analogy of a giraffe probably springs to mind, but here we should be careful of an oversimplified analogy. While giraffes are specialist browsers of upper-canopy leaves, they usually maintain their necks at a modest angle, and in the case of females the neck may actually be horizontal up to 50% of the time. Quadrupedally feeding sauropods may have had a very similar neck posture some of the time. There is, however, the possibility that the neck was held at a more acute angle when they were feeding on all fours, and bipedal or tripodal feeding would have meant that the neck was held in a vertical or subvertical position. In light of sauropods' coniferous diet, the skeletal morphology, pneumaticity and bracing of the cervical vertebrae are best explained as adaptations that served not only for supporting the neck when the sauropod was horizontal but also to effectively distribute the neck's weight when the animal raised its head to eat.

Brontosaurus and *Diplodocus* are the best-known representatives of the *diplodocids*, a subfamily that evolved some of the most extreme special-

Not all sauropods were adapted to high browsing, and low-level feeding evolved in at least three clades, the dicraeosaurids, titanosaurs and rebbachisaurids. Although the skull of the dicraeosaurid *Brachytrachelopan* is unknown, the titanosaur *Bonitasaura* (*top left*) featured keratinized, blade-like structures immediately behind the anterior teeth of the upper and lower jaw (*top right*) to chop through tough vegetation. The rebbachisaurid *Nigersaurus* (*bottom right*) possessed enormously wide, flat jaws that housed tightly compacted batteries of incisor teeth that were continuously replaced as they wore down and fell out (*bottom left*). All three forms were small and, for sauropods, had short necks, features that were adapted to a low-browsing niche where a long neck wasn't needed.

izations among the sauropods that were already adapted to high reach. In addition to their ultralightweight spinal columns and double supraspinal ligaments, diplodocids had short trunks and relatively short front legs. This reduced the forward body weight and brought the center of mass to just in front of the hips, while the double-nuchal system was aided by powerful dorsal muscles, attached to an especially wide, massive pelvis. Another key mechanism for upright rearing in diplodocids and other sauropods were the *caudofemoralis* muscles. These paired muscles, originating mainly along the proximal chevron bones and possibly the transverse processes on the underside and sides of the tail, inserted onto the upper rear *femora*, or thigh bones, and were a main force in pulling the animal into an upright position. When diplodocids assumed a tripodal position during longer feeding periods, the weight of the body was partially transferred to the mid-caudals of the tail, where another distinctive feature, *double-fused and paired chevron bones*, evenly distributed the weight. Finally, like all other sauropods, diplodocids and other diplodocoids possessed outwardly curving, banana-shaped claws on the hindfeet that could dig into the ground and (in the case of diplodocids and some others) especially enlarged, "logger's pike" claws on the first digit of the manus, which could help hold onto a tree trunk and provide further stability. All this added up to making it especially easy to rear up and stay that way.

Other features of some sauropods' skeletal morphology also support the idea that they could rear. The first is the presence of bifurcated neural spines in the above genera, which we don't see in other, presumably more quadrupedal long-necked forms, like brachiosaurids and omeisaurids. As pointed out in Chapter 4, vertical control in a dorsally braced, subvertical beam is most effectively controlled by a *double* cable rather than a single one, and a double supraspinous ligament is likely to have independently evolved among various sauropods to stabilize the neck in an upright feeding posture. A second is the height of the dorsal, sacral and proximal caudal neural spines in the three genera. If these sauropods were habitual quadrupeds, why are their spines so tall compared with those of other forms, whose longer necks would have needed even more support? Along with the wide and massive pelvis, which is much more robust than in other families, these features are best explained as anchoring points for the dorsal musculature, which was another main force in bringing the trunk and neck into an erect position. Finally, there are the proportionally shorter trunk and shorter forelimbs of the diplodocids. As discussed, this effectively shifted the center of gravity back to a point just in front of the hips, which greatly eased tilting the body backward. All these

Diplodocus and other diplodocids relied on a feeding technique called **leaf stripping**. When a mouthful of arboreal browse was seized in the jaws (*top*), the head moved down and back (*center*), where the conifer foliage was pulled past and between the forwardly directed and beveled upper and lower teeth (*bottom*), leaving the stems. The up/down head motion was aided by the extreme ventral position of the occipital condyle, compared to the more horizontal one in other sauropods.

skeletal features make a case for *Diplodocus* and *Brontosaurus* (as well as some other sauropods) being extremely capable, and habitual, upright feeders. Of the diplodocoids, the dicraeosaurids have all of the above features associated with rearing, but they are small and have relatively short necks. Possibly these bridged the feeding gap between low and medium browsing (see Chapter 10). Basal macronarians like *Camarasaurus* have relatively short dorsal, sacral and anterior caudal neural spines, and this may mean that these were generalist feeders, able not only to habitually browse quadrupedally but also to rear to reach higher food.

A second important skeletal adaptation for upright feeding lay in the position and angle of the pelvic bones in relation to the dorsal vertebrae. In some sauropod groups, the twin blades of the ilia (hip bones) were tilted back at an angle slightly up and away from the axis of the spine, a condition Paul has termed a *retroverted pelvis*. This slight but important modification enabled these sauropods to more easily shift their centers of gravity into a stable posture when they reared on two legs, and may have made it possible to walk slowly around a tree to maneuver into a better feeding position. The anterior prominences on the twin blades of the ilia of most large animals are also points of origin for strong tendon sheets that support the mass and weight of the abdomen, and an upwardly tilted pelvis would have conferred a mechanical advantage in supporting a weighty belly by easing the strain on the spine.

Paul has suggested that there may have been four basic feeding postures among sauropods in which bipedal/tripodal behavior occurred. In the more basal or primitive "**cetiosaurids**," whose necks were moderately long and upwardly flexed, and whose shoulders were level with their untilted hips, rearing was only occasional. **Shunosaurids** and **diplodocids**, which had chevron bones on the underside of the tail, used these, as well as their massive pelvises, hindlimbs and short trunks, to assume a stationary, tripodal position; in these types, immobility was determined by the fact that the pelvis was not tilted, which would inhibit hindlimb motion when the body was erect. **Camarasaurs** did not have sled chevrons, but their retroverted pelvises would have made it possible to rear bipedally to feed. In the case of some forms these may have walked slowly on their hind legs for brief periods. In the **euhelopodids**, a somphospondylan basal titanosaur clade, the tail is currently unknown, but if found, it could imply that both rearing and bipedality were possible. The last group, the hyperlong-necked **mamenchisaurids** and **omeisaurids**, had a potentially tremendous reach from a quadrupedal position, but the presence of both sled chevrons *and* tilted pelvises shows that tripodal feeding and bipedal capability were also possible in these forms. The wide-bodied **titanosaurs** may have had their own form of tripodality. In this case Paul's findings, although speculative and not currently based on rigorous biomechanical analyses, nevertheless seem to persuasively account for the osteology of these forms, and they could be tested through the process of finite element analysis (see Chapter 7).

EATING, SAUROPOD STYLE

With their heads in the trees, and almost in the clouds, how did sauropods harvest conifer foliage? By the Late Jurassic, sauropods' strong jaws bore well-engineered and highly specialized teeth. These took several forms that evolved to harvest different types of conifer foliage and other vegetation. **Robust cropping dentitions** generally resemble thick, sharp-edged spoons, whose crowns were wider than the roots. Found in the deep, massive skulls of "cetiosaurs" and camarasaurs, these retained the basic cropping function of the prosauropod teeth. Coming together to slice like a pair of gardening shears, their cutting surfaces moved diagonally and toward each other to interlock. This was probably best suited to cropping coarse foliage occurring in clumps and would have been effective in cutting the woody, tough stems that supported most conifer

leaflets. Camarasaurs also may have had a limited form of *oral processing*, or chewing. In their study of associated and separate teeth from *Camarasaurus*, James Madsen Jr. and Mary A. Carey (both DinoLab) inferred, based on the grouping of the teeth and the size, shape and pattern of lateral grooves and wear facets, that camarasaur teeth could to some extent grind or crush, as well as cut. The pronounced "shoulders," or corners produced by heavy wear on such teeth, give us still another clue as to how these sauropods collected foliage. These *might* have been produced by habitual side-to-side movement of plant material across the *lingual*, or inside, surface of the teeth, but it's probably more likely that such shoulder wear indicates that camarasaurs, as well as some other advanced sauropods, may have been forcefully pulling or tearing bites of foliage away from the tree.

Although not as specialized overall for upright feeding as the diplodocids, camarasaurs' bifid neck vertebrae do show a moderate adaptation toward this. Some species, such as *Cathetosaurus lewisii*, also possessed a tilted or retroverted pelvis (*right*) that was more specialized than that of its close relative *Camarasaurus* (*left*). This aided in easing the strain placed on the hips with paired, anterior and posterior laminae (*thick lines*) that transferred much vertical loading (*arrows*) downward to the tail.

Compressed cone-chisel dentitions are sharp and narrower, and this shape gave the shallower, less robust jaws of brachiosaurids and some titanosaurs a *precision shear bite* that was probably suited to *slicing* through foliage. This contrasts with the simple *orthal bite*, or less closely spaced tooth-to-tooth jaw closure, of basal and some advanced sauropods. A precision shear bite means that the teeth came into much closer occlusion, or contact, than in earlier sauropod forms. Because of their neck lengths, brachiosaurids and their close titanosaur relatives typically browsed at higher levels than camarasaurs, and this tooth shape may reflect more selectivity about the type of foliage they harvested. Titanosaurs (see Chapter 11) have teeth that were in some ways similar to the diplodocids' but with some macronarian characteristics, a feature that indicates their origin among the brachiosaurid-related forms. At least one ground-browsing species, *Bonitasaura salgadoi*, had shearing blades on opposing upper and lower teeth that could cut through tough vegetation.

As the name implies, **pencil-like dentitions** are cylindrical, more numerous and positioned at the front of the jaws into a closely parallel arrangement that can be termed a "tooth comb."

These occur in the diplodocoid sauropods, the *diplodocids* and their close relatives the *rebbachisaurids and dicraeosaurids*. Like their theropod cousins, all sauropods had replaceable teeth. When a tooth became worn down, it was pushed out by a new replacement that grew from a chamber inside the root cavity. This was taken even further in the pencil-like diplodocid dentitions, which, like a mammalian rodent's incisors, were continuously growing and open rooted. Open-rooted teeth imply a food-gathering activity that wears the tooth down quickly, and the abrasion angles on those of diplodocids are usually one of, or a combination of, two basic kinds. One is at a 90- to 95-degree angle to the main axis of the tooth, the other at a 50-degree angle to the tip, or *apex*, of the tooth. Diplodocids like *Diplodocus*, *Apatosaurus* and *Barosaurus* are most likely to have incurred this type of wear by using their teeth in two ways, which can be termed *leaf stripping* (**raking**) and **grasping**. *Leaf stripping* (or raking) was probably the most frequent food-gathering technique, in which the teeth were used to enclose a mass of conifer leaflets and then strip them from their stem by pulling the head down and backward; this is the action that would have produced the 90- to 95-degree angle of wear. **Grasping** was employed to pull off large objects like seed cones or pods from their stems. A third behavior, gnawing, was probably unlikely, however, because a recent biomechanical modeling test shows that this would have created serious stresses in the skull and premaxillary teeth.

Some workers, such as Bakker, Norman and Dodson, have come to basically similar conclusions at about the same time with respect to leaf stripping (or raking), while in 1994 Barrett and Upchurch proposed a detailed model of how leaf stripping would work. In this concept, the lower tooth comb would hold and stabilize the leaflet-bearing branch against the inner, or lingual, side of the upper tooth comb. At the same time, the upper tooth comb would exert downward pressure away from the base of the branch toward its tip, allowing the stem to slide away but separating and pulling off the leaflets.

Tooth batteries, a derivation of pencil-like teeth, are an extreme and fascinating specialization found in *Nigersaurus taqueti* and some others among the diplodocoid rebbachisaurids (see Chapter 10 for further discussion). As wide as they were long, the extremely wide upper and lower jaws of *Nigersaurus* contain a single tooth row at the very front of the muzzle. Although pencil-like in form, the teeth were much smaller, and instead of being backed up by just one, a "nest" of replacement teeth, already formed, were ready to take their places as each tooth was worn out. These created a single, abrasive grinding surface, anticipating the larger, wider ones that later Cretaceous hadrosaurs would evolve in the rear of their upper and lower jaws. There is speculation among workers as to their use. Some, such as Paul Sereno (University of Chicago), think that *Nigersaurus*'s relatively short neck and wide mouth are signs that it browsed, unlike most sauropods, on low-growing vegetation, but others, such as coauthor Wedel, think that the evenly spaced teeth might have functioned more like a comb, possibly for straining water plants or invertebrates, as with modern flamingos.

The **pencil-like chisel teeth** of titanosaurs (see Chapter 11 for a more thorough discussion) like *Rapetosaurus*, *Ampelosaurus* and *Saltasaurus* form a fifth category. Although rather elongate and rounded like those of a diplodocoid, they are also somewhat flattened and have a chisel-like tip produced by wear. The diplodocoid-like character of the teeth in advanced forms, as well as the fact that some titanosaurs' overall skull shape (*Rapetosaurus, Tapuiasaurus, Nemegtosaurus*) shows affinities with these sauropods, suggests that the group had evolved away from its basal macronarian ancestors to become diplodocoid-like in its feeding adaptations. In 2007, Upchurch and his associates noted that the wear facets on *all* sauropod teeth pose an apparent paradox— what caused this wear if there was little or no oral processing (chewing or grinding) going on? The conclusion the workers arrived at is that the wear reflects the large number of repeated jaw opening and closing actions—perhaps many thousands per day—related to the cropping of large volumes of plant fodder connected to bulk food processing. Tooth replacement rates in sauropods were high, and this fits with the heavy usage placed on them every day of the animal's life.

THE EVIDENCE OF MICROWEAR

Further clues to sauropod feeding behavior have come from another source as well, *dental microwear*. This usually appears as scratches, gouges and pits on the surfaces of an animal's teeth and is usually created by the foods it eats and tiny, inorganic particles of grit. These are often present as (1) grit that the plant eater may take in from the nearby surface when it crops food close to the ground, or (2) grit ingested when wind picks up and blows dust and debris onto upper plant surfaces. Microwear can also be produced by the movement of the animal's jaws and teeth as it collects the food.

How do food materials themselves create microwear? In addition to their sometimes hard or rough outer shapes and textures, the cell walls of most plants are supported on a microscopic level by *phytoliths,* crystallized spicules of inorganic mineral compounds that tend to form distinctive shapes in each plant, and some are larger and potentially more abrasive than others. If the conifer leaflets and twigs presumably eaten by an adult *Camarasaurus,* for example, contained a higher proportion of more abrasive phytoliths than those eaten by *Diplodocus,* it could help to explain different microwear patterns. This could be tested by conducting a systematic study to see whether some of the conifer leaflets we have discussed earlier contain larger phytoliths than others, and also by examining the teeth of the *Camarasaurus, Diplodocus* and other sauropods, which might confirm this by revealing consistent associations, if preserved, of phytoliths. It would be another way of knowing what types of vegetation were preferred by certain sauropod species.

The size, depth and direction of microwear markings, as well as their numbers and location on the teeth, can furnish information about both the qualities of foods (coarse, hard or soft) and the way the animal punctures, slices or chews these foods. Comparisons with the microwear patterns of living and some extinct mammals with known diets can help with sauropods, but only when it comes to patterns we see on incisor teeth, since sauropods didn't chew and had only cropping dentitions. In sauropods, as with other dinosaurs, teeth were continuously replaced as they wore down, and thin-section studies of these show that sauropods had the fastest replacement rate known in any vertebrate, anywhere from 62 to as little as 30 days. This means that the markings on each tooth represent *a very brief record* of feeding for that sauropod. In addition, since even hard tooth enamel surfaces gradually wear down with eating, *microwear only reflects the last few meals* of the animal before its death. Depending on the abundance of a particular food item and the season a sauropod species' food may have been available, we could expect

Analyses of sauropod tooth microwear show consistent patterns that usually fall into two catagories. Large scratches and pits (*top*) mean that there was coarser, more abrasive material in the diet, presumably closer to ground level, while finer scratches and pitting (*bottom*) indicate that the food plants were higher above ground and less apt to be covered by hard particles that produced scratching and pitting. This provides clues to the general type and height of foliage that formed sauropod diets.

In filling high-browsing niches for feeding, sauropods evolved two basic jaw and head shapes. Non-selective feeders had broad muzzles, as do cows (*top*), for cropping the largest quantity possible of calorie-poor but abundant browse with each bite, while selective feeders, although, like deer (*bottom*), having comparatively narrow muzzles that took in smaller quantities, made up for this by concentrating on more nutritious, calorie-rich arboreal foliage. Some, such as camarasaurs, were generalists, with muzzles intermediate between broad and narrow. Whether selective or non-selective, each main sauropod clade had distinctive teeth to deal with different foliage.

the type of microwear to vary within a sampling of preserved teeth from a particular species. In this case more is better: the bigger the sampling, or *database*, from teeth belonging to a species studied, the more likely it is that we can detect recurring patterns. If we have only a small sample of teeth, the microwear may or may not actually represent a real tendency. When we can correlate a recurrent pattern of tooth microwear with certain types of plants and cropping motions that are likely to produce these, this can help to construct feeding preferences and behaviors for particular sauropod species.

Because the skulls and dentition of North American Morrison Formation sauropods like *Camarasaurus* and *Diplodocus* were until recently among the best preserved and most well known, these in the past have been the focus of most workers. Utilizing scanning electron microscope (SEM) photos taken from the teeth of *Camarasaurus* and *Diplodocus*, Anthony Fiorillo (Dallas Museum of Natural History) and Paul Upchurch (University College London) and their colleagues in independent studies during the 1990s discovered patterns of pits, coarse scratches and fine scratches that, in spite of the small sampling, suggested patterns that appeared distinctive for each species. The larger pits and deeper, broader scratches found on the surface of adult *Camarasaurus* teeth indicated to these workers that coarser plant products that contained (or were coated by) relatively abrasive materials may have formed the diet for this sauropod. The smaller pits and finer scratches on those of *Diplodocus* adults suggest that its food centered on more delicate, less abrasive plant forms. Interestingly, Fiorillo also found that juvenile *Camarasaurus* microwear patterns were like those of adult *Diplodocus*. This suggests that there was a dietary overlap period, in which the young animals were apparently selectively eating vegetation with finer foliage than that which formed the diet of adult camarasaurs. Finer microwear patterning is what would be expected from relatively soft, herbaceous material, while *Equisetum* and similar plants could account for the pits, since these grew close to the ground and may have accumulated grit and other hard particles. *Equisetum*, like many other ground-level plants, derives much of its support from the mineral silica embedded in its cell walls, and this should also produce scratches.

The presence of fine microwear in juvenile *Camarasaurus* and adult *Diplodocus* is significant, because at first it seems to challenge the concept of diplodocids as primarily high-reach feeders. Since camarasaur juveniles were obviously unable to reach higher canopy levels and more abrasive materials would be expected closer to the ground, the similarity in scratch patterns could mean that adult diplodocids were actually taking in plants much closer to ground level as their main food source. Another possibility, however, is that the striation patterning produced by upper-level feeding on adult *Diplodocus* teeth, although very similar to that on juvenile *Camarasaurus* teeth, was created by tooth contact with seemingly similar but actually different plants. The younger, more tender lower-story vegetation that formed the juvenile camarasaurs' diet at this stage may have duplicated the patterning created by equally tender, delicate upperstory conifer leaflets that the adult *Diplodocus* may have stripped off. This seems to be paralleled among some ground-level-feeding modern primates that, like diplodocoids, practice leaf stripping. Here scratches outnumber pits, because instead of being compressed down to create pits, the grit particles are drawn across the tooth enamel as the leaves are pulled off the stem. This creates scratches instead.

SELECTIVE AND NON-SELECTIVE FEEDING

The issue of diplodocoid feeding behavior is a subject of recurring debate among sauropod specialists. It was recently reexamined in a comprehensive 2011 study conducted by John Whitlock (University of Michigan). In this, Whitlock compared the microwear observed from a small sampling of the teeth of diplodocoids like *Diplodocus*, *Apatosaurus*, *Dicraeosaurus*, *Nigersaurus* and *Rebbachisaurus* with that of macronarians like *Camarasaurus* and *Brachiosaurus*. His findings were different from those of Fiorillo and Upchurch: unlike these workers, Whitlock observed a great deal of large, coarse pits and gouges forming the microwear in the diplodocoids, and he felt that this was caused by the above genera (except for *Dicraeosaurus*) feeding close to ground level, on plants that accumulated a lot of grit. He also went a step further and combined these findings with a hypothesis derived from a joint study by Christina M. Janis (Brown University) and David Ehrhardt (Carnegie Institution), a second independent study by N. Solunias (New York College of Osteopathic Medicine) and S. Moellecken (University of California, San Francisco), a third by Helene Dompierre and C. S. Churcher (both University of Toronto) and finally a fourth by Iain Gordon and A. Ilius (both James Hutton Institute). These workers noticed a correlation between the muzzle shape of extinct mammalian ungulates (like camelids) and their diets. Here, anterior snout or muzzle shapes that were rounder and narrower in dorsal view generally seemed to correspond with species that were *selective* in their plant food preferences (usually less abundant, finely structured vegetation tending toward high nutritional value). These shapes could be compared to those of species with wider, squared-off snouts, which often belonged to *non-selective* feeders that often fed on abundant but coarse vegetation of low nutritive value, often found near ground level. In short, if you're a picky eater, you can have a smaller, narrower snout because the smaller width and shape let you selectively pick out the really choice stuff, but with a big, wide mouth you can grab off a lot more poor-quality chow, in bulk, to make up for the difference in nutrition—think of a daintily feeding deer versus a grass-mowing cow. Earlier, Matt Carrano (Museum of Natural History, Smithsonian Institution) and his colleagues noticed the same pattern occurring among narrow-snouted (lambeosaurine) and wide-snouted (hadrosaurine) hadrosaurs.

Based on snout shapes and his own microwear studies, Whitlock synthesized both findings to construct an innovative theory to explain feeding behavior in diplodocoids. The wide-snouted (and therefore non-selective) diplodocoids, because of the prevalence of large, coarse pits and other microwear features on their tooth surfaces, were eating abundant, comparatively gritty plants that grew close to the ground, presumably like wide-mouthed hadrosaurine dinosaurs and some mammalian ungulates. These contrast with round-snouted macronarian sauropods, which were generally mid- to high-canopy, selective feeders. Whitlock ruled out the possibility that diplodocoids were leaf strippers, even though he admitted that ground-level browsing left similar microwear patterns.

Getting back to what microwear really says about sauropod diets, this theory seems to conveniently fit with the "vacuum-cleaner" hypothesis mentioned earlier. So does it prove that diplodocoids weren't arboreal feeders after all, but were actually using their long necks in a horizontal position to feed? Actually, no. As we already touched on, pits and gouges from presumed ground-level feeding could also be explained by both upper-level wind-blown grit and tooth–

food contact with the hard coverings on the fruiting ovules of *Araucarites* and other conifers. The same large pits *also occur* on the tooth surfaces of moderately selective, mid- to high-canopy feeders like *Camarasaurus* and *Brachiosaurus* with rounded snouts. Leaf-stripping feeding behavior, which Whitlock rejected, was recently shown in an even more recent 2012 study by Mark T. Young (University of Bristol) and his associates to be not only a possible but also a highly likely form of feeding for diplodocoids. In this study, a computer-generated model of the skull of *Diplodocus* was put through a series of stress tests, simulating (1) muscle-driven biting or grasping, (2) leaf stripping and (3) bark stripping (gnawing). In the first two tests the skull bones experienced only very low levels of stress and were actually demonstrated to be "overengineered," or shaped and connected more strongly than necessary, to deal with these activities. The third, gnawing, created much higher stress levels in bones and teeth, making this behavior less likely. As a result of these tests, the known morphology of the skull of *Diplodocus* and its close relatives suggests that among these sauropods, leaf stripping and grasping of conifer fruits were the main ways that they fed. Taking this into account, the fine microwear patterns, together with pits reported in the 2011 study, are probably best explained by this behavior and the presence of wind-blown grit.

If this is true, how does arboreal browsing in diplodocoids (and other sauropods) fit with the analog of square-snouted, non-selective to round-snouted, selective known mammalian feeders? Although the Whitlock hypothesis seems logical, one important thing it doesn't take into consideration is the possibility of ***non-selective arboreal***, as well as ***non-selective ground-level, feeding***. No known mammal, living or extinct, was adapted to the former feeding strategy. It doesn't mean that diplodocoids never fed from the ground level, and in fact they may have done so when their preferred arboreal browse was seasonally scarce and nutritious ground-level plants were abundant. Some clades, in fact, produced species like the dicraeosaurid *Brachytrachelopan* and the rebbachisaurid *Nigersaurus*, which bucked the typical sauropod trend and became highly specialized, probably *non-selective low browsers* (see Chapter 10). Because some wide-snouted herbivores can—and do—browse as well as graze (demonstrated by Robert S. Feranec [New York State Museum] in certain extinct camels, and known in modern cattle), this flexibility in feeding behavior widens the animal species' feeding niche. That, in turn, leads to an advantage in survival and might have been one reason for the diplodocoids' great evolutionary success. While taking this into account, however, the various adaptations to low-, mid-, and high-canopy browsing still seem to fit best as the main feeding adaptation among these sauropods and others, since it coevolved with the development of high intake, very long gut retention and fermentation of plant foods. This was a sauropod innovation unique among vertebrates. Along with possible feeding flexibility, the success of diplodocoids (and other sauropods) may possibly have been based on concentrating on harvesting mainly a few types of highly abundant conifer foliage—what could these have been?

THE MISSING CONIFER

Although several of the Mesozoic conifer families are represented today by living species, one, the *Cheirolepidiaceae*, is totally extinct. This is very unfortunate, because some things known about them point to their being especially good candidates for sauropod food. For one thing,

Selective Feeders

Camarasaurus
(camarasauromorph)
large, robust, spoon shaped
for coarse cropping

Turiasaurus
(eusauropod)
heart shaped (for
cropping unknown
foliage type?)

Mamenchisaurus
(mamenchisaurid eusauropod)
medium, spoon shaped
(for cropping finer browse?)

Non-selective Feeders

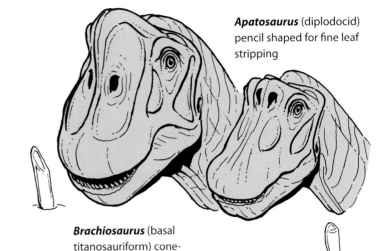

Apatosaurus (diplodocid)
pencil shaped for fine leaf
stripping

Brachiosaurus (basal
titanosauriform) cone-
chisel shaped for nipping

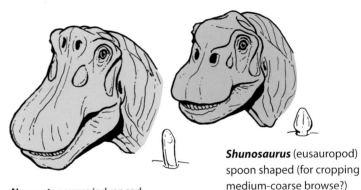

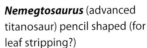

Nemegtosaurus (advanced
titanosaur) pencil shaped (for
leaf stripping?)

Shunosaurus (eusauropod)
spoon shaped (for cropping
medium-coarse browse?)

Selective and Non-selective Feeders

In spite of all being high bulk feeders, as they radiated to adapt to different conifers species and feeding levels sauropods evolved different head shapes and tooth morphologies to best handle a wide variety of these plants. Some, such as mamenchisaurids and turiasaurids, had relatively narrow jaws that took in fewer but more caloric leaves per bite, while brachiosaurids and diplodocoids had wide jaws for maximizing their bites of lower-quality browse. In this way competition was reduced among different sauropod clades. Later-evolving titanosaurs had skulls and teeth that, although macronarian, were very diplodocoid-like, perhaps an adaptation to both feeding strategies. Camarasaurs were probably generalist feeders, able to take advantage of a wide variety of arboreal vegetation.

Sauropods, depending on species and life stage, probably used a wide variety of plants during the Mesozoic, depending on the region they lived in and seasonal availability. Besides the cheirolepidiaceans, or "cheiros," like *Brachyphyllum* (**1**), *Podozamites* (**2**) and *Pagiophyllum* (**5**), these may have also typically included *Araucaria* (**3**), *Podocarpus* (**4**), *Ginkgo* (**6**), *Equisetoides* (**7**) and *Equisetum* (**8**). The last was probably a staple food for young sauropods of all types.

finds show this conifer family to have arisen and become established worldwide by the Late Triassic, just when the prosauropods and earliest sauropods were evolving. In comparison with other conifer families, the "cheiros" had the greatest diversity of growth pattern, habitat and morphology. Some types were tall trees with 1 m (3′6″) trunks that grew at least 23.4 m (76′9″) in height and had long life spans, attested to by the discovery of an in situ stand of trunks from the Late Jurassic of England, with individual trees that may have ranged from 200 to possibly 700 years in age. This meant that they were among the *dominants* in a long-lived forest ecosystem. Besides tree-like forms, other types were herbaceous or shrub-like, growing in low, dense stands in a tidal or coastal marsh setting. "Cheiros" were a part of many plant communities, ranging from almost monospecific/low-diversity assemblages in brackish (slightly salty) or hypersaline coastal environments to species-rich communities in *mesic* (moist), *riparian* (streamside) locations. They also apparently thrived under semiarid to completely arid conditions and in strongly seasonal climates, especially at low paleolatitudes. This would have made them tolerant of conditions at localities like the Brushy Basin and other Morrison formations.

Fossil cheirolepidiacean foliage occurs as two general types. One, *Brachyphyllum* or *Pagiophyllum*, bears leaves or cuticles that are scale-like or pointing outward and arranged in a spiral pattern; the other, *Frenelopsis* or *Pseudofrenelopsis*, has leaves or cuticles that form tightly around the stem and have a jointed appearance. Despite these differences, characteristics that all members of the family have in common are a distinctive type of pollen, separately named *Classopollis*, and less typically thick cuticles with sunken *stomata* (air exchange pores) with *papillae* (tiny projections) that extend over these. Other species may have had fleshy leaves and been deciduous.

Abundant, widely distributed and appearing in many forms, "cheiros" may have been a main food source that fueled much of sauropods' evolutionary radiation. The family appears to have become extinct in the Northern Hemisphere by the very Late Cenomanian/Early Turonian stages of the Cretaceous, while in the Southern Hemisphere it became severely restricted in diversity during the mid-Coniacian. Although it survived the mass extinctions at the very end of the Maastrichtian, it finally flickered out during the early Paleocene. This had seemingly drastic consequences for diplodocoids (see Chapter 11). Since they left no close relatives, we can't run fermentation experiments like the ones described earlier on extant conifers, horsetails and other plants to see how digestible they were, or how much energy they could have produced. We also don't know how quickly most "cheiros" grew or could regenerate after being cropped by a hungry sauropod.

RESOURCE PARTITIONING: HOW TO SHARE A SALAD

During their early lives, *sympatric* (living in the same area, at the same time) juvenile sauropods of all families were in direct competition with each other, as well as other herbivores, for the most nutritious vegetation of any kind that was available. As they grew in maturity (and in height), competition with one another and their genetically determined feeding morphologies may have channeled them into the feeding zones for which they were best suited. Sauropods also may have been instinctually "hard-wired," like some birds and mammals, to select those types of plants that were the most nutritious at a given stage of their development. Given the abundance and high nutritional value of low-growing *Equisetum*, as mentioned earlier, it was probably a main food source for most sauropod species as juveniles. The recently discovered skull of a juvenile *Diplodocus*, with its relatively narrow anterior muzzle compared with the much broader, more squared-off one typical of adults, suggests that when young this species was adapted toward browsing from less coarse, more calorie rich and, of course, much shorter food plants. As with modern African bovid ungulates, muzzle shape implies browsing and grazing adaptations: the dainty, more pointed mouths of smaller antelope like Thompson's gazelles grade into those with progressively wider ones (wildebeests) until one reaches the very wide muzzles of feeders like Cape buffalo. The two diplodocids whose skulls are reasonably known, *Diplodocus* and *Apatosaurus*, have anterior muzzles that by comparison are narrower in the first, more gracile genus and proportionately broader in the second, more robust one. Along with their average neck lengths and other postcranial morphologies, this implies that mature adults fed from different tree canopy levels, and possibly different conifer and other tree species.

Even though they probably formed their own group associations based on relative age, many juvenile sauropods may still have required proximity to adult feeding areas. Among modern herbivorous mammals like elephants, pandas, koalas and hippos, juveniles practice an instinctive behavior known as *coprophagy*, the deliberate eating of feces. The sterile digestive tracts of these babies at birth lack the gut flora that are necessary for each species in breaking down tough plant materials, and they can only acquire this by eating what has already passed through the adults' systems. This would have been a highly effective way for each type of sauropod juvenile to accumulate the right mix of microflora as it grew up (and grew taller) and to gradually begin acquiring more and more adult browse in its diet. If social separation of juveniles from adults gradually decreased with growing size and age, proximity to feeding adults may have also enabled older sauropods, by association and observation, to begin preferring certain types of browse.

How did such a diverse array of these kinds of dinosaurs share the limited food resources that may have been available to them? How could any single resource, in one regional environment, support so many gigantic herbivores? To visualize this situation, imagine a savannah the size of East Africa inhabited by not one but *several* types of elephant species, all requiring massive amounts of vegetation on a daily basis. There are limits to what even an abundant resource can provide, and modern communities of large mammalian herbivores can only coexist and maintain their diversity by utilizing different portions of a general food source and at different times, a condition known as **resource partitioning**.

The great number of grass-eating ungulates in East Africa's savannas, for example, is made possible by the fact that although all of them eat grass, all are in some way ecologically separated, some by each species' food requirements, others by their preference for different types of grassland: dry or swampy, open or wooded. Some ungulate herbivores, such as zebras, wildebeests, topis and Thompson's gazelles, all eat the same kind of grass but avoid competition by consuming *different parts* of the plant at *different stages* in its growth. This begins when zebras use their upper and lower incisors to nip off and eat the outer stems, which are too tough and lacking in nutrition for the three antelopes but are a food from which the zebra's hindgut system can efficiently extract nutrients. Topis are next, reaching for the more tender lower stems with pointed muzzles, followed by the wildebeests, whose squared-off muzzles are adapted to crop the horizontal leaves. After several days, the grass plant sends out new sprouts from its base, which the Thompson's gazelles, with their small, dainty muzzles, can easily obtain.

Another example of resource partitioning, among closely related species within the same area and for the same food, is shown by the herons of Okefenokee Swamp, Florida. In this situation, five kinds of heron may hunt for fish on the same tidewater shoal. While relatively short-legged green herons perch on the prop roots of mangrove trees, passively waiting for prey to come within reach, longer-legged, more active Louisiana herons wade out into the shallows to hunt. Snowy egrets feed farther out, skimming the water surface with one foot and often flushing out a fish. Larger and taller reddish egrets have the most complex behavior, first stirring the water to frighten the prey and then spreading their wings to offer the fish a shady, apparently safe shelter; when it darts into the shade, the heron strikes. The stately great blue heron, tallest and largest of the five, can wade into deeper waters harboring fish that would be too difficult for the others to reach. It's risky, of course, to infer too much about the feeding dynamics of sauropods, or other dinosaurs, from those of modern mammals on the East African plains, or herons in the bayous of Florida. To directly compare dinosaurs with large mammals or birds that appear to have similar ecological roles is in many ways like comparing melons and kiwifruit: the two were vastly different, and in the case of sauropods we have to use caution because there is no direct mammalian or avian analogy.

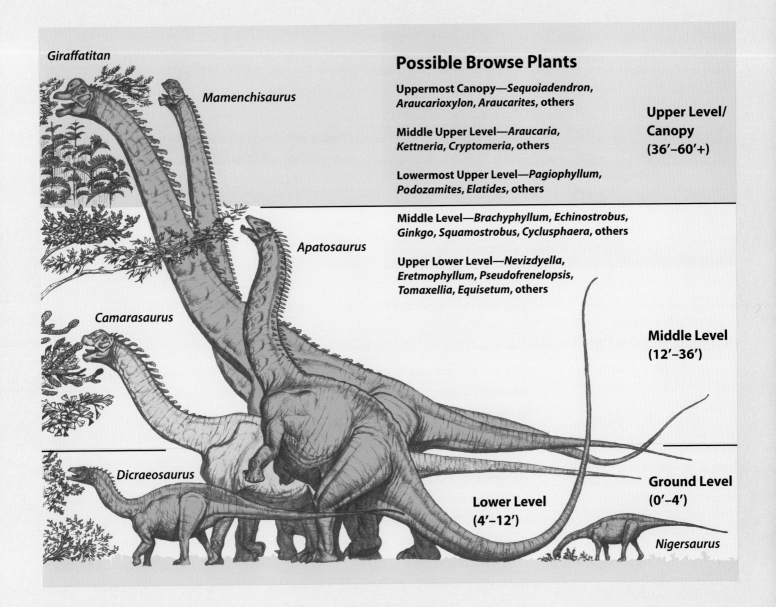

Possible Browse Plants

Uppermost Canopy—*Sequoiadendron, Araucarioxylon, Araucarites*, others

Middle Upper Level—*Araucaria, Kettneria, Cryptomeria*, others

Lowermost Upper Level—*Pagiophyllum, Podozamites, Elatides*, others

Upper Level/ Canopy (36'–60'+)

Middle Level—*Brachyphyllum, Echinostrobus, Ginkgo, Squamostrobus, Cyclusphaera*, others

Upper Lower Level—*Nevizdyella, Eretmophyllum, Pseudofrenelopsis, Tomaxellia, Equisetum*, others

Giraffatitan

Mamenchisaurus

Apatosaurus

Camarasaurus

Middle Level (12'–36')

Dicraeosaurus

Lower Level (4'–12')

Ground Level (0'–4')

Nigersaurus

Resource Partitioning

In regions like the Morrison and Tendaguru, competition for browse among various sauropod clades was mostly eliminated (in adults) by resource partitioning, with each form adapted to a particular type of browse. Small, shorter-necked species (*Dicraeosaurus*) were best suited to harvesting low to medium-height foliage. Moderate-neck-length, flexible feeders (*Camarasaurus*) used their robust, spoon-like teeth to pull off coarse browse at medium to medium-high levels, either quadrupedally or bipedally, while rearing-specialist feeders (*Apatosaurus*) could exploit a wide range of fine conifer foliage by leaf stripping in the higher medium to lower upper canopy. The fantastically long-necked and tall brachiosaurids (*Giraffatitan*) and the Asian mamenchisaurids (*Mamenchisaurus*) could reach farther still, accessing the uppermost canopy. In contrast to all these were short-necked, very small ground-browsing specialists like *Nigersaurus*, at right.

Scorpion flies, or mecopterans, which as fossils are known from the Early to Mid-Jurassic onward and were once much more diverse as an order, may have been important pollinators of some extinct conifers like cheirolepidiaceans. Unlike other conifers, these had female ovules that were poorly suited for wind pollination, and may instead have relied on the long proboscis of mecopterans for fertilization. If adverse conditions had affected either the pollinator or its host, the outcome might have been drastic for both and could have meant extinction for some cheirolepidiacean-dependent kinds of sauropods.

CONIFER CUISINE

One question of great importance in assessing the likelihood of conifers as a main food source for sauropods was this plant group's abundance and taxonomic diversity from the Mid-Jurassic until the Late Cretaceous, the time of the sauropods' radiation into diverse clades. The fact that sauropods were so different in their feeding ecologies throughout their evolution suggests that conifers, in turn, must have been not only diverse but also abundant if they were to have been a main, and adequate, food source. This is especially crucial when we think of environments like the Morrison Basin of the Late Jurassic North American West, in which several species and genera of these dinosaurs coexisted. Until recently, this huge geographic area, with its probably widely varying ecosystems, was described as mainly dry fern prairie broken up by occasional thin, patchy woodlands, restricted in quantity and distribution, which may have grown primarily near seasonal and permanent rivers. In reevaluating this scenario, however, Gee has now discovered that some Late Jurassic Morrison environments were not necessarily just semiarid fern savannas but instead were often locally dominated by a diversity of forest-forming trees, mostly conifers. These not only were species diverse but also formed widespread forests. Evidence for this comes from recent pollen sampling and fieldwork by Carol Hotton (National Center for Biotechnology Information) and Nina Baghai-Riding (Delta State University) and the findings of previously unknown conifer species based on fossil cones collected from eastern and southern Utah. All of these, including new fossil leaf-based floras, paint the picture of locally moister and much more extensive conifer forests and fern wetlands that coexisted with the Morrison sauropods, providing an adequate variety of browse to support them.

With the likelihood of extensive varieties and abundance of conifers throughout the sauropods' geologic history, other questions arise: What kinds were they possibly eating? How did such a wide range of species actually share the food resources available to them? All of the well-known Mesozoic ecosystems were marked by a highly varied assortment of large, medium and small dinosaurian herbivores. Of these, the sauropods would have formed a guild, or group, of closely related herbivores sharing similar lifestyles and food preferences and were primary plant consumers that were adapted mainly to browsing from trees. Some types, such as *Camarasaurus*, *Brachiosaurus* and *Barosaurus*, are known from closely related genera and species that lived in areas that are now England, Portugal and Tanzania, as well as other subcontinental areas of Laurasia and Gondwana. This suggests that similar sauropod guilds and ecosystems existed elsewhere around the Mesozoic world.

WHERE'S THE LEAF?

Before discussing this, we first have to honestly admit to a frustrating lack of evidence in determining definite conifer (or any other) plant food preferences in sauropods. As yet we have no actual boli (stomach contents), preserved pollen, phytoliths or even partially digested plant contents that can point to any specific conifer taxon in relation to any discovered skeletal remains. Another potential information source, coprolites, or fossil feces, should occur in abundance (as well as size), but here too we have to date almost nothing, an exception being the "Type A" titanosaur dung that comes from the Late Cretaceous Lameta Formation of central India and another large Cretaceous coprolite with definite conifer fragments (see Chapters 5 and 11). This contained both definite, but unidentifiable as to species, conifer and *graminiform* (grass) fragments. Adding to this "absence of evidence" is another daunting problem, one that comes from the science of paleobotany itself. This is the fact that, except under rare conditions, plant fossils are almost never found as whole organisms and are usually represented by disarticulated, fragmentary remains. Even when found intact, the micromorphological details and internal tissue patterns are often so poorly preserved that it's sometimes hard to determine the patterns that could indicate overall anatomy and phylogenetic relationships. This is why paleobotanists usually assign different taxonomic names to different fossilized parts of plants when they're found, much as ichnologists do with trackways: as strongly as you might suspect a fossil trackway was made by a certain dinosaur, you really can't assume it was unless you find that dinosaur's actual fossil foot bones inside the last footprints in the trackway sequence. Since this as yet hasn't happened, a distinctive type of trackway is given its own taxonomic name. So it is with plants. In the case of extinct conifers, the genus name *Pagiophyllum* actually represents a number of possible taxa, based on stems with a distinctive leaflet pattern, *Araucarioxylon* and *Sequoiadendron* are names denoting a type of fossil wood and *Araucarites* is a taxon reserved for cone scales with a certain morphology. Linking a particular plant species to a given sauropod species just isn't possible yet, but as long as we can accept this limit to our knowledge, we can nevertheless attempt some reasonable speculation about which *general* types of conifers they (might) have been eating.

In 2000, Upchurch and Barrett made a systematic and highly thorough study of the dentitions and skull morphologies of the known sauropod taxa. The dentition and neck length of basal sauropodomorphs (*Vulcanodon*, *Barapasaurus*) and eusauropods like shunosaurids (*Shunosaurus*), omeisaurids and mamenchisaurids (*Omeisaurus*, *Mamenchisaurus*), camarasaurs (*Camarasaurus*), basal titanosauriforms (*Brachiosaurus*, *Euhelopus*) and diplodocoids (*Diplodocus*, *Apatosaurus*, *Barosaurus*, *Dicraeosaurus*) were taken into account, as well as available microwear evidence and the probable physical limits of their vertical feeding ranges. Adult sauropods of each species had neck lengths that would have corresponded to the typical maximal heights of their food trees' vegetational mass. Another fundamental difference was the adaptation of dentition to plant forms. Moderate-level vegetation was utilized by

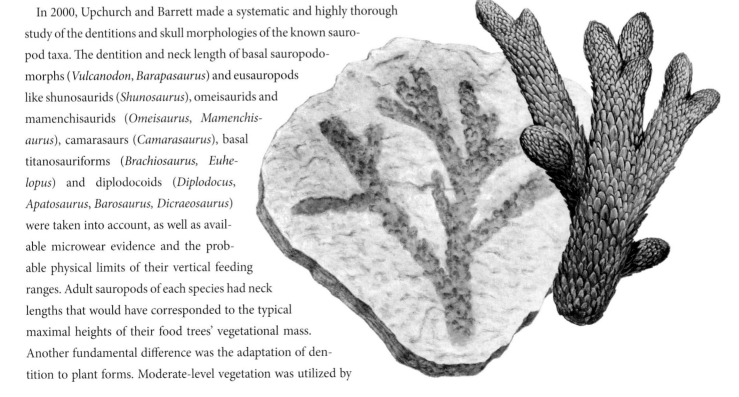

A fossil spray of *Brachiophyllum*-type foliage. This cheirolipiadicean may have been one of the major food sources for sauropods.

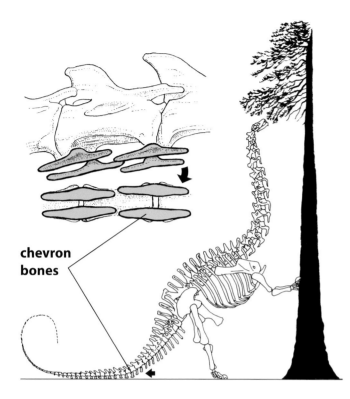

chevron bones

The fact that a sauropod's center of gravity was just forward of its hips, along with strong spinal leverage and a massively constructed pelvis, meant that it was easy for most sauropods to adopt a rearing pose and stay that way for long periods of time, sometimes slowly walking bipedally if necessary. In this they were aided by their "third leg," a muscular tail whose chevron bones (*blue areas*) in some clades transferred diagonal stress horizontally.

conifer feeders with medium-sized, spoon-shaped teeth ("cetiosaurs"; the eusauropod or basal macronarian *Jobaria* and others), while the bigger, tougher bracts and leaflets were the province of the later, large-spoon-toothed camarasaurs, whose necks made them respectively lower- to mid-canopy browsers (*Camarasaurus*). Very high, moderately coarse foliage was the feeding niche of the fantastically long necked *Mamenchisaurus, Omeisaurus* and other Asian forms. Some medium to tall foliage was best harvested by the precision-shear bite of the brachiosaurids and some later titanosaurs. Lastly, the finer, more delicate branchlets were the specialty of the pencil-toothed diplodocoids (*Diplodocus, Barosaurus, Brontosaurus* and possibly low-feeding *Dicraeosaurus*), which raked large quantities at a time from the stems. The presence of spines found in association with some modern conifer species suggests that in Mesozoic forms these could have been a deterrent against excessive cropping by these sauropods and other dinosaurs, which, like modern giraffes, may have had seasonal preferences. Other factors that may have influenced sauropod plant selection might have been the abundance and size of the leaflets, the physical accessibility of a given food tree, its growth form and even the presence of aromatic substances. Ultimately, adult sauropods living in the Mesozoic ecosystems avoided food competition with one another by evolving dentitions, jaw morphologies, neck lengths and behaviors that were suited to harvesting different conifer food sources, both from differing conifer species and from different areas of a food species' canopy. Some unique forms such as *Nigersaurus* and *Brachytrachelopan* became secondarily adapted to low browsing, probably from short conifers and other low-growing plants. Based on all this, we can attempt to reconstruct the following basic feeding ecologies for known groups of sauropods. One eusauropod clade, the turiasaurs, is currently known from such fragmentary material that as yet we just don't know enough about how they ate and what foliage they might have eaten, and so they are omitted here.

EUSAUROPOD ADAPTATIONS AND DIETS

Housed in rather short, broad skulls, the moderately sized spoon-like teeth of this basal clade of sauropods may have been best suited to cropping smaller leaves and moderately coarse cuticles. The medium to short (for sauropods) neck lengths suggest that they were adapted for harvesting browse from medium to tall trees, which may have included those represented by foliage types like *Pagiophyllum, Podozamites*, species such as *Ginkgo* and similar forms. The upturned upper muzzles of some cetiosaurs like *Shunosaurus* may be an adaptation for cropping browse with the head held subvertically, which, together with the short neck, might have reflected habitual rearing. *Omeisaurus* and *Mamenchisaurus* were endemic forms unique to Asia. They followed the eusauropod basic adaptation of moderately sized, spoon-like teeth mounted in somewhat longer, still broad upper and lower jaws, but these were mounted on very long necks. Along with this came high shoulders and massive scapulae to form an anchor for the muscles of the cervical vertebrae, which in one large form, *Hudiesaurus,* were bifurcated at the base of the neck for extra lateral control. Although mostly smaller bodied and less massive, this clade converged on certain later neosauropods, the brachiosaurids, in having large, robust shoulders and longer forelimbs. They could therefore reach into the high canopy to reach the tallest conifers represented by unknown but possible ancient *Sequoia*-like species and by wood types like *Sequoiadendron* and *Araucarioxylon,* whose assumed small cuticles and fruiting bodies may have been a main food source. The omeisaurids' retroverted pelvises indicate that these forms, unlike the bigger brachiosaurids (which, if ever, seldom did so), may have also habitually reared.

MAMENCHISAURID ADAPTATIONS AND DIETS

Mamenchisaurids' jaws were rather long and, in contrast to the first group, narrow. These, as well as the smaller teeth, imply selectivity toward a finer leaf or cuticle foliage source, which they reached with fantastically long necks, the longest of any known sauropod. In the largest species yet discovered, *Mamenchisaurus canadorum*, this was almost 17 m (55′), and like *Hudiesaurus,* it required the extra control of bifurcated neural spines at the neck base. Possibly because of this extreme neck length, some mamenchisaurids had only moderately enlarged, higher shoulders—the neck probably made a taller body unnecessary. The high-reach capabilities suggest that they were aiming, in the mature adults, for the tallest cupressaceaen and taxodiacaean conifer species, which may have been similar in height to *Seqouiadendron, Araucarites* and *Taxodium*.

MACRONARIAN ADAPTATIONS AND DIETS

The relatively medium-sized and spoon-shaped teeth of known **basal macronarians**, as with cetiosaurs, show a generalized cropping dentition. Those of some African types like *Jobaria* (although this may actually have been a eusauropod) and *Atlasaurus* (some workers consider this species not to be a macronarian, however, but a derived eusauropod) are set within rather narrow skulls, and together with their long forelimbs and robust shoulders (especially in *Atlasaurus*), this suggests a medium-high specialization toward a finer, possibly more nutritious foliage that allowed for comparatively more selectivity per bite. This was a similar adaptation to the mamenchisaurids but took place on a lower level. *Camarasaurus* has differential wear patterns on the most anterior teeth, moderately elongated glenoid fossa and simple orthal bite, and it had a jaw action that was suitable for harvesting a broad spectrum of vegetation. The fact that "shoulders," or shelf-like macrowear, are commonly found on the big, robust teeth suggests that a ripping or tearing action could have taken place during feeding, as opposed to the more precise, shear bite likely for brachiosaurids. Because of the potential (but probably limited) ability of its teeth to crush and interlock, camarasaurs were probably rather generalized feeders adapted to larger, coarser leaves and cuticles. They could have routinely harvested a large variety of coarser conifer and broad-leaved ginkgo foliage, as well as cheirolepidiacean and cupressacean conifers. In addition to this, *Camarasaurus*'s and other forms' spinal bifurcations, as well as their retroverted pelvises, would have allowed them to rear and walk around a tree. Besides foliage, eusauropods and macronarians could also have bitten into the female cones of araucarians, which, if they were similar to modern forms although not nearly as large (in the case of the modern Australian bunya-bunya, *Araucaria bidwelli*, football sized), would probably still have been packed with dozens of nutritious seeds. Cones and male reproductive organs of all conifers would have been valuable, although seasonal, high-protein food sources for all sauropods that could reach them. Resins, too, are sometimes rich in nutrients, and while these probably could not be obtained in any quantity to totally sustain a camarasaur (or any other sauropod), they could have been a supplemental source of carbohydrates if the bark was gnawed through with the powerful teeth.

DIPLODOCOID ADAPTATIONS AND DIETS

The diplodocoids radiated into three major known clades, the **rebbachisaurids, diplodocids** and **dicraeosaurids**. Rebbachisaurids, although a basal diplodocoid group, survived far longer than their sister taxa. This may have been because the greater diversity and distribution of the conifer taxa in South America and other southern continents during the Early to Mid-Cretaceous remained suitable for exploitation by an (at first) more general type of tooth morphology. Skulls and teeth are fragmentary (in the large forms *Limaysaurus* and *Demandasaurus*), but one small, short-necked lineage, represented by *Nigersaurus* (described earlier in this chapter), apparently bucked the general sauropod rule. This species became a *habitual low browser* of fine, possibly non-coniferous vegetation, although its relatively massive hindlimbs would have allowed it to rear. The generally small size of known rebbachisaurids, if they did browse arboreally, suggests a lower canopy feeding level that might have included shorter cheirolepidiaceans, horsetails and ferns. Dicraeosaurids, except for the North American *Suuwassea*, are known from the southern continents and generally small in size. Including the African *Dicraeosaurus* and South American *Amargasaurus*, their relatively short necks were probably an adaptation for specialist browsing in low-canopy settings in which they harvested browse in a distinctive way. Plants similar to those fed upon by rebbachisaurids may have been included in their diets. The upper and lower jaws of *D. hansemanni*, along with *Suuwassea* the species with the most complete skull, are slightly deeper than those of diplodocids like *Diplodocus*, suggesting a more powerful bite that could have been used to harvest leaflets or cuticles that required greater force to rake off. One fascinating

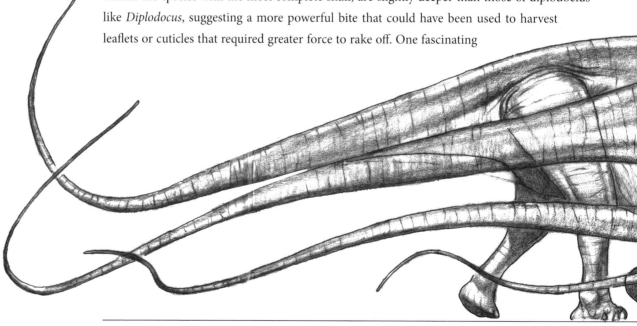

but unfortunately incomplete small form missing its skull, *Brachytrachelopan mesai* (described earlier), was a bizarrely short-necked dicraeosaurid that may have occupied in Argentina the feeding niche that the stegosaurs did in the northern continent—a generalized ground and low-canopy browser. Diplodocids were slender to massive high-canopy feeders whose often incredibly long necks in many ways paralleled those of the endemic Asian mamenchisaurids. These were the most highly adapted of all sauropods to bipedal/tripodal feeding and possessed teeth that were suited for harvesting conifer foliage, whose small, open pattern of branchlets and blade-shaped leaves would lend themselves to being harvested by a raking type of jaw action. Conifers represented by leaflet and wood types like *Podozamites*, *Pityocladus*, *Pagiophyllum*, *Brachyphyllum*, *Elatocladus* and *Protopiceoxylon* (a close relative of the modern *Keteleeria*, which survives in Taiwan) may all have been potential foods for these sauropods because of their leaflet pattern. Which ones were fed upon by which diplodocid at present can only be guessed at, but the neck lengths, jaw widths and teeth of this dinosaur group may hint at the kind of resource partitioning that was going on.

The long tails of the flagellicaudatan diplodocids, ending in an ultrathin "whiplash," are a characteristic feature of this family and have sometimes been the focus of speculation about the possibility of their functioning as "supersonic whips" to strike at faces and eyes of predators to deter their attacks (see Chapter 9). Could they also have played a role in feeding ecology? There is no way at present to know whether diplodocids (or any other dinosaurs) were territorial in regard to any aspects of their behavior, but it would make sense in the case of some sauropods in regard to defending feeding space. If a multi-ton animal rearing bipedally or tripodally to reach a favored conifer species needed to stay in this posture and location for a given length of time to browse the maximum amount, it would benefit from keeping competitive high browsers away. By lashing at other hungry adults of its own or other species with a long tail, a *Diplodocus*, *Brontosaurus* or *Apatosaurus* could defend and maintain an optimal feeding zone for itself until it was ready to move on to the next tree.

TITANOSAURIFORM ADAPTATIONS AND DIETS

This group comprised two (and possibly more) major clades, the basal macronarians such as the *brachiosaurids* and *somphospondylans (Euhelopus)* and the more advanced *titanosaurs*. Brachiosaurids, in spite of their ranging from small (*Europasaurus*) to gigantic (*Giraffatitan*) size, were high-canopy feeders with long forelimbs and big shoulder blades, indicating that they converged with euhelopids and omeisaurids toward this same niche. As with the latter two, their longer, somewhat broader jaws cropped huge amounts of foliage at a time, but instead of spoon-like teeth, these are equipped with compressed cone-chisel ones, meaning that they used a precision-shear bite. As titanosaurid skulls are becoming better known (*Tupiasaurus*, *Rapetosaurus*, others), the food-gathering portions show in some ways a convergence with those of the earlier diplodocids: pencil-like teeth restricted to the anterior jaws and "underslung" inferior orbital fenestra. This means that the *propaliny*, or fore-and-aft action of the jaws, may have been similar, but at the same time the jaws in these forms are deeper and more robust, housing muscles that probably produced a more powerful bite, closer to that of brachiosaurs and more basal forms. At the same time there is a distinctive "waist" or narrowness to the long skull when viewed from above, strongly recalling the appearance of advanced hadrosaurs like *Edmontosaurus annectens*. This could mean that (like the presumably low-browsing hadrosaurs) there were broad, lateral head motions, as well as clamping and pulling ones, going on. Although they are known from all the early protocontinents, the fact that the titanosaurs were more highly diverse in the southern landmasses (by the later Cretaceous, Africa, Madagascar, Australia and especially South America) may reflect the corresponding diversity and prevalence of many conifer species in these areas. Extinct close relatives of the modern monkey puzzles and other araucarians now restricted to the Southern Hemisphere may have been mainstays for the titanosaurs until the end of the Cretaceous, as well as the cheirolepid *Otwayia* in areas like Australia. On the other hand, titanosaur fossil dung from India, shown to contain angiosperms like early grasses and some other vegetation, suggests that some forms were accommodating themselves to this increasingly dominant type of plant.

GREENER PASTURES

Sauropods must have had a tremendous impact on the ecosystems where they lived. By thinning out the foliage of dominant tree species and trampling smaller understory vegetation, they would have stimulated growth and opened up areas for the takeover of plants that needed more light, increasing the potential range of browse for ornithischians and juveniles of their own kind. Sauropods were probably not, however, in the habit of pushing over trees to feed in the manner of African elephants, as some authors have suggested. If they were adapted to vertical feeding, there would be little advantage to expending energy in knocking down vegetation that was more easily accessed by rearing up, as well as damaging a limited food resource. It's also unlikely that, unless the browse was exceptionally plentiful, a group of sauropods would have stayed long in one place, having to move on after depleting what that locality was able to offer. Had there been extremes in seasonal rainfall, the pods would have migrated to find browse and water as the African plains herds do today, each species of sauropod following the growth cycles of their particular set of food conifers and other plants; during these marches, the highly nutritious horsetails, shorter conifers and less nutritious, but possibly plentiful, ferns could have sustained them. The equally plentiful dung that was produced, in turn, must have fertilized many kinds of plants and supplied some arthropods with a feast of partly digested plant matter. In time, the giant herbivores paid back to the land what they took: the death of a sauropod was a ponderous bequest that fed many scavengers and released precious nutrients back to the soil.

As the Tithonian stage at the end of the Jurassic period came to a close, sauropods had evolved into a worldwide spectrum of diverse genera and species whose primary (though not exclusive) adaptation was the exploitation of mid- to high-level conifer foliage as a main food source. They did this mainly by evolving enormous, high-volume hindgut digestive systems that continuously extracted nutrition by retaining forage for long periods. To support these, they also attained gigantic sizes and evolved tremendously strong, yet lightweight, skeletons in which the necks in particular were uniquely adapted to reaching a food source mainly inaccessible to other animals. Sauropod clades also evolved distinctive morphological jaw and dental adaptations that, along with differences in their necks, enabled species to partition conifer browse and avoid direct competition with one another. By any standard, they had become one of the most successful, as well as the largest and most widespread, herbivores among the dinosaurs.

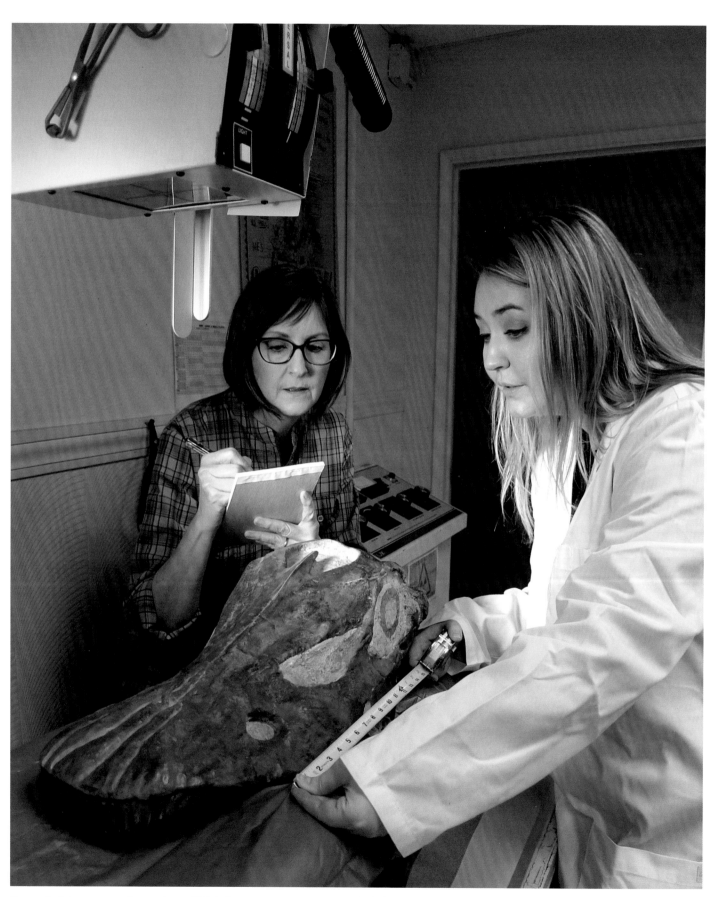

Two technicians mount the cranium of *Diplodocus carnegii* on a conveyer belt that will pass it through the opening of a CAT scanner, where secrets of its internal structure will be revealed.

CHAPTER SEVEN

A SAUROPOD IN THE LAB

As the 21st century progresses, a dazzling array of scientific technology has come to join the older traditional laboratory techniques in helping to measure, analyze and virtually model the possible appearances and life processes of ancient animals. For creatures as gigantic as sauropods, digital imaging has made many forms of study and testing available that were formerly denied by the huge size of the bones, while other recent laboratory techniques have opened up new possibilities for interpreting sauropod growth and biology.

TALES OF THE ISOTOPE. Around 150 million years ago, the Dry Season in the Late Jurassic savannas of western North America was punishing for all animals. Not the least of these were sauropods like *Camarasaurus*, which could endure only for so long without water; the only solution was to migrate to a water source. But where? The answer may lie in *oxygen isotopes*, which have recently been used to infer possible dinosaur paleobiology ranging from the fish-eating preferences of spinosaurs to the ability of some Chinese dinosaurs to endure cold. Dinosaurs replaced their teeth throughout their lives, and as they did so, they constantly incorporated oxygen from the water they drank into the mineral *bioapatite* that made up their tooth enamel. Oxygen occurs in different varieties (known as *isotopes*), the most typical being the "lighter" (each element has the same atomic number, but the isotope of each of these has a different numerical designation based on the varying number of neutrons) oxygen-16 and the "heavier" oxygen-18. Lakes, rivers and ponds all contain different blends of these oxygen isotopes, which leave their "signatures" in molecules that make up certain hard structures like tooth enamel. When Henry Fricke (Colorado College) sectioned a series of 32 *Camarasaurus* teeth collected from two Morrison localities, he found that there was much more of the lighter oxygen-16 variety in the specimens and less of the heavier oxygen-18, which is the predominant type found in the surrounding sediments. The teeth had the highest ratios of oxygen-16 to oxygen-18 ever found in the Morrison Basin.

Fricke reasoned that the camarasaurs must have obtained their water from higher ground than the basin. This is because as air rises, it cools and condenses, preferentially taking up the heavier oxygen-18 atoms to bond with hydrogen to form water molecules before the lighter oxygen-16 ones get their chance. When clouds form, they eventually deposit this load of "heavy" water as rain before that of the "lighter" water, and in lower elevations. The leftover, now more abundant "lighter" rain is what becomes a part of lakes, rivers and ponds in highland locations, and which would leave its isotope signature in the teeth of any animals drinking from them. In the Late Jurassic North American West, the closest high terrain then lay in a series of volcanic mountains west of the basin. If this is where the camarasaurs migrated to get their water, they were traveling about 300 km (186 mi) going from and returning to the sites where they eventually left their bones.

The migration was probably a seasonal one, an inference again based on the proportion of oxygen isotopes. Microsections of individual teeth, starting with the oldest enamel at the tip and progressing to the youngest at the root, revealed to Fricke a dip in the ratio of oxygen-18 to oxygen-16 and suggested that as the tooth grew out, the animal was on its way from the basin to the highlands, a journey of perhaps four to five months. The sauropod would have completed the return migration at some point shortly before its death. Although this hypothesis rests on both a very small sampling and assumptions as to the rate oxygen isotopes were absorbed into *Camarasaurus*'s tooth enamel, how this changed with temperature and how hot the Morrison Basin was at this time, it fits with a scenario of possible sauropod migration in this region to obtain water and better browse, similar to the great seasonal migrations of wildebeests and zebras on today's African Serengeti plains. While again based on a very narrow sampling and database, what makes this hypothesis testable is that similar microsections could be taken from the other archosaurians that shared the basin environment with *Camarasaurus*, like the small theropod *Fruitadens* and the numerous crocodilians like *Eutretauranosuchus (Goniopholis)*, both of which could certainly not have undertaken the migrations that their giant neighbor did, and should have a much higher proportion of oxygen-18 in their tooth enamel.

Isotopes may also have much to contribute to the ongoing debate about whether sauropods were truly endothermic like birds and mammals or whether they were **inertial endotherms**, huge but technically ectothermic animals whose large ratio of volume to surface area enabled them to lose body heat slowly, thus giving them the same metabolic advantages as an actual, smaller endotherm (see Chapter 12). This possibility arises from an isotope analysis method called **clumped isotope thermometry**, which focuses on the ratios of different isotopes that formed within minerals inside ancient rocks. Certain isotopes tend to form depending on

From studies of the absorption of certain oxygen isotopes during the final weeks of life, the teeth of *Camarasaurus* and other Morrison dinosaurs may yield the key to an understanding of the animals' migration patterns in their search for water and other nutrients.

whether a chemical reaction occurs at a higher or lower temperature: the lower the temperature at which the minerals form, the more the two isotopes carbon-13 and oxygen-18 tend to bond together, or "clump." By measuring how much of this resulting carbon dioxide was trapped in specific rocks as they were shaped at certain times in the past, workers like John Eiler (California Institute of Technology) can, in theory, determine how warm that locality was at the time and, with enough data, eventually construct a model of that region's paleoclimate.

Based on this approach, Robert Eagle (California Institute of Technology) felt that he could apply this same technique to the question of sauropod and other dinosaur body temperatures. In this study, 11 teeth from both Morrison age *Brachiosaurus* and *Camarasaurus* were sectioned and analyzed, and here the isotope clumping indicated that the bioapatite in the enamel of both species had formed between 36°C and 38°C (96°F to 100°F), about the same as mammalian body temperature. Again this is a very small database, but it is a finding that could be tested by comparing it with dental sections from a wider array of other Morrison vertebrates.

PALEONTOLOGY UNDER A LENS

Conventional 2D microscopes continue to serve well in histological research, but electronic microscopes, which generate images by bombarding the object for study with a cloud of electrons that bounce off and create a detailed, viewable image of a specimen, can reveal even objects as tiny as molecules in extraordinary detail. This technique is used when it's necessary to see actual textures and other surface details on bone and soft tissue, such as microwear scratches on teeth and the evidence of postmortem arthropod damage inflicted on the bone surfaces of the hypothetical camarasaurid carcass (discussed in Chapter 9).

Sauropod and other dinosaur fossils don't reveal their secrets easily, requiring lots of hard preparation and study. In the past this usually came as a result of comparative anatomy studies based on the external morphological details of the bones, but more recently **bone histology**, or bone tissue study, has revealed much about how these gigantic animals grew through the examination of ultrathin microsections from certain places on sauropod bones under a microscope (with images transferred to a monitor screen for easier viewing in real time) and comparison with corresponding ones in some modern archosaurs like crocs and birds, as well as reptiles and other less derived saurians. When sauropod and other dinosaur bone tissue has been examined, it has led to some exciting discoveries about how sauropods grew to adulthood, their possible life spans and even how we could determine their sex.

SECRETS OF BONE GROWTH

Living bone tissue is made up of *organic materials*, such as the complex protein **collagen**, organic minerals like **calcium** and **phosphorus** and inorganic ones like the compound mineral **hydroxyapatite,** which appears as crystals. Collagen, relatively soft, allows a bone to have suppleness and enough flexibility to avoid breaking under bending or twisting stress, while organic and inorganic minerals provide needed rigidity and support. To form and grow, bone, like any other body tissue, needs to be supplied with blood—not only because the living collagen needs food, oxygen and waste removal but because the organic and inorganic minerals are constantly being

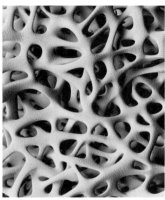

A microscopic close-up of inner spongy bone photographed through an electron microscope (*bottom*), and a microsection of bone taken with a lens microscope (*top*), revealing primary osteons (*circular shapes*) and Haversian canals (*large dark holes*). The very tiny dark spots are lacunae, which house the osteocytes that attract minerals to make the bone hard.

reabsorbed and deposited in response to the body's overall demands on the bone. Bones don't just sit there once formed; they are shaped and reshaped all throughout an animal's lifetime, existing primarily for support and anchoring muscles but also as stores of organic mineral reserves (like calcium), to be accessed when needed elsewhere in the body. Where they have to be hard and especially strong (on the outsides), bones have a dense *cortex*, while the middles (*cancellous areas*) and insides (*marrow cavities*) can be lighter, sometimes even hollow. The internal microstructure of bones can tell us a lot about how a sauropod lived. Did the animals grow slowly, like their reptile ancestors, or fast, like baby birds and mammals? When did they reach reproductive maturity, and how long did some species live? To find out, we have to leave the museum fossil prep room and go into a university or medical center lab.

All amniotes have comparable patterns in the way their bone tissue is laid down while they're juveniles and when they reach adulthood, and depending on what amniote we're dealing with and its stage of growth, these are highly distinctive. Such patterns can tell us how bone growth takes place, and they are revealed under a high-powered microscope as thin-sectioned slides. Let's look first at what's called *primary bone tissue*, the kind that's laid down when an animal starts growing. Bone cells, or *osteoblasts*, begin by secreting *osteoid*, which deposits *collagen fibers* and starts to embed *hydroxyapatite crystals*, in parallel formation, into the collagen matrix. In fossils the collagen is only rarely preserved, but the crystals stay and are evidence for the stages, and rates, of bone deposition. In the growth of primary bone, *endosteal* (*endo*, " inside"; *osteo*, "bone") *lamellar* (*lamina*, "layers"), rates are slow, and these collagen and hydroxyapatite layers show a tight parallel alignment, as well as separations or pauses, giving the microsection a striated (striped) appearance. Sometimes bone growth, like the annular growth rings of a tree responding to annual periods of wet and dry, also forms lines called *lines of arrested growth* (LAGs). These often occur when the animal's biorhythms are seasonally affected, and they are typical of ectotherms like reptiles. A second kind, *periosteal* (*peri*, "surrounding") *parallel fibered bone*, gets laid down at slightly faster rates, resulting in the fibers being closer together. The third, *periosteal woven* or *fibrous bone*, is dominated by a "woven" (interconnected), random pattern of collagen/hyroxyapatite fibers and generally shows fewer LAGs. Fast-growing living animals like birds and mammals often have a *combination* of lamellar and woven types called a *fibrolamellar complex*: the overall size and shape of the bone are determined by woven growth at the outer surfaces, but the canals needed for fine blood capillaries to penetrate these areas are partly filled in by slower-growing lamellar bone.

The above bone depositional types vary according to the animal's age and species, and under these conditions they require different patterns of vascularization, or blood supply. When the vascular canals become surrounded by newly deposited bone, they become *primary osteons*, and depending on the directions the bone has to grow in, these run in a *longitudinal* (parallel to the general direction of the whole bone), *circular* (making ring-like patterns around the center) and/ or *radial* (pointing from the center outward) pattern. In more rapidly growing bone, the circular and radial patterns combine to form interconnecting canals at separate points (and on the same horizontal planes) along the inside of the bone, becoming *laminar vascularization* (laminar bone). This sometimes develops into *plexiform vascularization* (plexiform bone) when short,

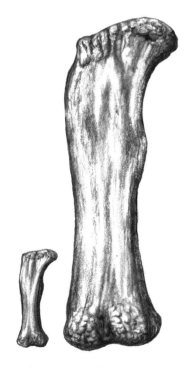

In her analysis of the cross sections of the femur and other bones of *Apatosaurus*, **Kristi Curry Rogers** found that by the age of 5, this sauropod had reached half its adult size, with full growth attained somewhere between ages 10 and 30—an amazing rate of growth similar to very large mammals, exceeded only by rorqual whales.

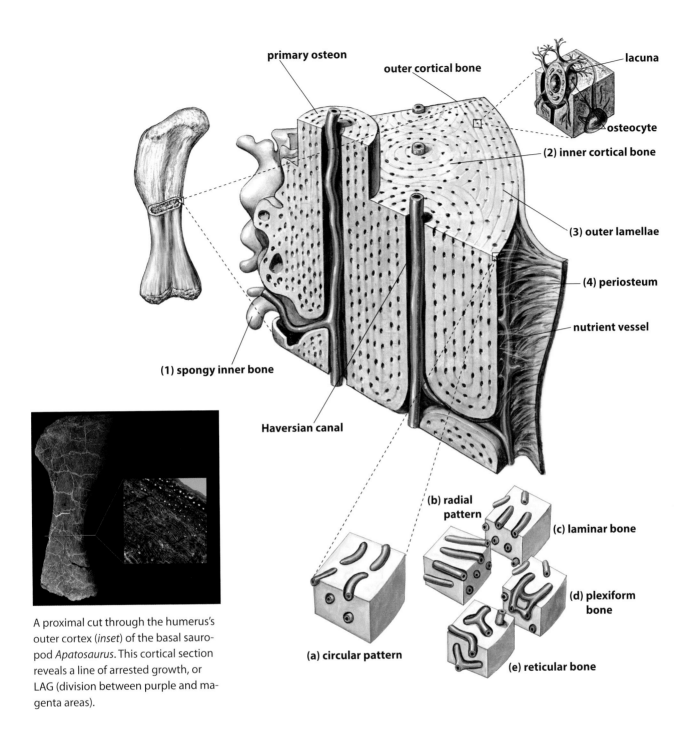

A proximal cut through the humerus's outer cortex (*inset*) of the basal sauropod *Apatosaurus*. This cortical section reveals a line of arrested growth, or LAG (division between purple and magenta areas).

Sauropod Bone Growth

A section from the femur (*left*) of the titanosaur *Neuquensaurus*, progressively enlarged to show (1) the spongy inner bone surrounding the marrow cavity, (2) inner cortical bone, containing primary osteons (growth centers within inner lamellae, or rings), Haversian canals containing nutrient vessels supplying osteocytes, or bone cells, housed within lacunae or spaces, (3) outer lamellae and (4) periosteum, or enclosing membrane. In addition to the longitudinal pattern shown, during moderate bone growth primary osteons can also take (a) circular and (b) radial patterns. In the fastest bone growth, primary osteons intercalate, or join together, forming (c) laminar bone, with connected circular and radial canals, (d) plexiform bone, with circular, radial and longitudinal intercalated canals, and (e) reticular bone, with oblique, irregular patterns.

longitudinal canals create further connections between the above. The fastest growth occurs when there's **reticulated vascularization** (reticular bone)—here the osteons form oblique, irregular canals. Reticular bone sometimes combines with laminar to grow bone not only fast but also efficiently, since the greatest space can be permeated by blood capillaries with both types working together. This type of fibrolamellar complex is typical of what we see in sauropod and other dinosaur bone tissue, as well as birds and placental mammals. As a bone continues to grow, a second stage of "remodeling" takes over and the primary bone is reabsorbed. Here vascularization becomes even more dense, becoming **Haversian bone**. This and fibrolamellar bone are very typical of many dinosaurs when we look at their bone sections. It tells us that, like highly derived living vertebrates, baby and juvenile dinosaurs grew very fast, and it indicates that they also had a fully endothermic metabolism at least during their early ontogeny.

FAST-TRACK BONES

So far, most studies have found that sauropod juveniles and adults commonly deposited abundant fibrolamellar growth. When dinosaurs, birds and mammals reach maturity, tissue samples show that this kind of bone deposition continues but usually slows down depending on the animal's age. Here, fast-growing sections may alternate with regions of slower growth and sometimes are marked by LAGs, but these aren't as common as they are in ectotherms like reptiles. When this happens in the outer (cortical) bone of sauropods, it often indicates mild slowdowns of bone deposition and reabsorption from fast to very fast, different from the steep slowdowns and LAGs seen in reptiles and other ectotherms. Taking into account factors like cyclical annual growth and seasonal stress from possible food scarcity, measurements from these markers, called **cortical stratifications**, can be used to make estimates of the sauropod's age at sexual maturity and at death. One important factor affecting this is "Amprino's Rule," a theory that the different patterns of bone deposition described earlier always happen within certain rates of growth, even in different taxa, individuals and parts of the skeleton. At one time it was assumed that if sauropods were like typical reptilian ectotherms, it would have taken them many decades, even a century, to reach their enormous adult sizes. Now we know, from microstructural bone similarities to birds and mammals, that such prolonged ontogeny wasn't necessary, and we can use Amprino's Rule to calculate from a given sample the minimum and maximum days an individual sauropod grew and when it became an adult. If, for example, we take a sample of laminar bone from a sauropod femur and compare this with the range of daily bone deposition rates for this tissue from the same bone in other vertebrates, we can use the highest and lowest growth rates in the extant animals to determine the approximate number of years a sauropod species took to reach sexual maturity, calculate how old it was at death and hint at its

potential longevity. In her 1999 histological analysis, Kristi Curry Rogers (Macalester College) analyzed several different limb bones from *Apatosaurus* individuals at four different size classes and stages of development, resulting in the first quantified study of sauropod growth strategy. Her findings demonstrated that *Apatosaurus* actually reached about half its adult size by as early as age 5, with fully adult growth attained between ages 10 and 30. Later research by P. Martin Sander (Steinmann-Institut für Geologie, Mineralogie und Palaeontologie, Reinische Friedrich-Wilhelms-University of Bonn) and Sandra Tuckmantel (Peqlab Biotechnologie GmbH) resulted in histological profiles for other sauropods, which matured at different rates: these findings showed that *Brachiosaurus* sexually matured at 40% maximum size, while *Barosaurus* may not have done this until 70%. *Janenschia* became reproductive at age 11 and maxed out its growth at 26. Longevity is another issue, since the studies only show how old the sauropod was at death, not how long an individual

By "wrapping" a computer-scanned skeleton in a virtual "skin" (*top*) and then calculating the volumes for each body section, a rough estimate of the total volume based on the skeleton of a sauropod-like *Giraffatitan* (*bottom*) can be approximated, but factors such as the variable density of muscle masses and visceral areas must be taken into account. Using this approach, sauropods may have been lighter than previously estimated.

might have actually lived. Old age and death may have occurred as early as age 30 (as is the case with the *Janenschia* sample) or could have been decades longer. Growth in a given species may also have depended on gender. Sander found in his histological sample from *Barosaurus* evidence for what he felt was sexual dimorphism in growth, based on one set of samples from this species that consistently showed rapid and continuous growth, while another showed slower, discontinuous deposition. He hypothesized that the highly remodeled bone in the second was likely a female's, with the growth slowdowns reflecting mineral withdrawals of calcium for eggshell formation.

IT'S A GIRL, AND SHE'S PREGNANT…

In 2001, Mary Schweitzer (North Carolina State University) and her lab associates received a fragment of *Tyrannosaurus rex* femur as a specimen for her microstudies from the Museum of the Rockies, and as is now the custom, the skeleton it came from was given a name, in this case "Bob," in honor of the field crew chief who actually found it. On first looking at it, however, Schweitzer's first words were "It's a girl, and she's pregnant." Her studies of living birds enabled her to immediately recognize that the lining inside the marrow cavity was preserved **medullary bone**, a special tissue that modern birds temporarily form to store calcium in preparation for egg

The use of CAT scanners (*top*) by Sam Noble/Oklahoma Museum of Natural History and other research teams has revolutionized the study of vertebrate internal skeletal structure (*bottom*). Through this process, delicate areas can be visually recorded in the finest detail without destroying the actual specimen.

laying. The *rex*'s medullary bone was almost exactly like that of ostriches and emus and showed that it, too, was a reproductive female anticipating the production of eggs. The confirmation of medullary tissue in a dinosaur, although a temporary condition, is a great discovery—if a data set of one or more distinctive skeletal features could be found *only* in association with medullary bone in sauropods, it would provide a means of sexing them and other dinosaurs. At present we can't do this.

WHAT DID THEY WEIGH?

An outcome of a sauropod's growth was not only its overall size but also its mass and weight—how did these change with age? A newly discovered principle that can be applied to this question is ***developmental mass extrapolation*** (DME), which is a scaling method developed by Gregory M. Erickson (Florida State University) and Tatyana Tumanova (Palaeontological Institut, Russian Academy of Sciences) that uses ***skeletochronologic*** (skeleton growth timing) data. Here we do a temporary switch from micro to macro. The idea is that if a linear measurement from an animal's skeleton (say, a femur) at a certain age scales consistently up as its body mass increases throughout its lifetime, and the body masses of more than just one individual at a given age are known, then an average based on similar measurements, taken at a certain stage during the animal's ontogeny, should predict its body mass (femur length is ideal because it scales in a persistent, predictable way throughout ontogeny). To apply DME to dinosaurs, we first take the cube of the femur's length for each known specimen representing a growth series and convert this into a percentage of the largest known adult femur. These percentages are then converted to fractions and multiplied by the estimated adult mass, which gives body masses for each ontogenetic stage represented by a femur. In sauropods this, of course, is highly dependent on estimating as accurate a typical adult mass as possible (see next section), but the validity of this method is verified by tests on humans and wild alligators, where the margin of error doesn't exceed 5%.

Based on studies by Erickson and Curry Rogers, dinosaurs as a clade seem to have had their own, distinctive growth trajectories. The larger the dinosaur species, the more rapid was the overall growth rate, with some species showing rates below, equal to and/or above those of birds and mammals. *All* dinosaurs grew at rates from *2 to 56 times faster* than living reptiles. This would have varied from taxon to taxon, but in the case of sauropods their growth trajectories generally approached those of modern whales, which are some of the fastest-growing placental mammals. Juveniles of all types grew incredibly fast, and this extended well into adulthood. Even at full maturity an *Apatosaurus* put on weight at the amazing rate of 14,460 g (31 lb 15 oz) per day, compared with 20,700 g (45 lb 10 oz) per day for a 30,000-ton (66,138 lb) gray whale, *Eshrichtius robustus*. All this growth was for the purpose of getting as big as possible to process maximal amounts of food.

WRAPPING IT ALL UP

In addition to juvenile growth rates, another persistent question about sauropods was their mass at adult sizes. The methods already discussed above and in the rest of this chapter are based on what's sometimes called the ***predictive regression approach***. You compare the lengths of a bone

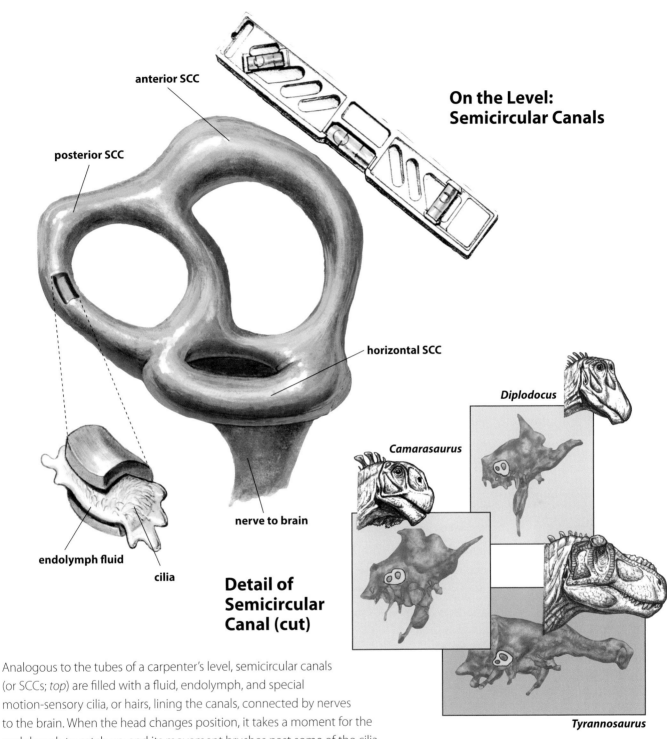

On the Level: Semicircular Canals

anterior SCC

posterior SCC

horizontal SCC

nerve to brain

endolymph fluid

cilia

Detail of Semicircular Canal (cut)

Diplodocus

Camarasaurus

Tyrannosaurus

Analogous to the tubes of a carpenter's level, semicircular canals (or SCCs; *top*) are filled with a fluid, endolymph, and special motion-sensory cilia, or hairs, lining the canals, connected by nerves to the brain. When the head changes position, it takes a moment for the endolymph to catch up, and its movement brushes past some of the cilia, which, depending on their location, tell the brain (*brown areas*) whether the head is tilted forward, backward, up or down. The somewhat short, stubby canals of sauropods (*Diplodocus* and *Camarasaurus*; *beige areas*) with their fewer numbers of cilia may mean that eye and head movements were less important to these herbivores than the SCCs of a carnivore like *Tyrannosaurus*, whose longer canals and greater numbers of cilia may have provided more nuanced information needed for quick head turns during predation. The horizontal SCCs of sauropods generally indicate the "alert" or default position of the head.

like the femur with those of known animals and assume the mass of the extinct species scales, or regresses from these points, in a similar way. A problem with this is that skeletal features can vary widely and may not necessarily correspond to mass in the same way in all animal groups. An alternative to this is the **volumetric approach**: create a drawing of the body, estimate how much volume it occupied, and multiply this by the predicted density. It's a reasonable alternative, but drawing a sauropod's body outline can be as subjective as that of a mass estimate based on a scale model. A twist on this is William Sellers's (University of Manchester) idea of doing a **convex hull estimate**, in which the separate, convex "hulls," or body sections (head and neck, torso, tail and limbs), from an animal's complete skeletal mount are scanned and then, with a software program, "wrapped" in virtual "skins" as tightly as possible. From the dimensions of these spaces the volumes and cumulative total mass are partially calculated. Why only partially? Sellers had discovered that in testing his technique on the skeletal mounts of modern mammals like wild boar and African elephants, whose typical adult weights are well known, the projected weights of his computer models were *consistently lighter by about 21%* than the actual ones. This is because the initial hull estimate didn't reflect the density and weight of inner muscles and internal organs, which contribute to the animal's mass, the missing 21%. When Sellers applied this method to Berlin's *Giraffatitan*, he got a mass of 23.2 tons, which included the added 21% (4.83 tons). This is extremely close to the estimate of 23.3 tons made by Taylor with other volumetric methods in 2009, a difference of only 177 kg. As with the predictive regression approach, there are potential pitfalls in doing convex hull estimates. The 21% factor that applies to mammals may not be as easily applicable to dinosaurs like sauropods, since the convex hulls of the mammals' limb segments are proportionately much larger than those of sauropods, while the density values used for mammals might not be applicable to sauropods, whose long but lightweight necks generally had a density of no more than half the legs. For sauropods (and other dinosaurs), realistic volumes depend on an accurate reconstruction of the torso, which can account for 70% of the total body volume in a huge species like *Giraffatitan*. Determining the shape, and as a result the volume, of a sauropod's torso can also be difficult owing to the distortion and preservation of the ribs.

ANATOMY LESSONS IN 3D

Just as 2D conventional microscopy has now been extended into the 3D world of electron microscopes, the process of taking X-ray photos has also evolved into a truly revolutionary process for seeing inside the macrostructure of sauropods and other dinosaurs. **X-ray computer-assisted tomography** (*tomos*, "slice"; *graphein*, "inscribe"), also known as a CAT scan, is a process in which multiple X-ray images of a specimen (like a skull) are photoscanned in sequence from end to end by a drum equipped with an X-ray transmitter and opposing digital image receiver/recorder. As the specimen rolls past it on a moving conveyer, the drum circulates entirely around it, and the resulting series of ultrathin cross-sectional photo images can then be patched together to make one continuous, virtual 3D image. Each kind of body tissue has its own degree of density (especially bone, which can be very dense or very thin), and these block the X-rays at different levels, which can be calibrated or "windowed" in a way that allows each kind of structure to show up differently. When the resulting scanned photo is produced into a single cross-sectional

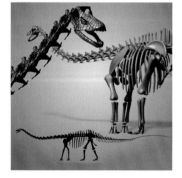

When correctly extrapolated from well-known skeletal elements, the process of parametric modeling can be used to reconstruct accurate digital facsimiles of entire skeletons, whose movements can be modeled to understand movement limitations and possible behaviors, such as this model of a *Diplodocus* skeleton by Turbosquid. **Kent Stevens** and **Michael Parrish** were the first to develop this in Dinomorph.

image, it's then combined with other sequential tissue scans from along the specimen's main axis to create a complete, virtual 3D *tomograph* that can be manipulated into any position desired. In something like a skull, the various structures can be digitally assigned with different colors to make them show up distinctly from one another. This technique has given us our first glimpses of the

actual spacial relationships and intricacies of internal olfactory chambers, vascular passages, endocranial shapes and other structures within the skulls of both modern and prehistoric animals, as well as completely accurate 3D images of unique fossil specimens. Although CAT has now been in existence for some time, Lawrence M. Witmer (Ohio University) and his laboratory have so far gone the farthest in producing interactive, highly detailed virtual anatomical online "atlases" of many vertebrates. These hold great promise in unlocking mysteries such as like the reason for some sauropods' unusual nasal structures.

CAT scans are also telling us a great deal about sauropod brains and sensory abilities. While the delicate skulls of these animals are rarely preserved intact and complete, the sturdy, comparatively thick *basicranium* section, or brain cavity, is one of the most likely parts to survive, and virtual endocranial images are yielding detailed information about brain architecture. Generally croc-like (in that the brain didn't closely fill the endocranial cavity), sauropod brains are, in contrast to these, vertical rather than horizontal in appearance, a result of the extreme shortening and tilting of the basicranium (especially in diplodocoids) that accompanied changes in the overall shape of the skull as it adapted to an upwardly oriented feeding posture. The shape and proportions of the endocranial cavity differ in various sauropod species, but in all of these the *pituitary body*, which contains the pituitary gland and other structures, is always huge, probably relating to the role this important endocrine gland played in regulating ontogenetic growth and other body functions. CAT scans also show that most sauropods had a complicated vascular structure and very large venous sinuses within their skulls, which left prominent grooves, recessed areas and foramina recorded by the bone surfaces. This may be evidence of a specialized system of cephalic veins and sinuses that regulated the great changes in blood pressure that probably occurred when the animals raised and lowered their heads.

SAUROPOD SENSES AND SENSIBILITIES

Of great interest regarding sauropod endocranial structure is the shape of the SCCs located in the region of the upper inner ear. CAT scans show that sauropods' SCCs were short and thick, with a large auditory vestibule, and are more similar to the basal condition of crocs rather than the elongate, slender ones of *Tyrannosaurus* and other theropods, which are more like those of birds. What does this mean? In living animals the shape, length and position of these generally

Missing skeletal elements (in this case, a left hindfoot of *Diplodocus*) can be reconstructed by first making a point cloud scan from corresponding elements on the other side of the body and then programming these as mirror images into a digital printer to create a full-size 3D copy, modeled in resin or another material. The printer can also be programmed to produce smaller or larger exact-scale replicas for study.

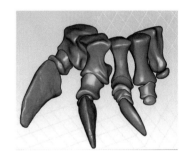

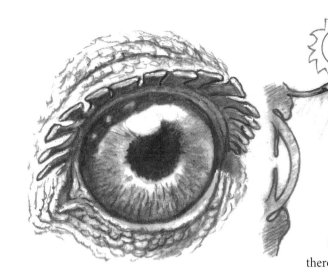

Like modern large ungulates (camel eye; *below*) sauropods would have required protection for their eyes from the sun's glare. These could possibly have taken the form of flap-like skin extensions (*above*).

correspond to hearing ability, balance, head posture and eye movements. In addition to unconsciously informing the animal of its basic vertical versus horizontal head position, SCCs sense turning movements and angular head acceleration, and their morphology (their length, curvature and degree of tilt) seems to correspond to relative speed and agility. A relatively long, slender canal has the potential of relaying more detailed and subtle information about spacial orientation than a short one because of the greater number of sensory hairs (cilia) it can contain, much like the way the retina of an eyeball that's packed with more optical cells (rods and cones) can perceive more nuanced information than a retina with fewer of these cells. This makes sense if we consider that bipedal theropods, which were faster and relatively more agile than slower-moving quadrupedal sauropods, needed a more finely tuned degree of spacial feedback because of their fast-paced predatory behaviors. The SCCs and the vestibules themselves also play a part in coordinating eye movements and gaze stabilization, also important to a predator. The relatively short, stubby canals of sauropods, especially the horizontal ones, may reflect less of a need for movements of the eyes and head, and a similar reduction of the SCCs in living whales has been linked to reduced eye mobility, potential eye movement and neck movement. On the other hand, a precise sense of verticality would be extremely important to an animal that had to raise its head, and perhaps its entire body, meters upward to feed. At this point the sauropod would now be bipedal, and its great weight and potential instability would require precision in maintaining its balance. This is where the large sacral ganglia (see Chapter 5) may have played a big role in coordinating the hindlimbs and tail after receiving initial impressions from the SCCs. Mediolateral (side-to-side) eye and head motions may have played less of a role in feeding behaviors than sagittal (up-and-down) head movements. Although the role of SCCs for determining these kinds of movements is still unclear, they do suggest head positions that seem to agree with the optimal amount of binocular vision an animal might be expected to need, based on eye position and skull length. Studies by Witmer show that *Camarasaurus*, for example, probably carried its head downtilted at an angle of less than 10 degrees, whereas *Diplodocus*, with its subvertical skull attachment to the neck, may have held its head at a much more sharply downturned 45 degrees (see Chapter 4 for more discussion of how camarasaurids and diplodocoids may have held their heads). The fact that these and other species show considerable differences in the shape and proportions of their SCCs implies that various types of sauropods probably had distinctly different typical head and neck positions.

Although the *optical lobes* from many sauropod CAT endocranial scans aren't unusually large, the bony orbits nevertheless are, suggesting that eye size, and consequently vision, was important. Like most archosaurs, they possessed bony *scleral rings* (internal eye supports) and likewise could probably perceive a full color spectrum, unlike the limited one possessed by most mammals. Their eyes may have been capable of taking in considerable detail and in tracking movements, but at this point there's no way to know. The *olfactory bulbs*, the portions of the brain that collect and centralize smell sensations, are, like the pituitary body, also usually large but short, in keeping with the caudal retraction of the nasal cavity and the same overall telescoping of the endocranial space. Although it indicates that this sense, like vision, was important to sauropods, the reason isn't at present clear; it could have made it possible to detect differences in the content and quality of browse, but it also might have been important as a first alert in warning of predators.

DIGITAL IMAGING IN VIRTUAL SPACE

As in other visual scientific applications, CAT and the ability to create ever more precise, detailed forms in 2D, virtual space have created a revolution in paleontology. Organic, complex shapes (like bones, which in the absence of the real specimen had to formerly be portrayed as multiple, drawn [figured] views of the same bone to be intuitively understood as 3D forms) are now increasingly portrayed as volumetric-looking shapes through programs known collectively as **computer-aided design** (CAD). These eliminate much potential ambiguity and can be readily shared on flat surfaces and electronic transmissions. The shapes and volumes of many sauropods' known skeletons, sometimes challenging to understand through more conventional photography and technical art, can be particularly accommodated by digital imaging and subsequent animation. A current technique is through a process called **parametric modeling**. Unlike the creation of a digital image by scanning and computerizing points from an object's surface and building a virtual 3D form by producing an increasingly refined image by connecting these points (such as the image of an actual fossil sauropod femur), in parametric modeling shapes like a femur are created from "the inside out" by establishing progressively more detailed measurements from the most basic ones, in combination with equally smaller and more refined geometric forms. These (the parameters) result in an idealized virtual 3D shape, like a conventional 2D drawing of a reconstructed bone versus the actual bone specimen. Just as in a 2D reconstruction, however, some interpretation takes place, and care must be taken by the modeler to base the shapes on original sources that are accurate. (This is a serious potential problem when using measurements that may be based on crushed or distorted bone specimens, and also when taken from 2D archival sources [photos or drawings] that themselves contain errors.) These shapes can then be animated as virtual forms in space by using CAD software such as Wavefront Studio and 3D-Studios Max. Parametric modeling was first successfully enacted in 1999 by the association of independent biomechanics researchers Chapman, Anderson and Jabo to depict a walking *Triceratops* skeleton, and the technique has been greatly refined by Kent Stevens (University of Oregon) in his ongoing Dinomorph programs and research. Although there is criticism among other workers regarding Stevens's choice of original material and subsequent interpretations of sauropod neck posture, Stevens's visually sophisticated Dinomorph programs allow us to see a species like *Apatosaurus* as a fully articulated, walking skeleton from any angle, from close or medium distance.

FROM VIRTUAL TO SOLID REALITY

A better-known technique called **digital surface scanning**, in use for creating virtual 3D forms, first gained widespread attention in 1993 as one of the processes that made possible the amazing animated dinosaurs in the film *Jurassic Park*. Here a solid scale model of a dinosaur like *Brachiosaurus* was laser-

In finite element analysis (FEA), each scanned point or group of points on a bone specimen or model (*Diplodocus* skull; *below*) is assigned a numerical value and color-coded. When during the process of computer modeling these values and colors change in response to simulated pressure or motion, these are digitally record and analyzed, indicating the amount of change optically (sequence of skulls; *above*) and numerically.

scanned from several angles to obtain a *point cloud*, or series of thousands of points that recorded differences in volume and general surface detail on a computer database. These allowed the film animators to digitally re-create the animal as a sophisticated "wire grid" or geometric form in virtual space, which, after being "clothed" in programmed, software-generated skin texture, color and highlight/shadow, could then be composited with live-action footage to create film sequences. This included the breathtaking scene in the movie of the *Brachiosaurus* rearing to crop foliage from a tree. In the world of real paleontology, digital surface scanning, in combination with CAT, is routinely used to create images of bone specimens in which the unique external features, such as distortions and damage, must be accurately recorded, as well as the overall shape.

This idea of volumetric imaging has been extended to the process of creating not just virtual but actual solid facsimiles, known as **digital 3D printing**, of original bones and other specimens in plastic materials on (theoretically) any scale. Here the surface of the object for reproduction is scanned and then converted into "sliced" digital cross sections (CAT images) that serve as a guide for the ultimate 3D reproduction ("print"). These are then converted, through the aid of an STL file formatting process, into a pattern of triangular facets that approximates tiny portions of the object's shape—the smaller the facets, the more detailed and subtle the translation of the surface features, such as a skull or other unique specimen. STL acts as an interface, or intermediary, between the CAD program and the printing machine itself, and it is sometimes used with other, scanner-generated input file formats like PLY or VRML (WRL) when a full-color "print" is desired. To create the actual reproduction or "print," the machine reads the design from the STL file and begins to deposit successive layers of photochemically reactive liquid, powder, paper or thin extruded sheet material onto a "build bed" or flat, horizontal construction tray inside a tank. The 3D shape is formed from the *bottom upward* as electronically scanned, exact space coordinates (now represented by the triangular facets) are sequentially laser-projected into each layer as it's laid down. Where the projected triangles make contact with each other within the material, they are programmed to cause the gel or other medium to chemically react (or through heat or light) and join at these points, slowly but precisely building up the solid form—somewhat analogous to faster-acting casting resin, which sets up and becomes hard with the addition of a catalyst, or chemical agent. Depending on the form's size, complexity and the material used,

Across the early protocontinents, limy mudflats bordering shallow seas formed ideal substrates for the preservation of sauropod tracks.

the process at present typically takes from several hours to several days. The resulting 3D shape needs no further treatment, and the unused material or medium is simply water jetted or sluiced away. Not only has digital 3D printing opened up wonderful possibilities of creating duplicates of one-of-a-kind specimens for researchers to share, but in theory it makes possible small- to large-scale models of sauropod and other dinosaur skeletal parts, models and other reproductions that could be used for exhibits and comparative studies.

DINING LIKE A *DIPLODOCUS*

What happened when a sauropod like *Diplodocus* used its skull for harvesting food? Digital imaging not only allows us to see inside a fossil's anatomical structures in a non-invasive way but also makes it possible to understand the way unusual body shapes functioned, through *virtual biomechanical modeling*. One derived technique, **finite element analysis** (FEA), was used by Mark T. Young (University of Bristol) and his associates in 2012 to evaluate the stresses and strains going on in the skull of *Diplodocus* in assessing its biomechanical capability when dealing with three different hypothesized feeding behaviors: (1) simple *occlusion*, or biting, (2) stripping leaves from a branch and (3) gnawing (or scraping) nutritious tree bark (see Chapter 5). In this case CAT scans were first obtained from an actual, almost complete *Diplodocus longus* skull, producing longitudinal and transverse cross sections, which, when composited together, revealed the shape and thickness of the skull's individual bones (the finite elements). An AMIRA software program was used to digitally correct bones that displayed breakage, distortion and other imperfections. From the resulting virtual, 3D surface model a grid version made up of 906,257 tetrahedral and hexahedral shapes was then produced with a SCANFE v. 2.0 program. These tiny geometric elements are warped, or change shape, when a particular force, or *loading magnitude*, is applied to them, and they are assigned a range of numerical values that increase or decrease when the shape is warped. When the virtual skull was made to model a particular feeding activity, the loading magnitude on a particular area (or areas) changed, causing the shapes to change and the values to go up or down. Each range of loading values was also given an arbitrary, symbolic color (blue for low, green for low to moderate, yellow for moderate to high and red for high) that visually indicated "hot spots" of functional stress, strain or deformation. The intensity of these was then evaluated to determine how well a particular part of a live *Diplodocus*'s skull would have acted under the stresses or loading magnitudes of different behaviors. What were the results of the biomechanical modeling tests? As it turned out, the sauropod's skull was not only strong enough but actually *overengineered* to deal with the forces involved in both seizing and biting foliage and in branch stripping. In gnawing or scraping bark, however, the loading magnitudes were much higher in certain areas like the tooth-bearing premaxillary and dentary bones, and it's unlikely that both these bones and the teeth they enclosed could have endured the stresses produced if *Diplodocus* had habitually relied on bark scraping as an important feeding strategy.

WHO MADE THE TRACKS?

FEA can also help in solving another question relating to sauropods: what kinds of environments did some types tend to live in, or prefer? If we could know this, we might someday be able to piece

Computer analysis of trackways and digitally reconstructed sauropod feet have enabled **Peter Falkingham** and his associates to make initial estimates of how certain types of sauropods locomoted over some substrates.

together whether titanosauriforms, for example, evolved to adapt to drier, more inland areas while diplodocoids hugged the coasts. Determining how they walked on different *substrates*, or ground surfaces like hard or soft bases, from the tracks they laid down is a starting point. *Brachiosaurus,* a macronarian titanosauriform, placed more weight on its manus, or forefoot, than the diplodocoid *Diplodocus*, which bore most of its weight at the rear of the body. Differences in centers of body mass and weight dispersal would result in differences in the depth of the impressions as the animal left its footprints and trackways. This bears on the question of whether some "manus only" and "pes only" trackways are correctly interpreted as behavioral, for example, "poling" with the forelimbs in shallow water or actual bipedal progression on the hindfeet. In 2009 Peter L. Falkingham (Royal Veterinary College) and his associates used FEA computer techniques after digitally creating *Brachiosaurus* and *Diplodocus* "virtual feet" to plot the depth of the "prints" that these sauropods' manus and pes would create, as well as the kind of impressions that could be expected from these on various substrates. Although the computer modeling experiments are still ongoing, at this point they suggest that rather than due to behavioral situations, the virtual "tracks" in the FEA analysis of weight distributions and foot surface areas show that there was more likely a range of substrates, from soft to firm, that would naturally result in "manus only" or "pes only" prints being left. Some trackway patterns that fall into catagories such as wide versus narrow can be correlated with particular substrates and localities. If certain types of sauropods (known through their body fossils and probable tracks) could be linked to such conditions at a particular time and place, a pattern of their adaptations to certain environments might emerge.

Falkingham and his associates were able to obtain in part a valuable theoretical database for their analysis from two previous studies (2003, 2006) by Donald M. Henderson (University of Calgary), who initially wanted to know how the bodies of live sauropods would have been affected when the animals were immersed in water, based on, at the time, new concepts of their mass. With highly pneumaticized skeletons and air sac systems that made their bodies buoyant instead of heavy when immersed in water factored in, Henderson's computer-generated models of *Apatosaurus*, *Brachiosaurus*, *Camarasaurus* and *Diplodocus* confirmed that in a freely floating state these sauropods would have to have been careful to avoid tipping over, like inflatable beach toys on a water surface. Based on its calculated center of mass, average adult size and limb length, each taxon had optimal depths at which it would have remained stable, and each had different abilities to float, with *Brachiosaurus* and *Camarasaurus* being able to "pole" themselves along the submerged bottom with their longer forelimbs, producing "manus only" prints under some conditions, while short-forelimbed diplodocids might have left "pes only" underwater tracks.

Although in this respect the finding runs counter to Falkingham and his associates' later conclusions about non-behavioral origins for such tracks, a second study by Henderson in 2006 specifically addressed the important question of *center of mass and body weight* distribution between fore- and hindfeet to determine the sauropod trackmaker's identity. When models of a virtual

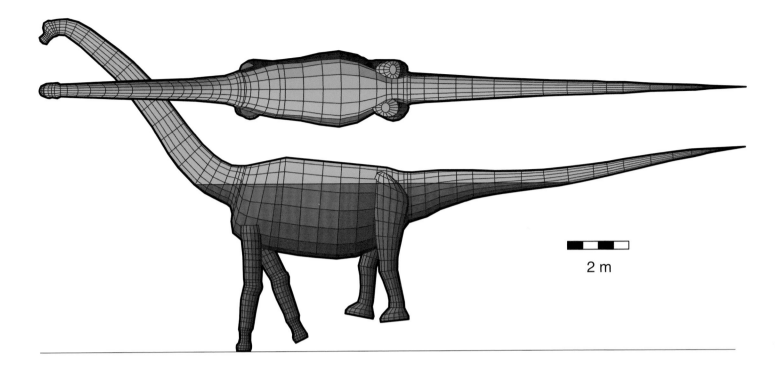

2 m

"walking" *Brachiosaurus* and *Diplodocus* (based on their body shapes, weight distribution and footprint shape) were enacted, he found that to stay within certain limits of stability and avoid the danger of tipping over while moving, each type of sauropod closely conformed to either a "wide-gauge" or "narrow-gauge" modeled gait. When either one of the two sauropod types was "forced" by the computer to walk in the way that the other did, it became unstable. This finding at first seems to help sort out the probable makers of the kinds of trackways found in certain localities, with the implication that the "wide-gauge" tracks, currently found from the Mid-Jurassic and until the end of the Cretaceous, were made by titanosauriforms like *Brachiosaurus* and their descendants the titanosaurs, while "narrow-gauge" trackways were produced by diplodocoids like *Diplodocus*. Henderson went on to propose, however, that, regardless of clade, probably *all* sauropods of 12 tons or more had wide-track gaits for the sake of stability, while narrow-track gaits may have been a basal gait for early types. This brings up a question: if wide gaits were a consequence of sheer size, did supergigantic, 30-ton-plus diplodocoids like *Supersaurus* adopt this, even though their center of mass/body plan would have predicted a narrow gait like the smaller *Diplodocus*? Would a tiny brachiosaurid like *Europasaurus* still have walked with a wide gait like giant *Brachiosaurus*, even though this wasn't necessary from the standpoint of its stability?

In spite of questions like these, it's clear that laboratory-based electronics, digital imaging and modeling, axial tomography, data analysis and other forms of computer technology have tremendously enlarged the possibilities of what can be learned about sauropods and other dinosaurs. From what at one time might have seemed like scattered, sometimes isolated sources of information, older disciplines now have exciting new tools that help these connect with one another as never before. Sauropods, formerly the resurrected denizens of natural history museums, now also have a home in the lab.

Based on his determinations of their density, body proportions and center of mass, **Donald M. Henderson** modeled various sauropods to learn how much stability they would have when immersed in different water levels while touching bottom with their feet. This. in turn, has aided in understanding and interpreting sauropod trackways.

Tracks left by sauropods may someday explain their habitat preferences.

Repeating a scene millions of years old, a hatchling titanosaur struggles to leave its once nurturing but now confining egg. Once it emerges, it has to make it to cover to avoid becoming snapped up.

CHAPTER EIGHT

THE NEXT GENERATION

"Some things start out small—very small. But sometimes the smallest things can make the biggest changes of all." This was the opening line in the overdub for Disney Production's 2000 animated film *Dinosaur*, and it could just as easily apply to the incredible development of an individual baby sauropod from its conception until it reached its full size years later. We're just beginning to understand how sauropods sexually matured, how they may have reproduced, how their young grew and the ways they may have survived—and more often didn't—in the Mesozoic world.

COMING TOGETHER. Day after day, inside the vast, warm body of a female *Ampelosaurus atacis*, a European titanosaur, nine undeveloped eggs were reaching maturity. In a male, kilometers away, a much greater number of sperm cells were almost in readiness for their long journey that would culminate in some of them meeting with the eggs to become **zygotes,** the beginnings of new sauropods. To date we have no fossils of any dinosaur reproductive organs, but in their absence it's reasonable to think that sauropods and other dinosaurs probably weren't too different from their living archosaur cousins (crocodilians and birds) in their basic equipment. In both of these, the female's **ovaries** and male's **testes** (or *gonads*, for each) are internally located directly underneath the kidneys, close to the spine. As hatchlings, female croc babies contain in their ovaries a large number of tiny proto-eggs, or **ovocytes**. Like later, fully developed eggs, these also have a supply of nutritive yolk. When the ovocytes reach an advanced stage, the ovaries (in birds, one remains rudimentary and non-functional) generate large, grape-like clusters of these yolk-filled spheres, each of which becomes the yolk of an un-fertilized egg before traveling into the paired *oviducts* (like the ovaries, in birds only one is functional; the other never develops). In contrast to the ovaries, the two testes produce thousands of sperm during a male's reproductive lifetime, and the testes, like the ovaries, enlarge during the breeding period to maintain production. As they develop to maturity, the sperm travel through a complexly convoluted system of small tubules, the ***vas deferens***, where they are held in readiness for mating. In most living birds and many reptiles, the male's and female's ***cloaca*** orifices, which serve as the only conduits for both eliminating wastes and reproduction, are simply pressed together during ***copulation***, or mating, resulting in the ejaculation of sperm-rich fluid into the female's oviducts. While the simple act of the male mounting the female and twisting his lower body as the female moves her tail to one side is usually enough to accomplish this, some male birds (ostriches and waterfowl) and crocs have a ***penis.*** This stays within a pouch just inside the cloaca until mating, when during sexual arousal it becomes engorged (with lymph in birds, in contrast to blood in mammals) and stiff. A penis means that a successful sperm transfer is virtually assured, especially when the mounting male is tall, heavy and unstable. In elephants, the 1.5 m (5′) organ (like the female's long vaginal opening) is far down on the *underside* and at the rear of the abdomen, which makes the angle of insertion and penetration easier. This is facilitated still more by the penis's S-curved shape, with tip pointing upward at an angle, and its great flexibility, powered by strong levator muscles. Capable of probing and extending on its own without conscious control, it finds the vulva and slips into the vagina with little or no movement from the male's main body. When the male climaxes, a flood of ***ejaculate*** or fluid with thousands of stored sperm is pumped upward from a muscular sac to meet the single ovum, which combines with a single sperm to become a zygote. Although we have no evidence, it's reasonable to think that male and female sauropods, because of their much greater bulk, height and potential instability, almost certainly evolved similar advanced, elephant-like copulatory organs and behavior rather than simple cloacal contact.

Apart from this, we can only guess at other aspects of sauropod sex and breeding behavior, such as whether mating was year-round or seasonal, and whether intraspecific fights between dominant males or specialized courtship displays between the sexes occurred. Hormones were the driving force here. These can produce huge morphological changes within a relatively short period of time, like the ferocious fanged snouts of male sockeye salmon or the extravagant antlers of a bull moose, grown fresh every year. Sauropods' eyes, ears, noses and other sensory organs were usually high off the ground, and because of the probability that they had color vision (see Chapter 5), this also opens the door for many possibilities like fancy head crests and bright breeding colors, similar to the beak and eye color changes that occur in birds when they're coming into mating condition. Did two brightly crested bull *Apatosaurus*, with their strong, boxy cervicals, engage in neck-battering contests like giraffes as they towered on their hind legs? Could a *Brachiosaurus* couple have flushed colors up and down the skins of their tall, vertical necks to advertise their readiness to mate?

What we do know is that dinosaurs, unlike mammals, retained the saurian (reptile and archosaur) form of reproduction known as **oviparity** ("egg birth"), or being *oviparous* (not to be confused with **ovoviviparity**, in which the egg is not only produced and fertilized but also retained within the mother's body during the embryo's development). This is a more primitive method than **viviparity** ("live birth"), or being *viviparous*, which characterizes almost all mammals and many lizards and snakes. In viviparous **amniotes** (land-living vertebrates in which eggs have an amniotic sac to cushion the embryo and an **allantois** sac to store wastes) the egg doesn't develop a shell, and the embryo and later fetus is retained inside the body of the mother until it reaches an advanced state of development before birth. Oviparous babies develop inside either a soft and leathery or hard-shelled egg, and for living saurians this may take many variations. During sea turtles' spectacular annual breeding congregations, in which an individual female may lay as many as 80–100 eggs before simply burying and leaving them, the fertile, soft-shelled eggs are left unprotected. The hatchlings make a brief, instinctive scramble to the surf, where many are lost to air, land and sea predators. As they grow out in the ocean, their chances of survival gradually improve with increasing size until, as adults, little can threaten them. Among the surviving archosaurs like crocodilians and birds, egg clutches are far smaller, but survival in hatchlings is aided by parental care behaviors like nest guarding and sheltering, as well as protection and feeding of young. This doesn't mean, however, that all extinct archosaurs (especially sauropods) may have had these nesting behaviors, or even nests like modern forms.

THE EGG, WITHIN AND WITHOUT

As the now fertilized dinosaur egg passed through the *uterus* (which, unlike the muscular, distensible structure of some mammals, was simply a specialized section of oviduct) and acquired its hard covering from the shell gland, the contents organized themselves into structures that would protect, feed and remove wastes from the sauropod-to-be. An **amniotic sac** and its clear fluid surrounded the embryo, creating a stable aquatic environment and a shock absorber. A **yolk sac** containing yolk, a nutritious mixture of proteins, fats, sugars and starches, channeled these nutrients through blood vessels into the developing primitive gut, the rear portion of which connected to the allantois. This sac stored the ammonia, urea and other wastes generated by the embryo,

Like modern whales, sauropods probably varied widely in their mating behavior. Some, like sperm whales, may have maintained "harems" during the breeding season, with the most powerful male fighting off rivals to maintain his exclusive right to breed, like these two *Brachiosaurus*. In others like gray whales, females may have had several breeding partners, with females choosing a male with the best stamina, vocalizations and other courtship behaviors. We'll probably never know.

The Mechanics of Mating

Because of their great size and height, male sauropods probably possessed a large, mammalian-like external penis instead of an internal, temporarily extrudable one positioned inside a pocket of the anterior cloacal wall, as in crocodilians and some avian archosaurs. Like those of elephants and whales, it would have been independently motile in seeking and penetrating the female's cloaca. On ejaculation from a bubo-seminal gland at the base of the penis, the seminal fluid would have been projected into one or both of the female's oviducts, which opened into the cloaca.

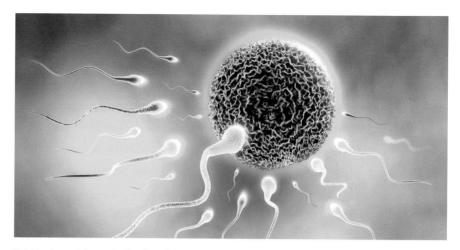

Within the oviduct of a fertile adolescent *Ampelosaurus* female, thousands of sperm converge toward a drifting ovum, or egg. The life journey of a sauropod is soon to begin.

During mating, it would be necessary for the female to deflect her thick tail to one side, allowing the male to lift one of his hind legs over the tail and balance on top of the female to allow the penis access.

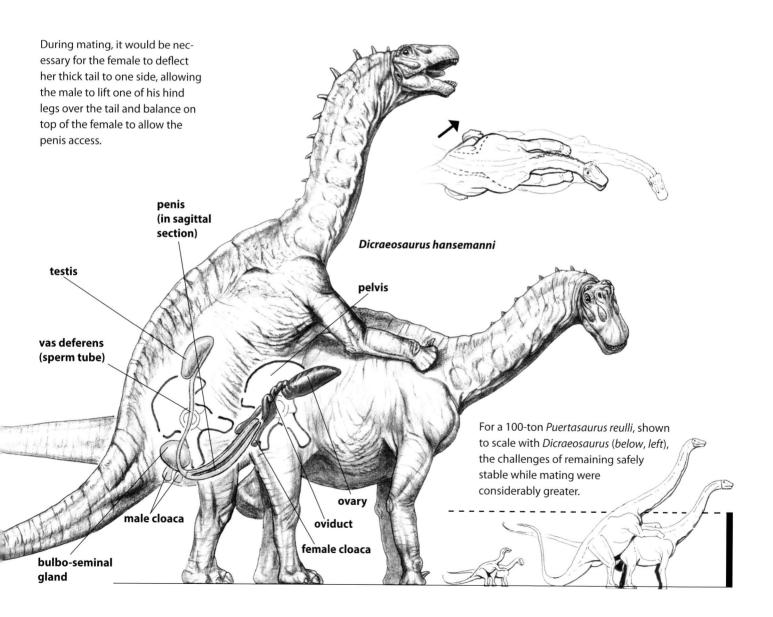

penis (in sagittal section)

testis

vas deferens (sperm tube)

Dicraeosaurus hansemanni

pelvis

male cloaca

bulbo-seminal gland

ovary

oviduct

female cloaca

For a 100-ton *Puertasaurus reulli*, shown to scale with *Dicraeosaurus* (*below, left*), the challenges of remaining safely stable while mating were considerably greater.

and expanded to accumulate these while the yolk became used up. Surrounding everything, the blood vessel–rich **chorion** and an outer gelatinous layer, the **albumen** ("white"), intially took care of diffusing oxygen and carbon dioxide gases into and out of the egg, while also admitting vital water. Soon after the egg was laid, however, the chorion and allantois fused to form the **chorioallantois membrane**, which took over from the albumen the job of embryonic breathing. The elastic albumen served, like the amniotic fluid, as an additional shock absorber if the embryo was suddenly bumped, and it also contained antimicrobial chemicals that protected against infection. Although the hard eggshell was comparatively thick (0.5 cm [0.196″]), it was nevertheless porous enough to allow gases and water to pass through. The shells of what were probably *Ampelosaurus* eggs are very porous, suggesting that the mother buried her clutch of 8–10 eggs under sandy soil or vegetation, where they would be protected from too much sun but still receive the oxygen and carbon dioxide release they required. Sometimes, as in modern birds and reptiles, the egg's shell formation went wrong, and abnormalities occurred, such as multiple shells. This happens when the muscular contractions of the oviduct temporarily send the already-covered egg back the way it came, where it gets a double or even multiple coating of shell; if so, the pores don't line up and are blocked, causing the embryo to suffocate. Multiple-eggshell pathology is the type most commonly seen in non-avian saurians, and this may be additional indirect evidence that some dinosaurs were more like these and less like birds in their egg production. The titanosaur sauropod eggs from South America have fewer pores than the European and Indian ones, a fact that contributes to the idea that, unlike their Old World counterparts, these were either barely covered or actually left unburied. Although this seems like an unlikely interpretation, it's supported by the fact that in sites like Auca Mahuevo, Argentina, the preserved eggs are surrounded by a matrix of mudstone. This would have resulted when nearby creek channels, swollen with rain, overflowed and buried the eggs in muddy silt. The hardened silt suffocated the embryos but also preserved the eggs as they sat in their sandy depressions, and it wouldn't have filled the spaces around the eggs if the nest had already been occupied by sand or vegetation.

The microscopic structure of dinosaur eggshell shows more variation than in any other amniote. Eggshell, like bone, contains crystallized minerals such as calcium to give strength and support, and the various prism-shaped microscopic structures have evolved to give different types of hard-shelled egg the support they need. The eggshells of crocs and most dinosaurs (except for those of theropods, which are structurally like those of their bird descendents) take the form of V-shaped, column-like groups of crystals that radiate from a bottom core or base. This form, known as *spherulitic*, is the way sauropod eggshells are constructed. Although most sauropod eggshells seem thick compared to modern eggs, they are actually comparatively thin, and this may have been a trade-off to allow for optimal gas exchange while offering the greatest amount of support. The outer surface of sauropod and other dinosaur eggs, like the hard-shelled eggs of ground-laying emus and some modern birds, have a texture, or *ornamentation*. In some titanosaurs the gray shells have an abundant, dense ornamentation of small bumps that sometimes coalesce into wormlike ridges. As with other nesters that depend on sand or decaying vegetation for incubation, this may have kept the vital pores unclogged by creating tiny air spaces around them.

NESTS AND EGG CLUTCHES

Sauropod egg clutches were laid, like those of sea turtles, in relatively shallow depressions that were probably mostly dug out by lunges from the mother's massive, clawed hindfeet, possibly assisted by the forefeet. The 8–10 eggs of a female *Ampelosaurus*, like those of other sauropods, when laid were spherical, about 25 cm (9.05″) in diameter and 5 L (1.50 gal) in volume. The nests at the Auca Mahuevo site in Argentina show that the female sauropod, in this case also a titanosaur, made an oval- or kidney-shaped shallow depression before laying her eggs. The Auca Mahuevo titanosaurs, unlike their European *Ampelosaurus* sisters, produced far more eggs per clutch, anywhere from 20 to almost 40. Before laying, the female probably bent her hind legs to get her rear body into a semi-squatting position: the hard-shelled eggs, after being expelled from the mother's body, would have broken on the ground or against each other if they had dropped any distance. It's also possible that female sauropods possessed a muscular, tubelike *ovipositor*, an extrudable extension of her cloacal wall, which reached the ground and deposited the eggs gently into the nest. A watery fluid secreted from this or from the wall of her cloaca may have lubricated the eggs as they passed through, and this would have kept the eggs from drying out until the clutch was covered.

Unlike those of other dinosaurs that show a definite pattern of adjoining or concentric rings of eggs, sometimes in layers, known sauropod egg clutches usually occur in groups of four to five in one to three layers. There's disagreement as to whether the adjoining eggs formed arcs (as the mother turned to lay more) or are random associations. Some Auca Mahuevo eggs are very isolated from the main clutch, and in living ostriches' communal nests these "sacrificial" or "trial" eggs are ones deliberately laid or moved out by the incubating female, making them more likely to be taken by nest-robbing predators, leaving the others to develop and hatch. It's also common for female birds (for example, chickens) during their first reproductive year to lay "trial eggs," often smaller than average, outside a nest.

TEMPERATURE IS EVERYTHING

The precise length of time it took for the eggs to hatch is hard to know, since the speed at which an embryo develops depends on the temperature. Ostriches and other birds maintain a fairly constant temperature of 40°C (104°F) when they are incubating, and under these circumstances a 1.5 kg (3.3 lb) sauropod egg—about the same size as an ostrich's—would therefore have needed about 60 days of incubation. But because of their sizes, sauropods couldn't incubate their eggs, and like modern crocs and alligators, they laid them in scooped-out depressions covered with sand or vegetation. Here the heat from sand or from the decaying plant matter provided warmth. In sea turtles, when the sand is cool—25°C (77°F)—from shade or rainy weather, incubation takes 65–70 days. If the nest is in a sunny location, egg temperatures can rise to 35°C (95°F), hatching the eggs in about 45 days. The mean temperature in nests of big archosaurs like the American crocodile (*Crocodylus acutus*) is 30°C (87°F) in June and goes up to 34°C (93°F) in August, when hatching occurs after a period of 77 days. Those few degrees are important, since in saurians temperature determines not just how soon the baby hatches but also its sex. Most vertebrates inherit sex chromosomes from their parents: two separate X chromosomes produce

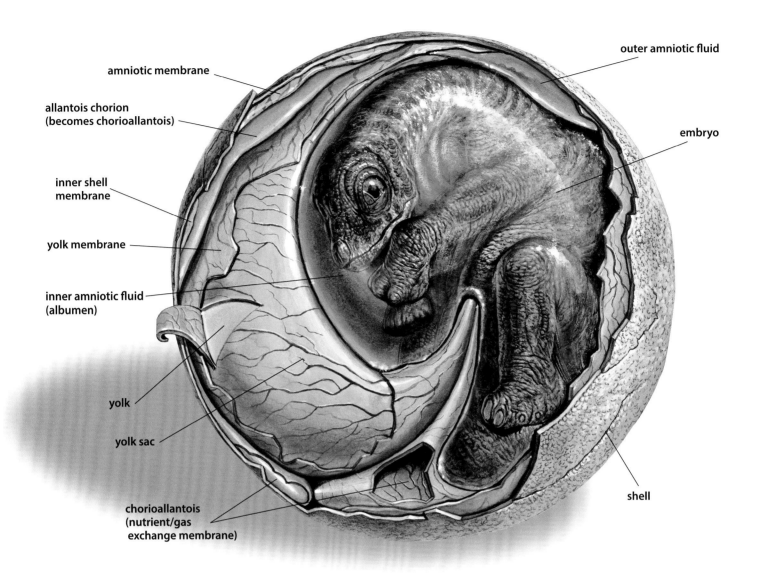

outer amniotic fluid

amniotic membrane

embryo

**allantois chorion
(becomes chorioallantois)**

**inner shell
membrane**

yolk membrane

**inner amniotic fluid
(albumen)**

yolk

yolk sac

shell

**chorioallantois
(nutrient/gas
exchange membrane)**

Within the Egg

A five- to six-week-old *Ampelosaurus* embryo, close to hatching. The overall structures of the egg were probably identical to those of its living archosaurian cousins, the crocodilians and birds. Baby sauropods, like hatchling turtles, may have emerged from the egg with remnant yolk sacs outside and under the belly to supply them with nutrition until they could find their own food.

a female, whereas an X and a Y result in a male. Saurians (and probably dinosaurs), however, don't and didn't have sex chromosomes—the incubation temperature determines the sex. In crocodilians, eggs that develop more slowly (at 30–31°C [86–87°F] or less) become *females*, whereas faster development (32–33°C [88–89°F]) results in *males*. Splitting the difference produces both sexes. In some croc species a temperature higher than 33°C produces males, but in others the opposite results, yielding so-called high-temperature females. Too high a temperature during the first third of incubation can kill the embryo or produce a variety of abnormalities, while those that are too low mean that the embryo doesn't develop. In the spherical cluster of

Except for the shell, at about 15 weeks, a developing sauropod may have superficially resembled this embryo of a snapping turtle.

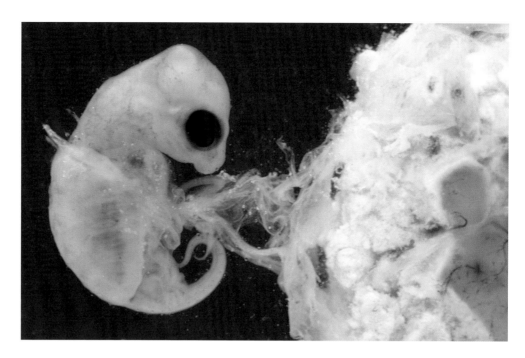

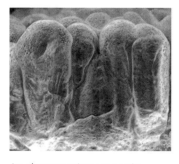

An electron microscope view of a sauropod eggshell in cross section, showing its columnar structure and a respiratory or breathing channel.

sea turtles' nests, the position of the egg in the clutch (a result of laying sequence) also can have an impact on sex, since eggs in the center will be slightly warmer and those at the edges a bit cooler. The known clutches of titanosaur eggs at Auca Mahuevo seem to be stacked on top of one another without any obvious arrangement except for sometimes being in two to three layers, and this may have evolved as a means of maintaining a more or less even sex ratio. Even the weather may have had an influence on sauropod hatching. Studies of sea turtles show that the middle third of the incubation period is a critical time period in sex determination, and a few days of cool, rainy weather can lower the nest's temperature from 4° to 6°; eggs well on their way to becoming female might end up as males. If the nests are swamped by high tides for too long, the eggs may die, since the eggshells can't extract oxygen from water and the embryos suffocate—this is what happened at Auca Muevo when freshwater streamlets during a flood overflowed their banks, killing the lives within before they could hatch.

A VOLCANIC NURSERY

Miles away from Auca Mahuevo, another titanosaur nesting site shows that some populations in northwest Argentina used the planet itself to incubate their eggs. At Sanagasta Geological Park, now extinct geysers and vents, some originally 4 m (13′) deep and at least 15 m (49′) wide, dominated a Late Cretaceous valley of the now Los Llanos Formation with intense geothermal activity. Here Lucas Fiorelli (Centro Regional de Investigaciones Cientificas y Transferencia Tecnologica) and Gerald Grellet-Tinner (previously cited institution, also Field Museum, Journey Museum) found 80 titanosaur egg clusters, as well as isolated shell fragments, in shale deposits rich in the crystals and minerals produced by this kind of heat. Each nest was located close to a vent, and the number of egg-bearing layers in the stratigraphy of the Sangasta Site indicates that, like at Auca Mahuevo, this occurred over a very long time. Unlike the Auca Mahuevo eggs, however, no embryonic skeletons have been found inside to allow identification of the species, but the tiny, rounded nodules surrounding the shells' pores are similar to those of the other site. This use of geothermal heat is practiced today by modern mallee fowl or brush turkeys (*Alectura lathami*, among others) of Australia and Micronesia. These and other megapodes all bury their eggs in large mounds of plant compost, monitoring the temperature with their sensitive beaks and periodically kicking off material to maintain a specific temperature. One in particular, the Polynesian scrubfowl (*Megapodius pritchardi*) of Niuafo, Tonga, routinely locates its nests close to the island's volcanic vents, and it's obviously a very old nesting strategy.

THE TREASURE OF AUCA MAHUEVO

Since the 19th century, large, spherical eggs, both intact and fragmented, have been discovered in Spain, Romania and southern France at the same localities and horizons associated with the skeletal remains of the titanosaurs *Ampelosaurus and Hypselosaurus,* but since no embryonic remains were found, these eggs could not be positively identified as those laid by these sauropods. Because of problems like this, paleontologists assign fossil eggs, like trackway types and body fossils, their own taxonomic names. Big, round eggs are called **megaloolithid** (*mega,* "big"; *oolithus,* "stone egg"), and the presumed sauropod eggs were named, among others, *Megaloolithus aurielensis.* Those from India are known as *M. baghensis* and by other names. In the mid-1990s, however, spectacular findings of thousands of Late Cretaceous megaloolithid eggs and nesting sites were made in Auca Mahuevo, northwestern Argentina (known as *Megaloolithus patagonicus*), as well as localities in Peru (*Megaloolithus pseudomammilare*), and many of these contained not only beautifully preserved embryonic bones but also fossilized integument, or skin. Although the early stage of development doesn't at present permit a species determination, the morphological features of the tiny skull bones and teeth now show that these were *nemegtosaurid* titanosaurs, most closely related to the Mongolian *Nemegtosaurus mongoliensis,* and they are the first sauropod embryos ever discovered. In these, as with most unhatched or unborn babies, the head and eyes were large, and (for sauropods) they had relatively short necks. Some characteristics also indicate that at this early stage in life titanosaurs clearly had skulls unlike those of their close relatives the brachiosaurids, a fact confirmed much later by the finds of complete adult titanosaur skulls. During ontogeny, dramatic changes happened to the nestlings' head proportions and features. As the skull bones grew, the *rostrum*, or snout, enlarged and became longer, while the external *nares*, or nasal openings, expanded and migrated ventrally from their primitive position at the anterior cranium, becoming located between and above the orbits, or eye sockets. A surprise is that in some embryos the teeth show some slight evidence of wear. Since their only food source was yolk prior to hatching, the reason may have been that tooth grinding could have taken place to create an occlusal or beveled surface to prepare the teeth for cropping plants as soon as the hatchling became free of the egg. Tooth grinding may also indicate that the jaw muscles had already started exercising in preparation for their vital role in obtaining food as soon as possible. Not as well preserved as the crania, the appendicular (limb) bones have no ossified ends but nonetheless have something very important to say. Unlike young mammals, in which the ossifying center and ends (the second known as *epiphyseal plates*) of the long bones grow *inward* toward each other, replacing cartilage and effectively ending further growth when these meet together and fuse, saurian bones begin to ossify from the *center toward the ends*. This means that there is no limit to the size the bone can grow, only that growth gradually slows with increasing age. It's one reason why the ends of sauropod limb bones, even as adults, have a simple, rounded look, almost like a cartoon bone. They didn't have the closely fitting, bony condyles or joint ends seen in most mammals, but instead had gnarly, **rugose** (roughened) surfaces capped by thick pads of first fibroid, and then thin hyaline cartilage at the surfaces. These cartilages could be continuously replaced with bone, which meant that the bone could keep increasing in linear size, as well as allometrical proportion (see Chapter 7). As a result, the bone got stronger as its sauropod owner became bigger. This shape and growth pattern is very similar to that found in living whales. These are the only mammals that have similarly rounded bone shapes, and they show the same astounding rate of growth projected for sauropods.

Just as exciting as what the tiny bones reveal is the first look we have of titanosaur skin. At the Auca Mahuevo site these are in the form of patches preserved as fossilized positive and negative impressions, or casts and molds, of the original skin. As opposed to the skeletal remains that are usually found flattened against the concave inner bottom of the shell, the *integument* is cemented onto egg fragments that are often scattered over the surfaces on which the nests are eroding. Instead of the typical overlapping scales found on living saurians, these skin impressions are non-overlapping *tubercle* (small to large, round bumps of similar size) patterns that take the form of rosettes (flowerlike arrangements of small tubercles surrounding a central, bigger one) and rows of larger tubercles. We know from finds at other localities that in some titanosaur species, adult remains are often associated with large, rounded *osteoderms* (**osteon**, "bone"; **dermis**, "skin") of bone that were originally embedded in the skin. Found in isolation without other fossil skin associations, these osteoderms may have formed a pattern of armor as in living crocs and alligators. On the other hand, very few osteoderms are ever recovered from titanosaur skeletal sites. This has led titanosaur specialists like Kristi Curry Rogers (Macalester College) and Mike D'Emic (Georgia Southern University) to think that they may have had another, previously unthought-of function, that of reservoirs of calcium and other needed minerals when these were seasonally unavailable (see Chapter 11). Evidence of this comes from the deep channels within the hollow structures, which were occupied by blood vessels that transported the minerals into the rest of the body. The osteoderms (although not present in the embryos), like other body features, probably grew and enlarged as the baby became an adult. Someday we may find a sauropod "mummy," like the ones known of hadrosaur and ankylosaur adults, with an almost completely preserved skin pattern.

NESTING COLONIES

The small clutch size and proportionate-to-adult-body-size egg dimensions of the European *Ampelosaurus* and the as-yet-unnamed nemegtosaurid titanosaurs from Auca Mahuevo imply that the individual females were following the strategy of producing several clutches per season. This is more like that of crocodilians, rather than one large one per female as is more common with sea turtles. At the Auca Mahuevo site, many individual clutches are found generally 2.7–3.0 m (9′–10′) apart, preserved in a widespread area composed of four stratigraphic layers, with only a

Sauropod eggs were round, and those of giant titanosaurs and others held about 5 L (1.50 gal) (*above*). Although the largest ever to exist, these, like all amniote eggs, were limited in size by the strength of the shell, whose thickness could not exceed the ability of oxygen and other gases to permeate through them. As with some living birds, some sauropod species' eggs were ornamented with distinctive patterns (*below*) that may have helped create tiny gas exchange spaces surrounding the shell.

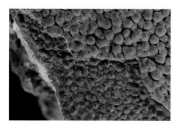

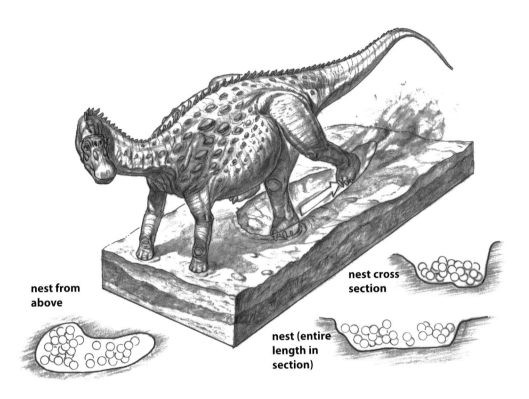

nest from above

nest cross section

nest (entire length in section)

Preparing the Nest

Sauropod nesting sites discovered at Auca Mahuevo and some other localities show that a shallow, crescent-shaped nest was excavated with lunges from powerful hindfeet. Following this, about 26–28 hard-shelled eggs were laid, loosely covered with soft, sandy soil or decayed vegetation, and then abandoned in the same way as a turtle nest. Because the eggs might have risked breakage if they fell, female sauropods may have possessed an extendable ovipositor (*above, right*), an extension of the cloacal wall, which gently released the eggs one at a time within the nest.

small amount of fragmented eggshell per clutch. From this we can reach three conclusions. First, the titanosaurs were practicing ***communal nesting,*** since many nests occupy one layer in a given area, far more than one female could produce in a single season. If they were like Australian freshwater crocodiles (*Crocodylus johnsoni*) or American alligators (*Alligator mississippiensis*) they might have been "pulse nesters," with females in a local nesting ground mostly laying their eggs at the same time, or "prolonged nesters," with a season lasting six months or more. Second, the females exhibited ***site fidelity,*** or the returning to the same general locality as a group, time after time, to dig their nests and lay eggs. From calculations based on the total thickness of nest-bearing strata, this colony could have used the site for a very long time, perhaps about 10,000 years. Third, because of the limited amount of eggshell fragmentation found associated with each egg and within the borders of each clutch, ***the hatchlings were precocial***. This means that on hatching the babies were either almost or completely able to survive on their own, like baby sea turtles or pheasants. It's the opposite of ***altricial*** babies, which are usually hatched or born blind, naked and needing feeding and protection, as in robins. In a fossil egg, as in a modern one, signs that hatching took place are that the uppermost portion of shell is missing and the interior

At Sanagasta Geologic Park, Argentina (*right*), discoveries of titanosaur nesting sites adjacent to the then contemporaneous volcanic vents show that some sauropods took advantage of geothermal heat to incubate their eggs, much as do mallee scrubfowl in Tonga (*above*).

is filled with the sediment that surrounded the egg. If they had been altricial, with each baby staying within the nest to receive parental protection and feeding, the empty shells would have shown much more crushing and fragmentation than if the hatchling had just cracked its shell, pushed its way out and then left. If we try to imagine a scenario with altricial sauropod hatchlings, problems come to mind: a huge adult sauropod would have been an awkward mother, as likely to accidentally squash as to protect her young. Like baby birds, young sauropods would need to be fed, and often. Feeding for themselves was probably an almost round-the-clock activity for adults, and the parent would have had little time to bring vegetation (presumably regurgitated) to the ravenous babies while avoiding other defensive parents protecting their nests. In additional support of the idea that baby sauropods were precocial and left their nests on hatching, so far no hatchling or older juvenile remains have been found at Auca Mahuevo or other nesting sites.

INTO THE WORLD

When baby birds and crocs are ready to break out, a tiny, triangular structure called the *caruncle*, or "egg tooth" (not actually a tooth, but keratin like your fingernail), forms at the tip of the beak or snout, and the hard point helps to apply concentrated pressure inside the shell to crack it. By using this, as well as pushing and straining with its body, head and neck, the baby breaks enough of the shell to emerge, wet and exhausted by the effort. The skeletal titanosaur embryos from Auca Mahuevo and other South American sites also have egg teeth. After hatching, each baby sauropod and its nestmates had to squirm upward (in the case of *Ampelosaurus*, through several inches or more of sand or decayed plant compost), to reach the surface of the nest. In gaining freedom, the babies ironically triggered the first threat to their survival. As with today's croc and turtle hatchlings, there were sharp-eyed predators on constant patrol that would have detected the sounds, odors and other disturbances at the surface of the nest and seen the emerging young (see Chapter 9). With a lack of any direct evidence, our informed imaginations must here re-create the baby sauropods' frenzied rush to survive. After struggling to reach the surfaces of their nests, those that were not snatched within the first minutes by opportunistic predators broke into an awkward, hyperenergetic amble, like little wind-up toys, toward vegetation cover. This was most likely to be composed of dense stands of various tall to medium-height forms of ferns, a type of plant universally common during the entire Mesozoic, which may have grown close to the nesting site in the form of either dry-adapted "fern prairie" species or moisture-loving forms in shady areas underneath woodland canopy. Another could have been horsetails, which also occur in tall, dense stands. We can safely imagine that the babies would have had some form of *cryptic*, or concealing, patterns and colors on their skins, like ratite or galliforme chicks, to break up their body contours and make them blend in with their cover, if they reached it. Even making it to the concealing umbrella of ferns or other vegetation was no guarantee of survival, however—death almost certainly lurked here as well, in the form of opportunistic reptilian and mammalian predators. Instinctively, the babies might have spread out to minimize their chances of being detected.

Having momentarily escaped being eaten, the sauropod babies' next vital need was to find food themselves, as well as water. Their small size and relatively short necks meant that any food would have to be close to the ground and abundant enough to supply the large amounts of calories and protein they needed for their continuous growth demands. New studies and reevaluations of Mesozoic plants by Gee show that of all the plants known to exist during this time, the highest in overall food value and most easily accessible to young dinosaur herbivores would have been the varieties of *horsetails* or *scouring rushes* (*Equisetum*, *Equisetoides* and others). Many of these are short (maturing at about 0.45–0.60 m [1.5'–2']) and highly abundant in moist areas, creating dense stands. They also grow fast—the tip, the most nutritious part of the plant, regenerates quickly when it's nipped off, and new plants spread in part by forming underground runners. *Equisetum* usually prefers moist environments near water, and this fact and the plants' own high water content would have provided a growing baby sauropod with what it needed to survive. A second possible plant food source might have been several species of succulent, high-calorie *halophytes*, members of a now extinct conifer family known as the *Cheirolepidiaceae*. These were common in many environments and, like some other conifers, might have had a high food value. Mushrooms and other types of fungi, growing, like *Equisetum*, in moist environments, not only could have contributed valuable dietary fiber but were probably, like today's species, rich in folic acid and other vitamins. Finally, it's also quite possible that had it encountered one, a sauropod baby at this stage of life might occasionally have snapped up any succulent insect it saw—arthropods are great sources of calories, fats and protein, the last up to about 13% per gram of body weight. Some of the earliest sauropodomorphs were probably to some degree omnivorous, and the young of later, advanced species may have temporarily retained this primitive food preference.

GROWING UP SAUROPOD

In spite of the odds against their survival, a new generation of baby sauropods had the strength of numbers, and time, on its side. If the clutches at Auca Mahuevo and other major sites hatched *synchronously* (at the same time) as do those of most living colonial sea turtles, several thousand hatchlings within minutes would have begun their instinctive scramble toward cover. Although the resident predators took a terrible toll, perhaps 60%–80% of these, each day a hatchling lived

At Auca Mahuevo, not only the tiny skeletons but also the skin of these titanosaurs and its distinctive patterns were preserved in exquisite detail.

improved its chances. Ecologists who study the importance of age in the success or failure of survival in a particular animal species have found that this takes the form of three basic patterns, symbolized by chart curves. In the first, there's high juvenile survivorship in long-lived species in which the young are protected by parents from predation and are very unlikely to die before they reach sexual maturity (elephants, humans in affluent, westernized societies). Here the convex curve stays level for a long time, since survival doesn't drop off until well after the animal attains adulthood. This contrasts with the second, mostly straight linear curve that represents a steady, constant risk of mortality for some species during their entire lifetime (medium-sized, moderate-lived mammals, like a deer). Finally, a third, concave-shaped curve is typical of animals that experience high juvenile mortality early in life (small, short-lived birds and mammals, like quail and rabbits) but, like the species in curve 2, then have a steady, linear survival rate once they reach a certain age and size. This is caused mostly by predation and tends to kill off individuals before they can sexually mature and reproduce. What can we deduce about the survival rates of dinosaurs, especially sauropods? Like elephants and some other living mammals, they were usually very large to huge and presumably fairly long-lived, but *unlike* these, they produced *abundant young.* This last characteristic makes them very different from the very large, long-lived mammals we know today. Living dinosaurian relatives such as crocs have high juvenile mortality rates like small animals, yet can grow large and live a long time like big mammals. Unlike dinosaurs, however, crocs take a long time to grow before they reach sexual maturity. Sauropods (and other dinosaurs) grew as fast as or faster than any living endothermic bird or mammal, and because they also sexually matured much faster and produced abundant young, they had a huge survival advantage over a big-bodied mammal. Because a sauropod wasn't like any modern animal, we can't make very simple comparisons or predictions, but it's reasonable to think that the huge number of young would have almost guaranteed that at least a few individuals would become old enough to perpetuate the species.

FAST-FORWARDING

Sauropods were dinosaurs that specialized in taking in large volumes of food at an almost continuous rate and letting it ferment in their vat-like gut systems to yield a steady amount of nutrition. This meant that getting to their huge adult sizes as quickly as possible was the end goal, and as a side benefit it would also make young sauropods less and less vulnerable to predation the larger they grew. If there was a survival advantage in having large young, like the much later big-bodied mammals, why didn't big-bodied sauropods do the same thing? The short answer is that long before sauropods had begun reaching large to huge body sizes, their ancestors had evolved and retained an ovoviviparous mode of reproduction: all known dinosaurs stayed with this. As evidence of this, some of the earliest known sauropodomorphs, such as *Massospondylus,* laid hard-shelled eggs. Then why not huge, and fewer, eggs? Kiwis, the living relatives of the giant extinct moas of New Zealand, actually produce one egg at a time that is gigantic in proportion to the adult bird. Kiwis are small, however, and although the egg is huge for their size, it's still a small egg. The larger an egg is, the thicker its shell must allometrically become (see Chapter 7) to provide enough support to prevent it from breaking or collapsing under its own mass. A second

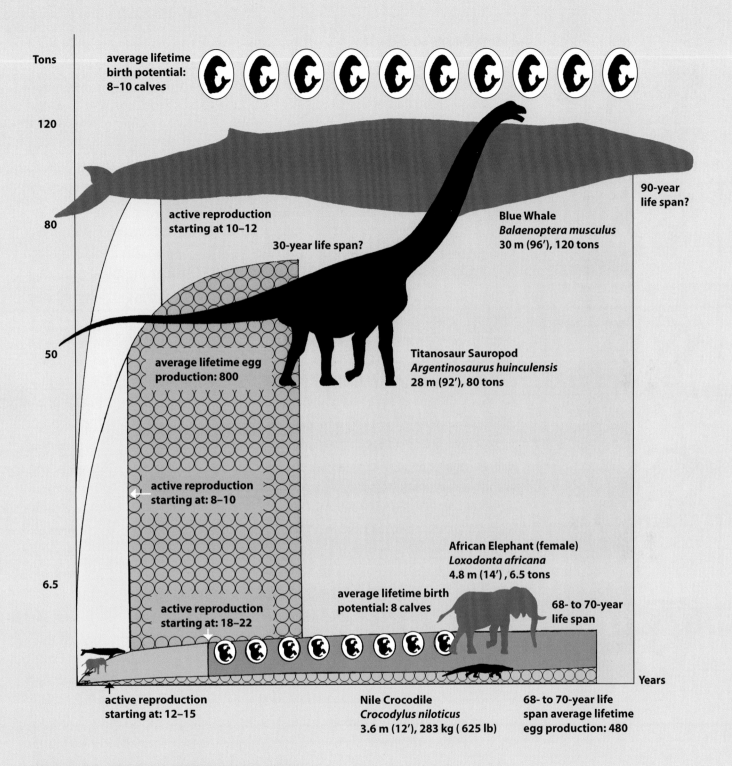

Tons

average lifetime
birth potential:
8–10 calves

120

active reproduction
starting at 10–12

30-year life span?

Blue Whale
Balaenoptera musculus
30 m (96′), 120 tons

90-year
life span?

80

50

average lifetime egg
production: 800

Titanosaur Sauropod
Argentinosaurus huinculensis
28 m (92′), 80 tons

active reproduction
starting at: 8–10

African Elephant (female)
Loxodonta africana
4.8 m (14′) , 6.5 tons

68- to 70-year
life span

6.5

active reproduction
starting at 18–22

average lifetime birth
potential: 8 calves

active reproduction
starting at: 12–15

Nile Crocodile
Crocodylus niloticus
3.6 m (12′), 283 kg (625 lb)

68- to 70-year life
span average lifetime
egg production: 480

Years

Maturity, Time and Reproduction

Starting off as 4–5 lb hatchlings, the biggest sauropods like *Argentinosaurus* attained and probably far exceeded weights of 50 tons by the time they reached full adult size at approximately age 20, a staggering increase by a magnitude of tens of thousands. Although their life span may have been over by about 30, in contrast to a much longer lived 65- to 70-year-old wild crocodile or elephant, sauropods and other dinosaurs attained sexual maturity long before adulthood, giving them a much higher lifetime output of potential offspring to offset high mortality rates. Elephants and other large mammals produce fewer young, but by nurturing babies with milk and parental care, these are more likely to survive to reproductive maturity.

p f pf l n
b
sq pmx
po
qj
mx
Hatchling j pe

Hypothetical steps

Adult

Names and abbreviations for selected bones: **b** = basioccipital; **f** = frontal; **j** = jugal; **l** = lacrimal; **mx** = maxillary; **n** = nasal; **p** = parietal; **pe** = postdental emargination; **pf** = prefrontal; **pmx** = premaxillary; **po** = postorbital; **qj** = quadratojugal; **sq** = squamosal.

From Baby to Adult

As a baby sauropod grew, its skull bones changed shape and proportions according to the demands that were placed on them. The bony nares (nostrils) retracted upward toward the top of the head, while the upper and lower jaws lengthened, deepened and became more robust to handle the loading stresses created by biting and pulling. As yet only hatchling (*top*) and fully adult (*bottom*) skulls are known in titanosaurs like *Tapuiasaurus macedoi*, but here hypothetical intermediate stages show how ontogenetic changes might have occurred.

problem is that the thicker the shell becomes, the harder it is for vital oxygen to pass into the developing embryo and for waste product gases like carbon dioxide to be released to the outside. The 1 L (1.05 gal) eggs of ostriches, moas and elephant birds are probably the largest that an egg can become and still allow the entrance of oxygen and release of carbon dioxide through the shell pores, and the eggs of known sauropods from India, South America and Europe were about the same size. This means that baby sauropods of even the largest species on hatching could be no longer than about 1 m (3.6′) from nose to tail and would have weighed no more than 5 kg (11 lb). For some of the most gigantic sauropods like *Argentinosaurus huinculensis*, which as adults may have reached lengths of 30–35 m (98′–115′) and weighed 75–80 metric tons, this meant an incredible size increase of five orders of magnitude from hatchling size, more than any known living or extinct amniote. So sauropods were stuck with having to produce very small babies in relation to their enormous optimal body sizes, and the only solution to this challenge was to grow fast. As with the annual growth rings in trees that tell age, some vertebrates produce a line for each year of their growth in their long bones. Many individuals are needed to construct a growth curve because the inner growth lines are lost to bone remodeling, but when a large enough sampling is available from individuals of the same species, a rough estimate of the individual's current age at the time of death may be possible. In sauropods these growth marks are rare and sometimes only occur during later **ontogeny** (*ontos*, self; *genesis*, origin), but findings suggest that some species attained their full adult size at an age of four decades. We also now know that juvenile sauropods grew up fast—very fast. Discoveries of fossil sauropod juveniles are rare, but well-preserved skeletal remains show that these, like those of young mammalian and avian endotherms, are highly vascularized and have lamellar bone, a sure sign of rapidly growing tissue that required a rich supply of nutrients. Histology studies indicate that sauropods also had *even higher* growth rates than other fast-growing dinosaur babies like hadrosaurs. This explains why growth lines or marks are rare in juveniles, since growth was nonstop and didn't pause to form lines until they reached an older age. How fast was fast? Calculations based on correlations of weight gain to skeletal growth in modern endotherms and applied to sauropods show that this was anywhere from an amazing 500 kg (1,102 lb) to *2,000 kg (4,409 lb), or 2 metric tons, per year*. This is in the range of juvenile whales, which also put on an astounding amount of weight and length as they become adults. One remarkable thing about both the breeding female and male *Ampelosaurus* is that they were most often adolescents, "teenagers" that were not even close to approaching their probable full adult ages or sizes, which may have been at least 40 years and approximately 13 m (45′) and 12 tons.

Sauropod hatchlings started life with short necks, big heads, eyes and nares and the probable ability to rapidly amble bipedally, a survival plus. The ontogeny of *Camarasaurus* (*this page and p. 189, bottom*), among the best-known and most well preserved species, documents this.

(1) 8- to 12-month juvenile
(2) subadult
(3) young juvenile

In living vertebrates, growth slows as an animal reaches adolescence, and at a certain point it generally corresponds with the onset of sexual maturity. In the sauropods studied this seems to have taken place during the second decade of life, about the same time as that of a large modern herbivore such as an elephant, but whose body growth curves peak at a much earlier age and size.

STRATEGIES FOR SURVIVAL

In the animal world there are generally two types of reproductive strategies, each of which is based on the best chances for a particular species' survival. If you're an animal that's small, short-lived and vulnerable to predation both as a juvenile and as an adult, your species will have a better chance of survival if you have large numbers of young at the same time as other breeding females in your area—we see this with rodents and other small mammals. When lots of your young appear at the same time, many may fall to predation, but even if only a few escape and live to reproduce, it will guarantee the survival of your species. If you can't generate a large number of babies at one time, have several smaller clutches during the breeding season, in different locations, so that they'll be less likely to be found and will then have a better chance. Your clutch or litter has an even *further* advantage if the young don't require a big investment of your time and care (now you can get back to producing another one) and are precocial or can survive on their own as quickly as possible. In modern biology the practice of producing relatively few, slowly growing young that are nurtured by one or both parents for an extended period of the juvenile's lifetime is known as a *K strategy*, in contrast to an *r strategy*, in which the female parent produces an overabundance of fast-growing, quickly maturing young. Although they're subject to high predation, the latter are apt to survive in enough numbers to continue the species.

Although reasonably complete juvenile skeletons are rare, specimens such as the baby *Apatosaurus* (*right*) and "Max" (*below*) give an idea of how young animals looked. On "Toni," a *Brachiosaurus* (*inset*), the large orbits and nares are juvenile features, indicating a relatively young animal.

If as an adult you're somewhat bigger and not as likely to be eaten (or to die from something else) very soon, you'll have a better survival chance, but your numerous, small young are still vulnerable. Early in their evolution, marsupial and placental mammals took the path of retaining the eggs, embryos and later fetuses inside the female parent's body to the point of an advanced state of development before the young were born. Bigger, more developed babies, of course, meant that there was now less room inside the body for a large litter, so selection occurred for a more limited number of young. As some mammals evolved to become larger and longer-lived, it was possible to offer their young more protection and care. More parental care, in spite of fewer offspring, made the offspring more likely to survive, and in the largest forms this corresponded to a delay in sexual maturity and the birth of even fewer young, whose even larger size and longer gestation periods meant that breeding happened less often. This strategy has generally worked well for big-bodied mammals, but the disadvantage is that when there's an extinction threat, there are fewer individuals of breeding age around that can produce offspring fast enough to maintain the population size during a die-off. The species may then go extinct. In animals that reproduce quickly and abundantly, there are more individuals to make the species bounce back if it hits bad times.

If we could travel as field biologists back in time to the Late Jurassic of western North America, Europe or Africa to study sauropod populations, what would we see? Having observed the abundant elephants and other mammalian big game herds so typical of *Animal Planet* programs, we might be amazed to see so few mature sauropod adults, while half-grown subadults and small juveniles would be very common. It would be the reverse of what we'd see in a herd of big present-day mammals, in which adults predominate in large numbers while there are only a few juveniles. This sauropod scenario would be the result of a reproductive strategy in which many young would be produced on a frequent basis, and reproduce early themselves, before reaching full maturity and size. Instead of two to three elephant cows in a herd of 10–15 giving birth to a single calf every year and a half to two years, we would see perhaps dozens of adolescent, or "teenage," sauropod cows laying multiple egg clutches collectively numbering in the hundreds or thousands each annual season. Unlike the elephant calves, which enjoy years of protection from their mothers and have an excellent chance of becoming reproductive adults themselves, the baby sauropods would be severely decimated by predators, with only a small number living long enough to mate and reproduce. For elephants and humans, a female must have about 2.1 offspring, on average, to balance the mortality rate and keep the population from declining. If sauropods were to likewise keep their numbers constant, a female would have to produce around *1,000 eggs during her lifetime*—500 times more offspring to keep the species afloat. Wasteful, yes, but it worked. Enough young survived, and long enough, to ensure that the race continued. Furthermore, because so many individuals were around to become reproductive at an early age, populations of sauropods that made up a given species could potentially replenish themselves fast if something occurred to threaten the survival of the entire species—a prolonged draught, a temporary die-off of plants that provided most adult foods, or the appearance of a new type of predator.

GIANTS IN THE MAKING

Do we know anything about the "gap years" in a juvenile sauropod's life, between the time it was very small and when it became a fully grown adult? At this point, there's not much, but we can speculate. One probability is that very small juveniles, aided by their cryptic patterns and coloration, kept, like baby crocs and alligators, a low profile by staying within areas of dense vegetational cover. If their main food source was *Equisetum* and possibly other nutritious plants, these would have naturally occurred in moist areas in or near these environments. As they rapidly became larger, their chances of survival would have gradually improved, but even as they outgrew the threat from resident forest predators (see Chapter 9), there were probably still others, lurking in the marginal open woodlands, that took their place. If luck was on its side, by age 3 a young sauropod would have reached the size of a large cow or draught horse. At this time the juvenile's size and need to possibly wean itself onto other, higher browse would have increasingly forced it into more open environments. In this stage, the original cryptic

On its way to becoming a full adult, a fast-growing *Camarasaurus* (*below*) underwent a change not only in absolute size but also in its body proportions and probably color and patterning. Baby sauropods were big headed and short necked and, like a hatchling alligator or ostrich (*above*), depended on concealing markings to avoid detection. The juvenile colors and crests and dewlap of the adult shown here are conjectural.

It was once speculated that the spectacular titanosaur nesting grounds of Auca Mahuevo, Argentina, were also nurseries, with adult sauropods furnishing regurgitated vegetation and protection against predators for the hatchlings. A number of factors make this scenario extremely unlikely if not impossible, and now it's known that the babies were on their own as soon as they hatched out.

patterning of streaks and dappling that mimicked filtered light on leaves might have gradually transformed itself into a *disruptive* one. In disruptive patterning, large, contrasting, irregular shapes divert a potential predator's attention away from a prey animal's characteristic shape (in this case a big body, long neck and tail) and make the overall contours less noticeable—like a human hunter's or military vehicle's camouflage. If greater size began to be a deterrent against an occasional killer, these older sauropod juveniles may have also started to exhibit counteroffensive behaviors, such as tail lashing, kicking with their powerful, clawed hindfeet and rearing to threaten with "thumb" claws, depending on the species. In addition, sauropods, often depicted in films like *Jurassic Park* and popular documentaries as cow-like "gentle giants," may in reality have had fairly nasty, aggressive temperaments when it came to deterring predation (as well as intimidating others of their own species). By analogy, brown bears (*Ursus arctos*), which spend most of their lives in open environments, are very aggressive compared to their fairly shy, retiring relatives the black bears (*Ursus americanus*) of forested environments. This difference may be at least due to the need at one time to have actively defended themselves, without the benefit of protective cover, from now extinct potential aggressors like dire wolves, sabertooths and short-faced bears, while in modern Siberia brown bears must occasionally face down tigers that attempt to prey on them.

With room to maneuver, an open environment might have made these tactics more likely to succeed. This also could have been the point at which sauropods might have started to become gregarious, sticking together for mutual defense: several pairs of alert eyes on top of tall necks would have picked up a predator's presence in enough time to mount a group defense. We know from discoveries of other juvenile dinosaurs, whose remains were uncovered in a group of animals about the same size and age at fossil sites, that this sometimes began when the young were still very small (a group of babies of the ankylosaurid *Pinacosaurus* were found buried together in Mongolia, apparently smothered in a collapsed sand dune). At the Mother's Day Quarry in south-central Montana, a group of diplodocid juveniles were uncovered that included several individuals that fell into two main size categories, one 75% (older juveniles) and the second 91% (subadults). This suggests that at least some sauropod types were gregarious, or, from an early age.

Assuming that at some point a given sauropod species was naturally gregarious, and that solitary individuals might later have begun to associate together to form what we'd call a herd (or pod), did older, larger juveniles ever mingle with more fully grown adults, and if so, at what point? In a discovery by Roland T. Bird (American Museum of Natural History) in the 1930s of trackways from Davenport Ranch, Texas, what seemed to be groups of small juvenile footprints were described to be in the center of much larger, adult ones going in the same direction. At first it was concluded that adults surrounded the babies to protect them as they traveled. The problem here, as pointed out by trackway specialist Martin Lockley (University of Colorado, Denver), is that the tracks of the some 23 different individuals show considerable overlapping, in such a way as to show that the small and large sauropods didn't pass through this location at the same time. In spite of this, there are correctly identified associations, as in the case of *Apatosaurus* and *Camarasaurus*, of fully adult bones with those of subadult or older juveniles that were in some cases 25% smaller. Not knowing the circumstances (the adult and juvenile bones could have occurred together through becoming separately deposited), however, we can't claim this as proof of juvenile–adult gregariousness. Sauropods were so different from one another and evolved over such a vast expanse of geologic time that although it's likely that there were many variations in their overall biology and behavior, in at least the large to gigantic forms that had to grow quickly juveniles probably didn't mingle with adults until they were much bigger, if at all. One problem with having very small young in the same group as adults would be the danger of an adult accidentally stepping on a baby. Another is that, since adults were probably always on the move to find yet more browse, it would have been a strain on juveniles, with their relatively shorter legs, to have kept up with the much longer strides of the adults. Finally, habitats that provided the right kind and amount of high browse for mature herd members might not have provided food or cover for young animals.

This doesn't necessarily mean that different sauropod age groups strictly separated themselves, however. As pointed out in Chapters 6 and 10, young sauropods probably weren't hatched with the microbial flora in their guts that they needed for digesting the adult arboreal browse preferred by each species. To acquire this, they may have deliberately eaten the feces of adult sauropods they encountered to obtain the microbes for their own systems, an instinctive behavior called *coprophagy*. If this happened, it means that juveniles couldn't have been totally segregated from adult habitats. Food preferences, although perhaps instinctive and controlled by the ability to physically reach for each species of sauropod, may also have been an acquired habit with juveniles as they grew up and informally associated with adults, eating what they ate. If they got this far, the pods of young sauropods were well on their way to surviving and approaching their reproductive years, and in making it to this point they could be considered true biological success stories. Seen in a different light, however, they were also multi-ton cans of quality animal protein sitting on the table. In the next chapter, we'll meet the can openers.

By chance, a wounded *Camarasaurus lentus* calf and its companion gain protection among nearby older pod members from attacking *Marshosaurus bicentesimus*. In spite of this, sauropods did not care for their young as do today's mammals, and the mortality rate of juveniles from predation was very high.

CHAPTER NINE
PREDATOR AND PREY
THE ANCIENT RACE

From even before the time they hatched until they died, sauropod dinosaurs were prime menu selections for a variety of predators. Depending on a sauropod's age and size, these predators took many forms, but the most common and adaptable were their carnivore cousins, the theropods, and during their evolution into giants, the sauropods evolved effective defenses. When a sauropod was finally brought down by predation, its huge body became food not just for theropod hunters but also for a wide variety of organisms.

DANGER OUTSIDE THE EGG. The threat to the baby *Camarasaurus lentus*'s lives began well before hatching. If it was anything like that of a modern crocodile or sea turtle, for a few minutes to half an hour or so before releasing its new lives the nest produced a faint odor. This was mainly from the smell of the egg contents clinging to the hatchlings and also from the metabolic gases and other wastes produced by the growing embryos inside the eggs. Discernible to humans in the case of crocs and turtles, and very much so to any local raccoon, monitor lizard, feral pig or dog that may chance by, it's an invitation to a meal. Croc and alligator mothers stick around the nest to defend their eggs, but in the case of turtles' unattended nesting colonies, only the sheer numbers of nests and quantities of eggs ensure that some will survive to hatch. If discovered, most of the eggs taken are nearest the surface, while the location of those deeper down buys them time until they're ready to hatch. Around the vegetated fringes of the sprawling sauropod nesting colonies at Auca Mahuevo and elsewhere around the world, there

Although mammals were generally small during the Mesozoic, some were large enough to actually prey on small juvenile dinosaurs. The powerful, badger-like *Rapenomamus robustus*, seen here killing abandoned theropod nestlings, would have used its probably keen sense of smell to search out and kill sauropod nestlings and very small young if it had the opportunity.

were would-be egg stealers. These probably included, as they do today at croc nesting sites, large predatory lizards like *Palaeosaniwa canadensis*, *Estesia mongoliensis* and others that resembled true monitor lizards (goannas or perenties). Although currently unknown from the sauropod egg sites, monitors are characteristic scavenger-predators of the modern tropics and subtropics. Keen of scent, fast, and powerful diggers, species like the Nile monitor (*Varanus niloticus*) haunt the riverbanks of east African rivers, where they take Nile crocodile (*Crocodylus niloticus*) eggs, as well as hatchlings, from unguarded nests. At this time there were also long-legged, fox-sized terrestrial crocs with short jaws like *Fruitachampsa callisoni* and *Araripesuchus wegeneri*, both of which may have haunted the nesting grounds. In competition with these were primitive, triconodontid mammals like *Gobiaconodon mongoliensis* and *Rapenomamus giganticus*, badger-sized, stocky animals with powerful teeth and limbs, well suited to the role of opportunistic predators. *Rapenomamus* is known to have been a killer of baby dinosaurs, since the fossilized abdominal cavity of one species, *R. robustus*, was discovered containing the remains of baby psittacosaurids, basal ceratopsians. At least one early boid, or boa constrictor–like snake, *Sanajah indicus*, was a definite sauropod egg and hatchling predator: its 3.5 m (10′) fossilized skeleton was found coiled around eggs of titanosaurids at a Cretaceous nesting site in Gujarat, India. Another possible large snake predator was *Dinilysia patagonica*.

As within the egg, on hatching baby sauropods were vulnerable to all these predators, but in making their dash to cover they would have attracted the notice of still more. Modern gulls and other opportunistic shore birds, with their high vantage points from the air, are among the first to see baby sea turtles break out. Although many known pterosaurs ("pterodactyls") may have been fish eaters, some larger forms might have snatched up and flown off with the puppy- to lion cub–sized nestlings in their jaws. Then, as now, there were 2.4–3 m (8′–10′) freshwater crocodiles *Eutretauranosuchus (Goniopholis) gilmorei* and other species, distant relatives of *Fruitachampsa*, which could have intercepted a baby or two before they reached cover.

Mesozoic woodlands and forests like those of today probably contained several types of small-area habitats, or *microhabitats*, compared to the larger, more homogeneous ones encountered in floodplains, deserts and other open environments. Some of these microhabitats have stable, less changing plant communities in warm climates and are often inhabited by their own set of predators. Here the hatchling sauropods, protected only by their cryptic patterns and inconspicuous size, would have been at risk from the resident boid snakes and triconodont mammals, as well as small to medium-sized theropod dinosaurs.

MEET THE THEROPODS

The theropods, the sauropods' saurischian cousins who had for the most part remained dedicated carnivores, were the greatest threat to sauropods of all ages and sizes. All of these had a highly developed sense of smell and excellent vision and hearing, and they probably included species that hunted at night as well as by day. Depending on the areas of Laurasia and Gondwana and the periods in which they lived, there were **guilds** of both generalist and specialist predators, all capable of killing juvenile and adult sauropods. A *carnivore guild* is a term borrowed by behavioral biologists from the medieval word for a group of local craftspeople or professionals that once specialized in making certain products without directly competing with one another. In this case it means a localized group of carnivores that are all after flesh but have evolved certain physical and behavioral adaptations that make certain types best suited to capture particular prey. Among modern animals an example of this would be among African big cats: smaller, light-weight cheetahs are specialists in running down fast prey like Thompson's gazelle, ambush hunters like leopards take bigger antelope, while the largest cats, the lions, routinely tackle herbivores

Both before and after they hatched, baby sauropods were vulnerable to small opportunistic predators such as monitor-like lizards (*above*) and snakes like *Sanajah* (*left*).

as big as Cape buffalo. There's no direct proof for it, but based on the variety of theropod species known from a particular time and place, it's reasonable to think that theropods also formed guilds. Because carnivores both large and small are and were opportunistic, or willing to try different behaviors to achieve this end, their roles weren't rigid and sometimes could and very likely did blur together. Small and large theropod species might have been like modern wolves and Harris's hawks, which aren't always solitary hunters and may cooperate in groups to bring down prey. During the Mesozoic era there were several, ever more biologically sophisticated groups of theropods that arose and gradually replaced one another. Some of these stayed small, but others produced medium-sized to giant hunters that have no equivalent among today's mammalian predators. This made things very tough for sauropods, since no matter how big they grew, either as individuals or as species, they had to face some kind of potential theropod predator.

Many of the smaller, chicken- to turkey-sized **coelurosaur**s were all fast, alert and equipped with clawed, grabbing hands, as well as sharp teeth, ideal for overcoming small prey, anywhere from insects to small vertebrates. Either solitary hunters or possibly gregarious like modern crows, smaller coelurosaurs might have descended on a nesting colony when the hatchlings' breakout occurred, each killing and running off with its prey. Moving on up the size range were **ceratosaurids** and **abelisaurids**. Although these also evolved into giant carnivores, some were small to horse-sized, like *Elaphrosaurus bambergi*. Theropods of these types and size ranges were the ones likely to have patrolled the margins of forests and woodlands, looking for any victims. At least one large species, *Aucasaurus garradoi,* may have preyed on adult titanosaurs as they laid their eggs, as well as hatchlings, since a partial adult *Aucasaurus* skeleton was found at the nesting colony in Auca Mahuevo. Other, later theropods took the form of **dromaeosaurids**, popularly known as "raptors." Of these, the wolf- to polar bear–sized deinonychids like *Deinonychus antirrhopus* and *Utahraptor ostrommaysi* may have been the most deadly, since finds of several individual *D. antirrhopus* skeletons with large prey like *Tenontosaurus tilletti* suggest that at least some were pack hunters like wolves, using their specialized, sickle-shaped first digit pedal claws to tear or disembowel larger dinosaurs. Dromaeosaurs, like modern African hunting dogs, might have preferred flatter, more unobstructed terrain like floodplains, where their speed might have given them an advantage. Finally, there were the large to enormous potential hunters of older juvenile and adult sauropods that included the **carnosaurs,** typified by *Allosaurus fragilis* and *Acrocanthosaurus atokensis*, as well as the bigger ceratosaurids, abelisaurids and huge **coelurosaurian avetheropods** like *Tyrannosaurus rex*. Carnosaurs, because of their massive size, were possibly best suited as ambush predators like tigers or leopards, hiding just within the edges of taller, vegetated areas. As with the smaller forms, their probably ratite-like feathered

skins would have featured streaks, blotches or other disruptive patterns to conceal their size and matched with their cover to let them make a sudden rush or sprint to capture unwary prey. Some medium-sized to large theropods, especially if they were pack hunters like dromaeosaurs, might have stuck more to the open habitats, where they would find more of a caloric payoff from large, more easily spotted prey, including solitary sauropod juveniles.

Just as the sauropods became adapted to herbivory, early theropods had begun to develop carnivorous specializations of their own. For small, Late Triassic forms like *Coelophysis bauri,* this took the form of teeth that could both puncture and also tear, thanks to small serrations along the convex side of the larger, upper teeth. Employed together with a series of rapid bites, the slender theropod could inflict a series of lacerations on its prey that caused loss of blood, weakening it to the point where it couldn't defend itself—a decisive bite was then all that was probably needed to finish it off. *Coelophysis* was probably one of the theropods that preyed on early sauropodomorphs like *Anchisaurus polyzelus* and *Ammosaurus major,* which lived in semiarid, rift valley habitats with occasional lakes.

A theropod predator's teeth and killing techniques were generally adapted to work in one of four ways. Many large forms followed *Coelophysis,* with **laterally compressed teeth** designed to pierce and slash, like *Allosaurus* or *Giganotosaurus (Carcharodontosaurus) carolinii.* Although the teeth in these forms were modest in size, the arms were sometimes robust and powerful. This was especially the case with *Allosaurus,* the most common large theropod of the Morrison Formation in western North America, in which the first manual digit bore a huge, curved claw. Capable of clenching and holding, its hand could hook and hold prey close within reach of the flexible neck and powerful jaws. *Ceratosaurus nasicornis,* although it had small forelimbs, possessed a huge, deep skull and **enormous, bladelike teeth** for deep penetration. Seemingly rare

The sauropods' most important predators were the bipedal theropods. These came in a great range of sizes and specializations, ranging from small, quick types like *Coelurus* (*above*), *Deinonychus* and other medium-sized to large dromaeosaurs, to much larger allosauroids and abelisaurids such as *Allosaurus* (*opposite*), and finally ending with gigantic avetheropods like *Tyrannosaurus* (*below*). All theropods preyed heavily on both juvenile and adult sauropods, both singly and probably in packs.

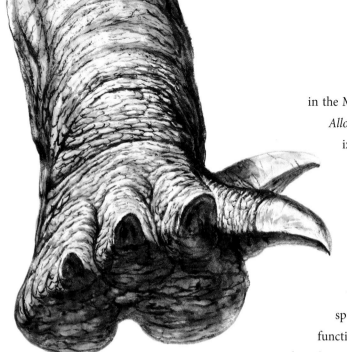

Present in all sauropods except for some titanosaurs, the innermost first (thumb) digit bore a claw, held up off the ground to keep it sharp. It could be rotated downward and would have been an effective weapon.

in the Morrison Formation fauna in comparison with the generalist hunter *Allosaurus*, *Ceratosaurus* may have been a solitary stalker that specialized in killing small to larger sauropods. Like *Allosaurus*, its relatively short hindlegs suggest that it wasn't fast, and it is likely to have been an ambush predator. In the third adaptation, the jaws probably did all the work. Abelisaurids like *Abelisaurus*, *Aucasaurus* and *Carnotaurus* had **short, stout teeth** that, along with the broad upper and lower jaws, were intended to clamp and hold. In contrast to the allosauroids, known abelisaurid forelimbs were small and atrophied, totally useless for grappling, but their long hindlimbs gave them more speed when they took off after prey. The possible piercing and holding function of the teeth, together with the broad skulls in these forms, may have been an adaptation to kill by clamping around the *trachea* (windpipe) and major blood vessels. In the fourth variation, bite strength was taken furthest by giant coelurosaurian avetheropods like *Tyrannosaurus rex* and similar forms, which had **long, massive teeth that were D-shaped anteriorly**, rather than laterally compressed, in cross section. Instead of slicing like steak knives, these were intended to tear out chunks of flesh. Backed by the huge, solid skull, these teeth could crush with tremendous force through bone as well as flesh.

It's also possible that many, perhaps even most, theropod predators had septic bites. The copiously dripping saliva within modern Komodo dragons' (*Varanus komodoensis*) mouths contains a potent broth of at least three different kinds of microbes, all of which begin to multiply and fester after they've been deposited by the huge lizard's lacerating bite. Within days the bitten prey develops *septicemia*, or blood poisoning, and either dies from this or is finished off by the dragon. A septic bite could have worked for theropods as well. Some dromaeosaurs, like some modern rear-fanged snakes and Gila monster (*Heloderma suspectum*) lizards, may have even been capable of injecting or pushing venom into a wound with grooves in some of their upper teeth. Venom injection in this case was perhaps more of a specialization to immobilize a fast-moving prey before it could escape. A slow sauropod wouldn't have made this necessary, but even a minor septic bite could have been lethal to the largest sauropod adult.

PREDATOR VERSUS PREY

To be successful and reasonably long-lived, a predator must become good at making cost/benefit judgments. It learns quickly, and often painfully, about whether it can overcome the animal it wants to eat, and this knowledge may come too late. If an experienced, adult prey has any reasonable defenses like claws or the ability to inflict a serious bite, the predator risks injury or even death every minute a kill is delayed. Among modern and extinct mammals, there are many proofs of this in the form of broken jaws, broken teeth and other injuries. These often show that if the accident didn't kill outright, it maimed the predator for life, impairing its ability to hunt and causing it to starve. A large, healthy animal in its prime takes longer and is harder to kill than a smaller, younger one, as well as one that is old, sick or crippled. Most modern predators try to make their kills with the least amount of effort and to minimize risk of injury or death.

Because we have nothing alive today that comes close to a small or giant theropod, we have to imagine what they might have been like as hunters by looking at some modern and extinct forms. The only medium-sized, terrestrial, bipedal theropod relatives that habitually hunt on the ground today are the South American seriema (*Cariama cristata*) and the African secretary bird (*Sagittarius serpentinus*). Both these birds include small vertebrates in their diet, which they kill by pursuing their prey and then stunning it with their feet or biting with their beaks. This would have probably worked just fine for small theropods preying on sauropod hatchlings, but for older individuals a quick bite to the throat, after gaining control of the head with their grasping hands, might have been required. Because larger theropods' bipedality made them potentially unstable, these may not have risked the use of ripping with one clawed foot while balancing on another, but instead relying on their forelimb talons and teeth. To minimize possible injury to themselves, medium to giant hunters may have used an "attack and retreat" technique, as theorized by Larry Witmer (Ohio University) for some extinct Cenozoic "terror birds." These seriema relatives were as large as 181.4 kg (400 lb) and had huge, deep beaks with hooked tips like hawks and eagles. In attacking prey, they might have thrust their powerful heads forward to make a ripping bite into the prey's flesh and then retreated momentarily, repeating this until the animal was killed or weakened from blood loss. As described

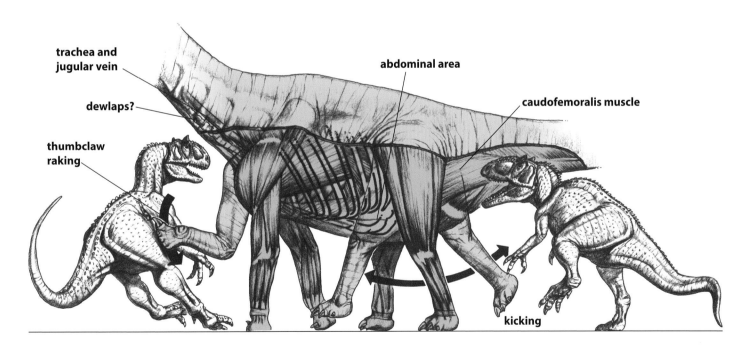

Vulnerability and Defense

Because no healthy adult sauropod could be killed outright, a theropod or its pack probably sought to first injure it in a key place like the underside of the rear belly (*black lines*), where a lacerating bite might cause massive bleeding or disembowelment. Other vulnerable places were the caudofemoralis muscle (important for power and relative speed) and the base of the neck. To counter this, a sauropod could kick forward and backward if a theropod came close enough to attack the belly or tail base, and lash out and down with its thumbclaw to protect the trachea and main blood vessels (*arrows*). Known to exist in some mummified hadrosaurs' skin, some sauropods may have possessed large, loose dewlaps at the neck base, which, as in Shar-Pei dogs, might have made a bite to the neck ineffectual.

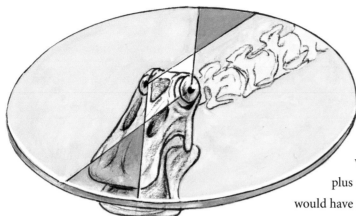

Although they were located at the sides, the fact that sauropods' eyes were positioned near the top of the head above the muzzle resulted in overlapping fields of view. This was particularly true of at least one small, ground-level browser, *Nigersaurus* (*above*), whose eyes were the most dorsally located of all. This gave them considerable visual overlap and, in the absence of a long neck, may have evolved to help detect nearby predators.

for *Coelophysis*, a slash-n-bite approach would have also worked well for toothed theropods. In imagining a hunting theropod we should also think of another archosaur relative, a crocodile. Although they might seem clumsy and water bound, they are masters of patience and stealth, investing long periods of time to bring themselves close enough to ambush their prey. When they do, they lunge forward with amazing speed, grabbing prey with powerfully muscled jaws. Any predator with these qualities, plus grasping hands, lightning-fast jaws and bird-like, bipedal agility, would have delivered a devastating attack.

Juvenile sauropods of all types were prime victims because of their superabundance and relative helplessness until they became older (see Chapter 8), supporting a wide spectrum of small to huge theropod species. In the well-known Morrison Formation *Allosaurus* may have been the most common large predator, with medium-length teeth that made it capable of killing larger sauropods but with legs and size that allowed it to be fast enough to capture juveniles, a generalist predator. If the rarity of its skeletal finds is an accurate indication, *Ceratosaurus* was much less common. Equipped with huge, bladelike teeth, it may have been a solitary hunter of larger sauropods, as was possible with the large *Torvosaurus*. At this point we can't be absolutely sure what theropod species might have been likely to have preyed on a given type of sauropod, but new finds in Tendaguru, Tanzania and Lourinha, Portugal, of different but similar species may provide a better picture of interspecies relationships.

Did sauropods have any defenses at all against theropod teeth and claws? Having coevolved over many millions of years with one kind of theropod or other that wanted their flesh, sauropods couldn't just be enormous, passive eating machines. As adults, *Anchisaurus* and other early to later prosauropod sauropodomorphs were well able to defend themselves. The tall, habitual rearing postures that they evolved to access arboreal browse and their strong, stoutly clawed forelimbs could inflict terrible damage, similar to the way a modern giant anteater (*Myrmecophaga tridacytla*) can rip into a dog or jaguar. These could also hook an attacker to bring it within range of the long hind claws, which then could disembowel in the manner of a cassowary or large kangaroo as it reared up on its strong tail. The long muscular tails of prosauropods and later sauropods were also important weapons. In spite of their speed and maneuverability, the great weakness of theropods was that they were obligate bipeds. A well-timed, well-directed blow from a long, massive tail could break a leg or at least easily knock them over and upset their balance while running or standing. As theropods grew larger with time, the consequences of this became more serious. While a small theropod like *Coelophysis* could right itself quickly and resume an attack or break off, a huge, bipedal form the size of *Allosaurus* or larger risked a broken rib, or worse, from a fall. At the very least it would have taken several seconds or minutes to get back up, which might have given its sauropod prey enough time to wheel around, rear up and come down on its forelimbs to crush its opponent's rib cage. In many sauropods, depending on age and size, the mass of the tail was equal to or greater than the mass of a typical adult theropod's entire body, making the tail a potent force.

While a theropod's Achilles heel was its bipedality, for a sauropod it was its slow speed. Medium-sized to large prosauropods were probably capable of a fast bipedal sprint, but early in their evolution the trade-off for primitive sauropods' increasingly large size and stability became their inability to run. Big species' long strides helped them cover more ground, but probably no sauropod species ever topped 40.2 km per hour (25 mph). Although young juveniles were virtually defenseless, older adolescent and adult sauropods almost certainly used their long, massive tails as primary weapons, and probably their powerful hind legs to kick backward and forward at an attacker. With their throats and heads largely out of reach, a moving sauropod's most vulnerable area may have been in back of the thigh: here a bite to the long *caudofemoralis* and flexor muscles, the main powerhouses for walking, would cripple the prey, almost immobilizing it and making it impossible to rear. A well-directed kick from a huge clawed foot would have covered this area and knocked out a big, unstable, bipedal predator if it came too close. All sauropods had huge areas for muscle attachment on their pelvises, but in one species, the macronarian *Brontomerus* (literally, "thunder thigh"), these were enormously expanded. An angry adult or even an older juvenile could kick backward or forward with great effect.

Some sauropods' tail *tips* also aided in their defense. In addition to having long, heavy tails, some Asian species such as the basal eusauropod *Shunosaurus lii* had short, dorsal spikes atop a bony club, while the African *Spinophorosaurus nigeriensis*, like stegosaurs, had four, laterally paired spikes, each of which was double pointed. Others, like the fantastically long-necked *Mamenchisaurus* and possibly the closely related *Omeisaurus*, may have had tiny but hard and bony clubs that could have been well coordinated to strike and injure a theropod's eyes and teeth. Although diplodocoids like *Diplodocus* and *Apatosaurus* had no spikes or clubs, they possessed unusually long, slender tail ends that they may have snapped like a whip, possibly lacerating an attacker's face, or at least creating a startling sound. With or without these structures, the long, supple end of the tail could have been directed at a theropod's body and head, either knocking out teeth or hitting the eyes. Today's crocodilians and some lizards likewise use their tails as weapons by lashing them sideways.

Protection against theropod attackers later *may* have been in the form of defensive armor. Some titanosaurids evolved lumpy, small to large osteoderms, vascularized and porous like the ones in modern crocs and alligators, which were imbedded in the skin and often found in isolation or in close association with a skeleton. That so few are usually found has suggested to some workers, however, that these probably didn't form a defensive covering of armor as in ankylosaurs. The osteoderms, because they were well supplied by blood, instead may have helped a titanosaur to shed excess heat by bringing warm blood to the skin surface and outside air, and to store needed minerals (see Chapter 11). In spite of this, anything to harden the target would make this an adaptive defense, since even some hard lumps embedded in the skin in strategic places would be much more difficult for teeth and claws to penetrate. There were also other weapons: bizarrely long, sharp dorsal neck spines are known in at least one dicraeosaurid diplodocoid, *Amargasaurus cazaui*. These were directed up and backward, suggesting that they might have been used to stab or skewer a predator that tried to attack at the shoulder or base of the neck. Modern giant sable antelope (*Hippotragus niger*) and Arabian oryx antelope (*Oryx*

Harris's hawks, here attacking a jackrabbit, are one of the few birds of prey to sometimes hunt cooperatively, suggesting the possibility that some nonavian theropods may have done the same to bring down sauropod giants.

dammah), with their long, backward-sweeping horns, are known to use this technique against lions, which can be stabbed when the antelope suddenly extends its neck backward during an attack. Because sauropod juveniles perhaps early on grew up as isolated individuals and only later in similar-age groups, defensive behaviors would at first have been purely instinctive and only learned from the example of others when older. One very early protection could have been similar to the defensive "freeze" seen in fawns when they sense danger. Combined with their probably cryptic coloration and patterning while hiding among ferns or other forest cover and a possible lack of scent, this might have helped the juvenile evade discovery by a predator.

SENSES FOR DEFENSE

To detect predators, sauropods probably had highly developed senses. All of the known forms had large orbits, or eye sockets, suggesting that big eyes and keen vision were important in giving advance warning of predators. That these were mounted on tall necks must have helped, giving most, except for low browsers like the small dicraeosaurid *Nigersaurus taqueti*, the advantage of height like today's giraffes in seeing attackers from far off. Although other species' orbits are located high on the skull to give as wide a visual scope as possible, those of *Nigersaurus* are among the highest in placement on the skull. With the eyes mounted on the sides, this created an overlapping visual field of at or close to the maximum 360 degrees. Like some modern ungulates, the pupils in some species might have been shaped like horizontal ellipses rather than being typically round, as in most mammals, to gather in as wide a visual field as possible. Prey animals with this arrangement, although usually nearsighted, are extremely sensitive to nearby small movements. Hypersensitive movement detection would have been vital to a small, low-browsing sauropod species whose relatively long, subhorizontal neck made a tempting target. Since color vision exists in most reptiles, it's probable that sauropods and many other dinosaurs also retained this. Color vision would have possibly aided in distinguishing predators, playing a role in assessing vegetation edibility and determining a potential rival or mate's readiness to fight or breed. In arboreally feeding sauropods, eyes may also have been shaded by long, keratinous upper eyelid filaments, like an ostrich's or mammalian ungulate's, to block out sun glare and protect against scratches.

Because the olfactory bulb, the anteriormost extension of the brain that processes odor, shows up as being well developed in known sauropod CAT scans and endocasts, it's likely that smell may also have aided in predator detection. Eusauropods and macronarians had large, moderately retracted bony nares that may have housed special air-moistening or moisture-conserving organs. Although there might also have been a structure within the bony nares that played a role in odor detection, at this time we don't know.

Sauropods could also possibly have used *sound* as a means of warning each other of the presence of a predator. Avian archosaurs have a unique sound-producing organ called the *syrinx*, which actually produces the sound, while crocs, like birds, have a *larynx*, or voice box. Somewhat like that of a mammal's, this produces sound. The indirect evidence of skull structures in hadrosaurs and other dinosaurs indicates that they were capable of amplifying sounds, while some dinosaurs' internal ears, in the form of a *tympanum* and *columella* (or *stapes*, inner ear bone, equivalent to a mammal's stapes), were adapted to hearing low-frequency sounds. With their long tracheas, capacious lungs and probably strong voices, most sauropods were potentially capable of creating deep, resonant calls, as well as other cries that could have alerted others in the pod of a hunting theropod. Such sounds might also have been instinctively understood by juveniles that were within hearing range, prompting them to adopt the defensive "freeze" mentioned earlier. If some species had convoluted lower tracheas like modern trumpeter swans, these would have increased sound resonance even more, and it's possible that the large nasal chambers of some macronarians might have housed some type of cartilaginous amplification structures.

Another possibility is that sauropods may have used *infrasound*, the creation of very low frequency sound waves, to warn of predators, as well as to communicate in other ways, over long distances. Infrasound has been known for many years as a way in which larger whale species communicate their presence to each other across the vast distances of the oceans, but recent studies by Peter Wrege (Cornell University) of the Elephant Listening Project and other research groups show that elephants, too, use this technique to contact herd members when they are separated. An elephant creates seismic vibrations (*Rayleigh waves*) as a deep rumble in the range of 14–35 Hz, traveling through the ground, as well as the air, in all directions at about 248–264 km (154–164 mi) per second; these have a range of up to 48–60 km (30–40 mi). How do they produce it? As large-bodied animals, elephants can summon up a lot of air volume in their huge lungs, which is forced across the larynx, whose muscles modulate the pitch and frequency of the sound. Elephants also have a looser *hyoid* (throat bone) structure than most other mammals, which allows for greater flexibility of the larynx, making it easier to produce and control the resonance of low-frequency sound. The sound is amplified by their long trunks and a *pharyngeal pouch* at the base of the tongue, which add even more resonance. In elephants sound is detected not only with superlarge *auditory ossicles* (sound-conducting bones) in their ears but also through layers of specialized cells in the skin of their feet called *Pacinian corpuscles*. Also found at the trunk tip, these cells are deformed by sound vibrations, and when this occurs, a signal is sent to the brain. Elephants and their distant relatives the "sea cows" (manatees and dugongs) are unique among living mammals in having a more saurian-like *cochlea* (an inner ear structure), which may facilitate a greater sensitivity to low sound vibrations and frequencies. There's no evi-

Based on field excavations of several differently aged *Mapusaurus* individuals found in one locality, theropod specialist **Phil Currie** has advanced the strong possibility that this giant carcharodontosaurid may have hunted in packs, allowing it to bring down sauropods that might have included gigantic *Argentinosaurus*.

dence that sauropods had this or any other of the above structures, but their large bodies, with the aid of their extensive air sac systems, long, resonant tracheas and probably more saurian-like ears, might have made it possible for them to have produced and sensed infrasound vibrations.

All these defense methods were probably most effective against a solitary attacker, but they may have broken down when theropods hunted in packs, with some hunters diverting a sauropod while others concentrated on a vulnerable area. There is increasing evidence to show that not only the specialized dromaeosaurs, some of which were huge in size, but also the giant carnosaurs and tyrannosaurids may have used group-hunting techniques when the prey was enormous and the stakes were high.

THE MYTH OF SIZE

One common claim regarding sauropods is that in the case of adults, their huge size made them more or less invulnerable to being attacked, as is often cited in respect to rhinos, hippos and elephants. A potential prey animal's large size in relation to a predator, however, may or may not be a deterrent to being attacked. It doesn't stop solitary weasels and shrews from killing rabbits and rodents many times their size, since the weasel uses a well-placed bite to the back of the rabbit's skull, usually paralyzing or killing it instantly. Medium-sized predators can and do habitually combine forces to bring down prey larger than themselves, as do wolves in the case of moose and lions with Cape buffalo (*Syncerus caffer*). Orcas, or killer whales (*Orcinus orca*), sometimes in packs attack blue whales (*Balaeonoptera musculus*) that are more than 3 times their length and 9 times their weight. The largest known land mammal predators that may have used group-hunting techniques were the Eurasian cave lion (*Panthera spaelea*) and South American dirk-toothed cat (*Smilodon populator*). These were huge felines probably weighing as much as 385 kg (850 lb), but even today much smaller modern lions in large groups may, even though rarely, take down individual African elephants when desperate. Probably big enough to tackle multi-ton mastodons and other giant prey, these extinct cats, as big as they were, would still have been dwarfed in size compared to the huge dinosaurian theropods. The fact that the latter were gigantic carnivores suggests that this was an adaptation toward taking down prey at least as big as, or bigger than, they were. This brings us to an observable fact throughout vertebrate evolution: *as the body size of a prey species increases, so does that of its major predator species.* Throughout the later Cenozoic era, or Age of Mammals, large body size in both canids and felids seems to have kept pace with increasingly larger potential prey that lived in the same times and places. In North America and elsewhere, both early dogs (*Hesperocyon gregarius*) and cats (*Pseudaelurus aeluroides*) started with small, fox- or ocelot-sized hunters who were suited to overcoming small ungulates (oreodontids) and other prey, but when prey species grew larger (cervids, or deer, as well as horses and camels), so did these types of predators, evolving into forms like *Canis dirus* and *Panthera atrox*. Their greater size enabled them to bring down the bigger game. A documented modern example of this is the gradual size increase of some populations of the coyote (*Canis latrans*) after it migrated eastward into the New England sector of the United States to replace the local extinction of its larger relative, the gray wolf (*Canis lupus*), many decades earlier. In the East, gray wolves were a major predator of the white-tailed or Virginia deer (*Odocoileus virginianus*) until humans exterminated wolves there around the mid-19th century. Coyotes, which west of the Mississippi River average about 15–18 kg (35–40 lb) and until then had hunted animals no bigger than a jackrabbit, now had available a much larger prey species. A larger animal is a much better investment of a predator's time and energy because of the bigger caloric payoff compared with a small one. Natural selection in New England coyote populations has favored bigger individuals with more robust jaws and teeth for taking down deer.

Theropods, the main predators of herbivorous dinosaurs, similarly show a size increase throughout Mesozoic time as an adaptation to bringing down ever larger prey. At the Sheep Creek and Como Bluff sites in Wyoming, two fossil localities that represent some of the lowermost and therefore oldest levels of the Morrison Formation of the American West, we find *Apatosaurus* (or *Brontosaurus*) skeletons with femur lengths measuring about 1,750 mm (6′), indicating a body length just under 21 m (70′). Bones found just past the top of Como Bluff, however, about a million years later, are from much larger brontosaurs. These femurs measure at about 2,000 mm (6.5′) and belonged to truly huge *Apatosaurus* that measured more than 25 m (85′). The earlier brontosaurs, tentatively assigned to the genus and species *B.* (or *A.*) *excelsus*, had evolved into a much larger species, *A.* (or *B.*) *ajax*. Other sauropods at the Morrison, *Camarasaurus lentus* and *Diplodocus longus*, show a similar size increase, becoming *C. supremus* and *D. carnegii.* Corresponding to these changes in sauropods are changes in the size of theropod predators like *Allosaurus fragilis*. The femur measurements of these carnivores at 850 mm (2′9″) remain fairly constant during most of the Morrison, but near the top of Como Bluff, the same site that produced the gigantic brontosaurs, their femur lengths increase to 1,200 mm (3′11″). This huge new species, *Saurophaganax maximus*, was keeping pace in size with its sauropod prey. A less common Morrison theropod, *Ceratosaurus nasicornis*, also underwent a size increase, apparently evolving into *C. magnirostris*. In the same way, by the Early Cretaceous in North America the super-allosauroid *Acrocanthosaurus* evolves to tackle *Sauroposeidon*, South American *Argentinosaurus* becomes potential prey for *Giganotosaurus* and later *Carcharodontosaurus* steps up to take advantage of the enormous *Rebbachisaurus* and *Paralititan* in North Africa. All this shows that getting bigger is no guarantee of escaping predation.

GANGING UP ON SAUROPODS?

While, as with modern carnivores, theropods would naturally have preferred smaller, weaker, less experienced and more easily subdued victims, the driving force of hunger can cause a predator to sometimes take risks and attempt new behaviors. These may or may not be rewarded by a successful hunt. As we've seen above, one solution is to get bigger yourself. This, of course, makes your chances of taking prey your own size, or larger, that much better, but what about the real prizes—insanely enormous sauropods like *Argentinosaurus*, *Puertasaurus* and others that may have weighed in at 50–75 tons, or even more? For this you want helpers, and owing to the risks involved, these should ideally be reliable and experienced pack members. Among the most habitual modern group hunters (wolves and lions, for example), pack members are closely related, often from the same litter, and this ensures a degree of cooperation and trust based on familiarity. At this point it's important to distinguish between *communal hunting* and *cooperative hunting* among vertebrates. In **communal hunting**, each carnivore in a group attacks as an individual and doesn't necessarily coordinate its attack with other group members, but the collective attack helps each hunter's chances. In **cooperative hunting**, the individuals act as a coordinated team, using a fine sense of timing, intuitive planning and reliance on each member's particular skills to bring down prey. Cooperative hunting depends on a rather high degree of intelligence and socialization among members. This, in turn, is usu-

ally reflected by the size of the brain, and the brains of some extinct and modern canids show enlarged areas that relate to socialization. At present there's debate among paleontologists as to whether theropod dinosaurs practiced communal or cooperative hunting. Some, like Larry Witmer, feel that large theropods may have practiced communal hunting but are skeptical, from what they see in brain endocasts, as to whether they were capable of cooperative techniques. Others, like Canadian paleontologist Phil Currie (Royal Tyrrell Museum), think cooperative hunting may have been possible. His opinion is partly

Giant avetheropods like *Tyrannosaurus* (*above* and *opposite*, *below*) dispatched their prey by the sheer, massive power of their bite, in which thousands of pounds of jaw pressure drove teeth that crushed and cut out bone as well as flesh.

based on the occurrence, at different geologic times and localities, of fossilized skeletal remains of different-sized individuals of the same theropod species found in close association. These included up to 12 juvenile, half-grown and adult *Albertosaurus (Gorgosaurus) libratus*, possibly representing a familial group of an adult and her offspring. Another possibly gregarious species, *Mapusaurus roscae*, is a large carcharodontid allosauroid from the Middle Cenomanian of the Late Cretaceous, Argentina, that was about 11.5 m (38′) long and weighed about 5 tons. This was a *sympatric* (living at the same place and time) theropod that shared the semiarid, open floodplain and riverine forest habitats of *Argentinosaurus huinculensis* (30 m [100′]) and other titanosaurids. If all these theropods were actually members of a family, they may have stayed close to help one another hunt cooperatively.

Can a fossil brain cavity tell us whether groups of theropods were smart enough to hunt and kill huge sauropods? Unlike a mammalian brain cavity, the skull of a saurian unfortunately doesn't closely mold to the actual, precise form of the brain itself. It only suggests the size and rough shape of more general areas like the *cerebrum*, the area corresponding with motor skills and cognitive ("thought") processing, the *cerebellum*, which governs coordination, and the *brain stem*, which controls automatic body functioning. Although here we can't study the contours of cerebral lobes that, as in mammals, would indicate the potential for social behavior, it doesn't necessarily rule out the possibility that theropods could effectively cooperate. Among living archosaurs, Nile crocodiles (*Crocodylus niloticus*) have been observed to work together to dismember killed animals, as well as jointly moving prey from one location to another, but there are no records of cooperative killing. The Harris's hawk (*Parabuteo unicinctus*), which preys on lizards, small birds and hares, usually is a solitary predator but sometimes hunts cooperatively, one of the few birds currently known to do so. Involving two to seven individuals, usually a mated pair and their offspring, the hawks use more than one technique. This may include each bird swooping onto a prey from different positions close to the ground, a single individual driving prey toward other waiting birds and, finally, a form of relay pursuit, in which a victim is harried by one hawk until that bird tires, whereupon another takes up the chase. Harris's hawks also guard their prey from theft by other predators and scavengers. Some small avetheropods like *Sinovenator changii* and *Saurornithoides junior* had brains with cerebral hemispheres almost as large as that of many modern birds. If they had been as intelligent as Harris's hawks and corvid species like crows and jays, it's certainly possible that these, and other theropods, might have had cooperative hunting as a part of their behavior.

By the time of the Lower Cedar Mountain Formation in Early Cretaceous Utah, a very robustly built 300 kg (600 lb) "raptor," or dromaeosaurid, had appeared, *Utahraptor ostrommaysi*. If this big, powerful predator had, like its smaller, wolf-sized cousin *Deinonychus* and the fox-sized *Velociraptor* in Asia, practiced group hunting, it would have been a terror to big brachiosaurids like *Cedarosaurus weiskopfae* and *Abydosaurus mcintoshi*, as well as the larger ornithopods. Furnished with an avianlike brain and overlapping, binocular vision, large and small dromaeosaurs were a revolution in theropod design. As pointed out earlier, they were not only well armed but also fast, and once a prey had been spotted and singled out, there would have been little chance of outrunning them. The size of big genera like *Utahraptor* might well have been an adaptation for killing large dinosaurs, including sauropods. If so, great size would have been of little help to the prey. Although the large arteries in a sauropod's body were deeply buried and the skin was thick, wounds from the sickle claws and sharp teeth would have eventually weakened the animal through rapid blood loss.

In other parts of North America, the rising seas of the Early Cretaceous gradually *transgressed*, or flooded, areas that had once been forest and open woods, creating new coastlines that now extended as far north as modern-day Oklahoma and middle Texas. The regions just north of the coast now became extensive floodplains, swamps and marshes crossed by broad rivers. It's here, along riverbeds like those of the Paluxy and Purgatoire near Glen Rose, that you can see pothole-deep trackways left by huge sauropods, probably from basal macronarians like *Paluxysaurus jonesi* and the enormous, 40-ton *Sauroposeidon proteles*. The trackways sometimes show as many as 23 individual animals walking together as a group, and the prints are sometimes overlaid by big, three-toed tracks. These prints may have been left by a huge carcharodont allosauroid of this time, *Acrocanthosaurus atokensis*. Although its arms were short and relatively weak, it had a thicker, more powerful neck and a longer, deeper skull than previous allosaurs. Together with its size (11 m [35′]) and tall, well-braced dorsal spine, *Acrocanthosaurus* was especially well suited to attack the biggest sauropods. And it may have hunted in groups. Some trackways from sites like Glen Rose and Davenport Ranch show that the sauropods were being followed by as many as 12 theropods at a time, and when the tracks of the sauropods make a turn, the theropod tracks go in the same direction. This, of course, can't tell us whether *Acrocanthosaurus* hunted cooperatively or not. The carnivores, like the herbivores, might have had to take the same route because of some physical constraint like water on one side, and although they were in a group, it doesn't mean they were a hunting pack. Shortly after these trackways were discovered by Roland T. Bird (American Museum of Natural History) in the 1930s, one set of sauropod-theropod prints was interpreted as evidence of a dramatic chase and attack on the sauropod by a lone hunter. The latter's prints seem to increase their stride (speeding up) and then skip a step as they veer over to the sauropod tracks, suggesting to some that the theropod briefly crossed over the sauropod's path to the right

An upside-down *Camarasaurus* pelvis from the Cleveland-Lloyd Quarry shows tooth scrapes from the feeding of large theropods on the right rear ilium. Because it was the origin of especially large, thick thigh muscles, this area received special attention from alpha predators when the flesh was being consumed.

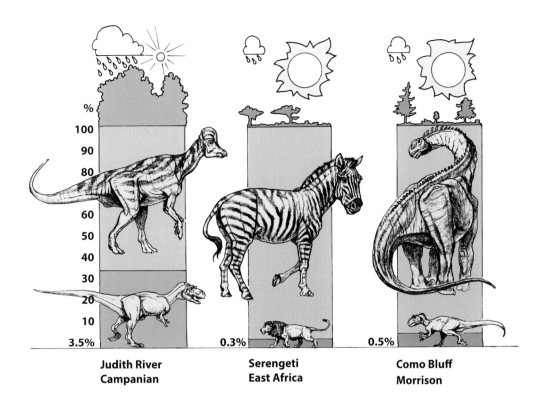

%
100
90
80
60
50
40
30
20
10
3.5%

0.3%

0.5%

**Judith River
Campanian**

**Serengeti
East Africa**

**Como Bluff
Morrison**

Preliminary studies of Mesozoic ecosystems show that the dinosaur prey-to-predator ratio may have differed according to the amount of vegetation cover available, which in turn was influenced by the amount of rainfall and aridity. In the well-vegetated Campanian-age localities like Judith River the cover meant that predators were more successful and could exist in a higher proportion of numbers to their prey, whereas, like today's Serengeti, the sparsely covered Morrison localities such as Como Bluff made hunting more difficult, and predators made up a smaller proportion because they needed more prey to sustain themselves.

and tried to grapple with the base of its intended prey's tail. Thrilled by this exciting scenario, these interpreters didn't consider that it would have been impossible for the hunter to have gotten over and around its victim's tail while running, as ichnologist Martin Lockley (University of Colorado, Denver) and others later pointed out. So no chase here, but the possibility still exists that a future trackway discovery may show a real act of predation.

MESOZOIC NUMBERS GAME

Ecologists are interested in understanding the dynamics of living ecosystems, and this includes predator–prey relationships, which can be complex and periodically change over time. If sauropod dinosaurs were a main prey source for theropods throughout the Mesozoic era, what can we learn, if anything, about what characterized this? A good place to start is by considering relationships between modern predators and prey, keeping in mind what differences may have existed between these and dinosaurs. One concerns the relative abundance of carnivorous predators to herbivorous prey, since this has a direct effect on the evolution of both. If, based on new findings from studies of skeletal anatomy, microscopic bone tissue and other areas of research, dinosaurs were endotherms like birds and mammals, it then makes sense that the abundance of large, predatory theropods in proportion to sauropod prey, such as that characterizing the Morrison Formation, should be generally similar to that of the dominant mammalian predators and prey of modern ecosystems.

Even *Carcharodontosaurus* or other huge alpha predators that made a kill (here *Paralititan*) might be challenged or even driven off by a smaller theropod like *Afrovenator* if it was outnumbered.

One natural present-day ecosystem with a large spectrum of habitats, plants and animals is the Serengeti plain in Tanzania, East Africa. Here species of very small (Oribi antelope, *Ourebia ourebi*) to huge (elephant) guilds of generalist and specialist herbivores are complemented by guilds of similar types of carnivores, ranging from shrews to lions. Many other species occupy feeding roles in between these extremes. In the Serengeti, as with other ecosystems dominated by endothermic vertebrates, so-called mammalian *apex predators* are much less numerous than their prey, averaging only 1/10 of 1% or less—their prey populations are almost 1,000 times greater than theirs. This is because, unlike crocs and other *ectothermic* vertebrate hunters, *endothermic* mammals need far more food to sustain their higher metabolic needs. A 45 kg (100 lb) leopard, for example, would need about 454 kg (1,000 lb) of meat per year to sustain itself, representing about 90 prey animals. A 45 kg ectothermic croc, on the other hand, would need only 45 kg, or 1/10 as much, representing only about nine individuals. A predator's energy use is also determined by its body size: a lion needs more calories to maintain itself than a leopard. So the higher the metabolic needs of a predator species, the bigger its prey base needs to be. Its numbers will always be fewer than the numbers of its prey. If we find several top predators in an ecosystem, either modern or ancient, this means that there has to be an increasingly broad-based pyramid of herbivores both abundant and diverse enough to sustain them. In the Serengeti the top predators are lion, leopard, cheetah, hyena and wild (Cape hunting) dog. During the Early Tithonian stage of the Late Jurassic, there were several large top predators that lived in the various Morrison ecosystems, including *Allosaurus, Ceratosaurus, Marshosaurus, Saurophaganax* and the megalosaurid *Torvosaurus*. Modern big-bodied predators such as those of the Serengeti generally (although not always) obtain most of their food from large prey, and this would also have been true of the Morrison theropods. Although there were large ornithischians like the thyreophoran *Stegosaurus ungulatus*, the polacanthan *Gargoyleosaurus parkpinorum* and the ornithopod *Camptosaurus aphanoecetes* available, most of the large herbivore diversity and body mass lay among the sauropods, such as *Diplodocus carnegii, Apatosaurus excelsus, Barosaurus lentus, Camarasaurus ssp.* and *Brachiosaurus altithorax*. Although theropods of the above-mentioned species certainly were major predators of any juvenile sauropods they came across, their sizes primarily reflect an evolutionary adaptation to taking down the big adults. These huge theropods could also not have maintained their populations (enough to leave fossils) if there hadn't been enough diversity of sauropod prey. This would suggest that in the Morrison and other ancient ecosystems, the predator-to-prey balance was similar to modern ones like the Serengeti.

When, however, Robert Bakker (then at Yale University) calculated the ratios of tyrannosaurid theropods to hadrosaurs and ceratopsians of the Judith River Formation, he found that this didn't all add up. In this Late Cretaceous time and Alberta locality, where big hadrosaurs and ceratopsians formed the theropods' main prey base, his findings showed that the predator/prey ratio was actually *greater* than that of the Serengeti. Here, modeled estimates indicate that predators amounted to 3.5% of their prey populations. This was *more* than the ratio of predators to prey in the Serengeti and some Indian wildlife reserves, where the ratio averaged 1% or less—3 or 4 times less than in dinosaur ecosystems. Why would this be?

There are several possible, related reasons. One is that climate, aridity and habitat quality play an important role in predator/prey ratios. The seasonally dry, shortgrass savannah of the Serengeti is very different from the environmental conditions that existed in the Judith ecosystem. In the Judith, soil, pollen and sedimentology studies indicate a lusher, wetter, year-round dense forest, which would have favored the ambush techniques of big tyrannosaurids like *Albertosaurus (Gorgosaurus) libratus*, contributing to a higher kill rate. In this ecosystem, where the summers were less dry and the forests thicker, there could be bigger populations of herbivorous dinosaurs that, in turn, could sustain greater numbers of predators, which the habitat conditions made more efficient and more common. Although the modern African savannahs support huge seasonal herds of game, the lack of suitable cover makes for fewer kills. This is aggravated when the Serengeti becomes seasonally dry each year, and the main big-bodied prey species like zebras and wildebeests migrate to greener pastures for a few months. It leaves resident territorial predators to hunt small prey, and more frequently, or go hungry. The result is that, to offset these disadvantages, a much bigger prey base becomes necessary for lions and other mammalian top predators to make a living.

Another is that the Serengeti is not the same now as it once was. Although one of the most natural and unspoiled large ecosystems on Earth, it has changed within relatively recent geological and historic time. While African faunas were not as severely pruned by the extinctions that occurred at the end of the last (Late Pleistocene) glacial period, they still lost some big-bodied mammalian predators. These included the saber-toothed cats *Megantereon nihowanensis* and *Dinofelis (Dinobastis) barlowi*, whose prey included juvenile (and possibly adult) modern and

A major kill by a large theropod like *Allosaurus* (*center*) would have attracted smaller opportunistic species such as *Ornitholestes* (*bottom left*) and tiny *Podokesaurus* (*bottom center*).

Two *Allosaurus fragilis* work over a highly decayed carcass of *Barosaurus lentus*, victim of a flood.

extinct elephants, as well as probably rhinos and hippos. Sabertooths are likely to have taken big antelope, zebras and buffalo as well, and if their presence is factored into the late prehistoric Serengeti, along with still extant big carnivores, it would increase the predator/prey ratio in the predators' favor. The faunas of today's world, as diverse and intact as they might seem to us, are what naturalists call *zoologically impoverished,* the result of the major extinctions of the *megafauna,* or large mammals, during the Late Pleistocene. With large carnivores like the sabertooths gone, the ungulate populations of some species, without their top predators, may have ballooned to unnatural numbers (for example, bison in North America and springbok antelope in South Africa). In East Africa, game species are diverse and their populations large, but the remaining predator species that survived the extinctions became fewer in diversity. This, as well as the greater challenge in hunting prey in an open environment like the Serengeti's, has helped to keep the ratio of predators to prey unnaturally lower than it might otherwise be. Adding to this situation are humans throughout historic time. Populations of lions and leopards declined from the activities of Europeanized ranchers and herders that killed these predators to protect their livestock. Because the Serengeti ecosystem extends far beyond the boundaries that were originally set up to enclose it, the migrations of both prey and predator populations taking place outside this area are now interfered with by many human activities. Unregulated sport hunting of lion and other predators until recently also played a role in reducing carnivore numbers, as well as tourist industries that disturb the predatory species' natural hunting. All this has probably kept carnivore populations in the Serengeti and other well-studied modern ecosystems from attaining natural levels of 3.0% to 3.5%, staying closer to 0.1%.

What about the different Morrison ecosystems? These were enormous, probably incorporated diverse habitats and lasted for millions of years. One locality that Bakker studied was Como Bluff, Wyoming, located in one of the latest horizons where the supersized brontosaurs and allosaurs were found. In this Late Jurassic time frame the climate and habitats were much more like the Serengeti than the later Judith. Summers were hotter and drier, often creating monthslong droughts in which hundreds of juvenile and adult sauropods and other herbivores either migrated or congregated around dwindling water holes before they dried up. The dead and dying provided feasts for giant allosaurids such as *Saurophaganax*, as well as smaller theropods like *Ornitholestes hermanni* and *Coelurus fragilis*. Although dense in places, some of the surrounding woodlands would have been more open with less cover for theropods, lessening the advantage of surprise and making successful kills less frequent. The seasonality of the rains also resembled the Serengeti's, and when the dry season arrived, the sauropod herds, like those of wildebeests and zebras, probably migrated to other areas where there was still abundant browse. If the dinosaur predator/prey theory was correct, this would mean that the predators were not only less successful but also less common. When Bakker crunched the numbers this time, the theory held: at Como Bluff, the predatory theropods made up only about 0.5% of the total dinosaur population.

One thing that has to be kept in mind is that while Bakker's finding seems like a confirmation of this theory, Como Bluff is only *one locality* among the many Late Jurassic habitats that existed within the long time span of the Morrison Formation. In his comprehensive 2003 paleoecological analysis of *all* of the then known Morrison localities, John R. Forster (University of New Mexico) came up with an overall predator/prey ratio (including non-dinosaurs as well as dinosaurians) of 8.6% for all geochrons sampled for the entire formation. As James Farlow had pointed out earlier, Forster concluded that a broader consumption of prey species (in the case of dinosaurs, ornithischian as well as sauropod) would result in a *lower* predator/prey ratio than if the theropod had concentrated more on just one species. If this was true, a very diverse guild of dinosaur predators would *seem* to have a lower predator/prey ratio than it really did. Depending on the geochron, Forster found that dinosaur-only predator/prey ratios in Late Jurassic western North America ranged from 4.4% to 11.0%, with an average of 6.44%.

A consistent problem that always crops up in paleontology is *preservation bias*. This is the fact that, among the thousands or millions of animals that ever lived belonging to a given species, only a very small number are preserved as fossils. The hard body parts of animals that lived in habitats like lowland areas, where water action and sedimentation favor fossilization, were more likely to be preserved than those of animals that lived in upland areas, where burial was less likely. In short, fossilization is not an equal-opportunity situation when it comes to preservation. This means that there's the possibility that more predators were actually around at Como Bluff than are represented by skeletal remains. This is why it's so important in paleontology (and other sciences) to obtain many samples and build a large base of comparative data. When an overwhelming amount of data supports a theory, it then starts to become confirmed as a fact, as with the theory of natural selection and evolution in the natural world. Returning to Como Bluff and the Morrison ecosystem, only more studies and data derived from sampling will absolutely confirm the theory that dinosaurian ecosystems followed the same principles as those of mammalian-dominated ones, and that these applied to the relationship between sauropods and their theropod predators. For the moment, however, it looks at least possible.

TINY TORMENTORS

While it didn't result in their deaths, sauropods and other dinosaurs were also food sources for bloodsucking insects. Preserved, intact individuals in Burmese amber show that by the Cretaceous there were extinct, mosquito-like tanyderid and chironomid biting flies. There were also

Preserved in Mid-Cretaceous amber, biting midges like this one and bloodsucking flies probably made life miserable for sauropods and other dinosaurs during the warm, humid, rainy months.

biting midges and ticks. While at this time there's no direct proof of actual parasitism, the fossilized skin from titanosaurs and a few other sauropods is relatively thin, and their huge warm bodies would have made them tempting targets. Sauropod skin in known types takes the form of angular-shaped, close-fitting polygonal tubercles. To some degree this might have offered somewhat more protection than other dinosaur types that had more unprotected skin surrounding the scales, but by then biting flies had already become specialized enough to probably overcome this. Mosquito-like forms, if they could get past the hard pavement, only had to probe as deeply as the outermost surface capillaries to obtain blood, while the bigger tabanids used their shorter stylets to lacerate past the capillary level and create a blood pool to suck from on the surface of the skin. The parasites not only took; they gave as well: the midgut wall in one midge from an amber specimen bears flagellate protozoans and cytoplasmic viruses, showing that these insects and probably their giant victims were hosts for a variety of diseases. If found, analysis of preserved blood from a future sauropod find could cast light on what plagued these and other dinosaurs.

BAD PREDATOR, GOOD PREDATOR

Just as prey species play a direct role in limiting predator populations and influencing species diversity, so do predators with prey, weeding out the individuals that are weak, less alert or in some other way more vulnerable to being attacked and successfully killed. While prey populations and diversity are limited from "the bottom up" (factors like the abundance of food and water in a given area, what kinds of feeding niches are available and how much competition goes on for these), they are also limited "from the top down" by predation. This constrains the populations of prey species, which, without the culling effect of the predators, would overpopulate and destroy their own food bases, as well as keeping unfit individuals within the gene pool. Without enough predation, herbivore species reproduce and overpopulate so much that they may eat up their food plants and begin to starve. Undernourished animals are also more prone to diseases, which can then spread to healthy individuals. Abnormalities that contribute to an individual being less fit are sometimes preserved. If these animals are successful in breeding and passing on their defects, the species as a whole becomes weaker and becomes much more vulnerable to extinction. The theropods were great destroyers of sauropod species of all sizes, but it was they who kept the sauropods fit and strong.

DEATH OF A GIANT

As already mentioned, prosauropod and sauropod adults in prime condition weren't pushovers. If adults did become prey, the ones most likely to be targeted were probably old, sick or somehow physically impaired, possibly owing to an injury from another, unsuccessful attack. Resident predators are familiar with the characteristics of their prey, sometimes extending to individual herd members, and with their keen observational skills are quick to notice anything that makes an animal more vulnerable. If adult sauropods were gregarious and inclined to come to each other's defense as with modern Cape buffalo, a herd's defensiveness might have compensated for any individual's weakness. On the other hand, if a given sauropod species was socially more loosely structured and didn't stick together, it would have been easier to separate a more vulnerable adult or large subadult. Group predatory tactics would have definitely increased the chances of a kill over a single individual's efforts. Nevertheless, dispatching an adult sauropod might have been a long, bloody process. Theropods would have attempted to make wounds to vital areas like the abdomen, rear leg and tail base, or simply by inflicting enough slashing or cookie-cutter bites to make the prey weaken from blood loss. The sauropod would rely on rearing and lashing out with its thumb claws, kicking or effectively using its tail. A kill could have taken several hours, but probably no more than a couple of days, owing to the likelihood of the prey sickening from the theropods' septic bites. If the vulnerable neck underside with its vital arteries and windpipe were ripped open, it was all over. The sauropod would have gone into shock, and the carnivores could begin tearing into its abdomen to get at the liver, heart and other soft organs. Unlike the relatively clean kills practiced by big feline carnivores such as lions, which bite down onto the nose or throat to cause suffocation, theropod kills were probably a lot more like those of Cape hunting dogs and Komodo dragons, in which the prey literally dies from being eaten alive.

Once the internal organs were consumed, the next areas to be eaten were probably the main muscle masses. In some preserved sauropod pelvises, tooth scrapes, perforations and other marks show up around the posterior ilia and femur, the places of origin and attachment for the big retractor muscles that powered the sauropod's rear legs. Because they had so much meat, these were choice areas, and if the hunters were in a group, these might have been quarreled over, with dominant members getting their share first, as in Komodo dragons and mammalian pack hunters. As it began to feed, a theropod might well have used one of its hindfeet to steady itself while it used the other against the carcass as a brace, so it could tear into the skin with its jaws. Powered by the strong neck, it could yank hard until a mass of meat was torn off, after which it would bolt the chunk down its throat, being unable to chew. Unlike mammals with their more precise dentition, this would have made them messy feeders, and by this time smaller theropod types like *Ornitholestes* would have surrounded the giant predators. These would have waited for a chance to snatch and run away with what they could, like jackals surrounding a lion kill. Although there's no direct evidence for it, some pterosaurs, too, could have swooped in to grab small morsels, as do modern vultures. With the amount of flesh available, a single hunter or group probably would have stayed close to the kill to eat regularly and defend it from being scavenged or taken over by bigger theropods of their own species or others. This could have lasted days to a week or more depending on the size of the sauropod and the theropod predator or predators. Most of the large species like *Allosaurus* or *Torvosaurus* might easily have been able to put away more than 82–91 kg (180–200 lb) of meat during one feeding, compared to 23 kg (50 lb) for a lion or 9 kg (20 lb) for a wolf. Since a successful kill in modern top predators often happens as little as 1 out of 20 tries, an individual typically gorges or eats as much as it can, especially if it senses that another,

As a thunderstorm breaks, a lone *Allosaurus fragilis* and two crocs take advantage of a pair of *Torvosaurus*'s unguarded *Diplodocus carnegii* kill. Even the largest *Allosaurus* would have been unlikely to have been able to kill an adult sauropod like this by themselves, requiring others to bring it down either communally or cooperatively.

more powerful predator may try to take over the kill. Numbers and aggressiveness can win out in these cases, and like a larger hyena pack running off a smaller number of lions, so a group of *Allosaurus* might have driven off a much bigger single or pair of *Torvosaurus*. Eventually the top predators, having eaten all the soft tissue they could easily obtain, would have drifted off, leaving the carcass to the waiting host of small to medium theropods, small land crocs, varanid-like lizards and opportunistic pterosaurs, which within weeks would have stripped the gigantic frame of its last shreds of flesh. Even if a sauropod lived long enough to die of natural causes other than direct predation, it was still subject to being consumed by big and small theropods, as well as smaller carnivores, by scavenging. If a group of *Allosaurus* were anything like today's Komodo dragons, their keen senses of smell would have brought them swiftly to the giant corpse as soon as they detected the first whiff of decomposition. Only if a dead behemoth were washed by flooding far from land would it end up beyond the reach of most carnivores. Here the pressure from steadily expanding gases in the guts would make its bloated belly at some point burst with a sound like a cannon shot, filling the air with unbelievable stench. Proof that this happened comes from occasional finds of otherwise articulated sauropod and other dinosaur skeletons with a blown-out rib or two, spots where the gases violently escaped.

FROM THE EVIDENCE

But this wasn't the end, only the beginning. Here we've entered into the territory of a formerly little-known science known technically as ***taphonomy*** (*taphe*, "grave" or "marker"), the study of what happens to the body of an organism from the point of death until it's fossilized. Taphonomy, in turn, is actually part of a more embracing discipline called ***forensics*** (*forensis*, from the Latin *forum*, implying "pertaining to law" or "public debate"), normally the techniques involved with the methodical gathering and analysis of evidence in matters like a crime, but here scientifically applied to a death in the natural world. These interconnected approaches can tell us how a dinosaur, ancient human or other organism died, what happened to its body after death and how it was preserved, hours or millions of years later. In the case of a giant sauropod carcass, a time-lapse film sequence might make this process seem like a slow-motion explosion: depending on how death occurred, things happened dramatically fast but then gradually slowed down more and more until conditions became stable, awaiting the processes of geologic sedimentation and erosion to cover and then uncover the body. Instead of just digging up and removing bones, modern paleontology now uses many methods and disciplines, such as taphonomy, to understand the paleobiology of an extinct animal while it's still in place.

Paleontologists are often particularly interested in the taphonomy of an individual prehistoric animal because this can yield clues to the life processes of its species. To reconstruct the postmortem history of a dinosaur or other giant vertebrate, many subtle clues must be discovered, studied and recorded. When this is done properly, we can expect to produce a reasonable account of what happened. In the case of a sauropod, these findings are sometimes similar to ones from observations of elephant carcasses that result from the deaths of these animals in game parks, as well as from undersea whale carcass localities, known as *whale falls*. As a result, the natural processes that typically occur following the death of a modern giant animal can be compared with the data from the excavation of an ancient one. In whale falls, the decomposition of a whale in the sea happens in stages, like those observed in land carcasses, with a succession of carrion feeders. First, the flesh is removed by large species like sharks, followed by smaller fish. After this, the body is colonized by both a range of marine invertebrates that feed directly on the carcass and bacterial mats that develop on it. When, after months or years, only the skeleton remains, specialized invertebrates, like bone-dwelling mussels and polychaete worms, along with bacteria, take over to exploit the rich store of fats and oils found in whale bones. Although this is a marine community, as we'll see, there may have been similarities to the exploitation of sauropod bones at a carcass site.

Let's suppose that the prey killed by a theropod or group of theropods was an older (age 5–7) juvenile *Camarasurus*, a common sauropod species that occurred in the Morrison geochron. Having found it, an imaginary group of paleontologists are now conducting a dig. Millions of years after the predators left, could forensic techniques cast light on its cause of death and what happened to the corpse afterward? For our hypothetical *Camarasaurus* skeleton, found eroding out of a dry wash and mostly complete and articulated, this starts with an overview of the body's pose and postmortem condition. In several places, the pelvic bones, anterior caudal vertebrae, some cervicals and both anterior and posterior surfaces of the disarticulated femora and humeri showed large curved scrapes, gouges and puncture holes. Surrounding the intact pelvis were five small to medium-sized theropod teeth, with rounded, smooth roots, that were assigned to *Allosaurus* after comparison with museum specimens known from this species. One very large tooth, which was broken off at what would have been the gum line, more closely resembled that of *Ceratosaurus*. Since there were no other objects like rocks or pebbles of similar weight and size in the *bedding plane* (the level in which the *Camarasaurus* was found), these could not have been carried to the site by water action or some other means. Theropods periodically shed their older, worn teeth, and these usually show rounded ends due to root absorption. Teeth were replaced continuously by new ones during feeding or predation, and so the isolated teeth were exciting proof of carnivore activity at the scene. But did the *Allosaurus* (or *Ceratosaurus*) actually kill the juvenile *Camarasaurus* or only scavenge from it? And if the sauropod *was* killed, which one did it? The answer came within a year and a half after the dig began. When the articulated cervicals were being prepared at the museum, two left cervical ribs, nos. 10 and 11, and one rib on the

The skeleton of a young *Camarasaurus*, cleaned of almost all edible tissue by vertebrate and invertebrate predators and scavengers, will soon be gradually covered by silty floodwaters, preserving its forensic story millions of years into the future.

right, no. 12, were found to be fractured. The ventral surfaces of the vertebrae corresponding to these also showed curved scrapes, but these were larger than the ones that characterized the damage to the pelvis. In matching the typical sizes and curves of the theropods' teeth to these, it was apparent that the giant sauropod had been bitten near the base of its neck by teeth matching those of *Ceratosaurus*. The bite was hard enough to break through the cervical ribs and scrape the centra. A bite like this would have been fatal, since the *trachea* (windpipe) and major blood vessels would have been severed, causing death within minutes. This points to at least one *Ceratosaurus* as the actual killer of the *Camarasaurus*. Because we don't find similar-sized tooth scrapes on the ilium and other body areas that had the choicest muscle masses, it's likely that the larger predator(s) was (were) soon driven off the kill by a pack of *Allosaurus*, which did leave their tooth marks on the pelvis. No actual dig has yet revealed a "smoking gun" or direct evidence of a direct kill by a theropod or theropod hunting group on a sauropod, and so far we only have indications of scavenging like tooth gouges and scrapes on bone. Such findings as these would be an amazing discovery, but with the application of modern and probably future techniques we may someday actually find just such evidence of predation.

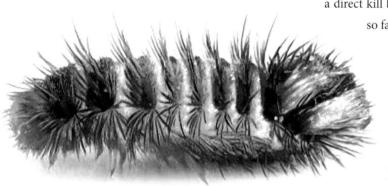

CALENDAR FLIES AND SKELETON CREWS

Even before the theropods and smaller vertebrates abandoned it, the now exposed, skeletal wreckage of the sauropod became a focus of attention from invertebrate scavengers. In one area near the distal end of the left shoulder blade a small patch of folded, preserved skin was found. This was exciting enough because of the information it gave about skin patterning in *Camarasaurus*, but within what may have been a gash caused by theropod claws or teeth several empty fossil pupa cases were found. Although not identical, these greatly resembled those of one of many species of blowflies or flesh flies, such as *Calliphora vomitoria* or *Sarcophaga haemorrhoidalis*. Flies, drawn to the chemicals emitted early during decay, are usually the first invertebrates to visit a corpse, often within minutes, laying their eggs on any exposed, soft tissue. When the egg hatches, the tissue provides food for the larva, or maggot, which then develops a rigid casing, or *puparium*, around itself. From this emerges the adult fly. Because the timing of each of the development stages in living species is well known, from this it can be inferred (*if* the life cycles of the ancient flies were similar to those of living ones) that for at least 14 or so days following death there was still soft tissue, taking into account such variable conditions as humidity, climate and season. In forensics, flies can act as calendars to point to the approximate time of death.

At or about seven weeks after death the carcass was visited by other, bone-modifying arthropods. These invertebrates specialize in eating dry tissues and produce the tiny pits, rosettes and borings on the surfaces of the bones. Like blowflies and other invertebrates that colonize corpses, these have life cycles that leave traces at predictable times and rates. Among the best known is the dermestid beetle (*Dermestes maculatus*), commonly used in museums to remove dry tissue from skeletons. Although no actual body fossils of this species date to the Late Jurassic, the ro-

Blowflies and flesh flies (*bottom*), followed by dermestid beetle larvae (*middle*), were present at least as early as the Late Jurassic as scavengers on sauropod and other dinosaur carcasses. These were followed by keratin-eating tineid moth larvae, which then as now left distinctive puparia (*top*) in which the adult moths developed.

settes and shallow pits on the bones matched those of damage to modern ones by dermestids so closely that the presence of this form, or a closely related one, was almost certain. After feeding for about four weeks, the dermestid larvae form puparia (from which the adult beetles emerge) by chewing into dry tissue and bone, creating their characteristic structures. In addition to the damage from the dermestids, some of the pedal phalanges had tiny traces, resembling borings, at the distal ends. This was the first evidence of activity from small moths similar to modern tineids, *Ceratophaga vastella* or *Tinea deperdella*, whose larvae may pupate inside bone but which actually feed on keratinous structures like horns or hooves, as they commonly do on ungulate carcasses in Africa. The presence of these pupae near the *Camarasaurus* pedal phalanges suggests that they were feeding on the tough skin, claws and pads of the feet.

The skeleton, which during excavation the team members had determined to be about 85% complete, lay on its right side, the one in contact with the original *substrate*, or ground surface, that existed at the time of the dinosaur's death. The bone surfaces on this side were mostly smooth and even, in contrast to those of the left, which in almost all cases were deeply fissured and showed evidence of flaking and cracking. This is a sign that the bones of the left side had been exposed to the drying effects of sun and air. From comparisons of this weathering to the kind seen in studies of modern elephant carcasses under similar conditions, it was possible to estimate the amount of *subaerial exposure*, or time the sauropod's skeleton had spent out in the open. In the case of the *Camarasaurus*, the amount of weathering took place anytime from about 9 to no more than 24 weeks (six months) later. The differing condition of the skeleton's bone surfaces suggested that the body had not been transported far from where it was killed. The close association of the limbs indicates that ligaments may have tightened and kept these bones, and most of the vertebrae, together while the carcass was drying out before final burial. After this time, the thick, cross-bedded layers of sandstone overlying the skeleton gave evidence that several of the bones were first scattered, and later completely buried, with coarse sediment by a series of monsoonally caused floods, followed later by fine sediment from quiet stream action. This had come in stages at the close of a long period of dryness, possibly a drought. The size and bedding level of the sediment grains provided clues to the relative speed and power, or *fluviatile energy*, of the moving water: water moving with higher fluviatile energy or velocity can move bigger-sized grains than a gentle current, which can only push fine sediments. The coarse sediments surrounding the articulated parts of the skeleton and larger, isolated bones were located on a lower level or bedding plane than the finer sediments, showing that the flooding happened first; if it had been the other way around, the finer sediments would have been swept away by the previous, more powerful water action.

During its lifetime, the *Camarasaurus* was just one of a community of sauropod genera and species that had reached an amazing diversity of forms, and these, in turn, were the prey for an equally impressive guild of theropod hunters and scavengers. Throughout time both sauropods and theropods had a mutual and profound effect on each other's evolution, and the race between the hunted and their hunters would continue into the eons ahead.

From the Mid-Jurassic period until the very end of the Cretaceous and beyond, Earth's original protocontinents underwent separation and constantly formed new ecosystems, vast stages for the development of sauropod and other dinosaur evolution.

CHAPTER TEN

AROUND THE MESOZOIC WORLD

Fossil evidence shows that throughout their long time on Earth, the sauropods' *bauplan* was a highly successful one, able to adapt to the wide variety of climates, environments and other species with which these dinosaurs coevolved as the continental landmasses separated. Although our knowledge of most Mesozoic ecosystems is at present very patchy, it's clear that when they were present, sauropods were "keystone" species that were a vital part of the environments in which they lived. The ecological portraits presented here, although speculative, offer some hypotheses as to what may have influenced major sauropod-dominated ecosystems.

THE VIEW FROM SPACE. We're in near-Earth orbit, high enough to get an overview of how much the lands that once made up the vastness of Pangaea have separated since the Late Triassic. By this point in the Middle Jurassic, what will one day become Asia, Europe and North America have completely separated from the parent supercontinent. South America and Africa are still tightly connected, while India and Antarctica (along with the island of Madagascar) still form a large land mass of their own. Australia is at first attached to this but later separates, remaining well within the south polar latitudes. The overwhelming dominance and aridity of the enormous deserts that characterized the inner vastness of Late Triassic Pangaea have now begun to give way to the influence of the oceans and their wind patterns, creating maritime and continental climate conditions that are more temperate and more familiar to us as visitors from the future.

Each continental mass, on becoming isolated from the others, becomes a potential laboratory for new species of *flora* and *fauna* (plants and animals), with some being **endemic**, or found nowhere else. Asia, formerly a part of the supercontinent of Laurasia (Europe–Asia), has now been separated the longest. Its plants and animals, as one would expect, are now the most distinctive because of isolation, as compared with those of Northern Gondwana (South America–Africa) and Southern Gondwana (India–Antarctica–Madagascar–Australia), which have been connected for much longer. Under these conditions, populations of a widespread species can still interbreed until final geographic separation occurs, at which point these populations will start to evolve into separate species. What was once Laurasia has now separated into an archipelago of large to small islands stranded between the bigger landmasses of North America to the west and Asia to the east. These islands are the nucleus of what will one day become Europe, but before coalescing and merging with Asia millions of years later, some will at times be close to North America and others closer to Africa. As in the bigger landmasses where dinosaur populations between these areas were in contact the longest, some former Laurasian species are also very similar, in some cases almost identical. By the Early to Middle Jurassic sauropods had evolved their basic feeding strategy and *bauplan*, or body structure: growing tall to reach and harvest an abundant, arboreal food source and processing it within a huge gut system that slowly but steadily yielded nutrients. By fast-forwarding and stopping at some key times and locations around the Mesozoic world, we can see how this sauropod *bauplan* evolved and adapted to very separated, very different ecosystems on the planet's shifting landmasses. In visiting the sauropod and other dinosaur faunas of these **geochrons**, we have to remember that each time/locality is only the tip of a vast geographic puzzle. We've made a good start in putting together a few sections, but we're now faced with the challenge of fitting in newly discovered localities and faunas that will help us see the big picture of sauropod evolution.

Proto-Asia

(1) Turgai Strait
(2) Primordial Tethys Ocean
(3) Primordial Pacific Ocean
(4) Dashanpu

During the Bathonian to Callovian stages of the Mid-Jurassic approximately 165 million years ago, Proto-Asia was a smaller landmass, separated from other continental areas by the Turgai Strait to the west and the primordial Tethys and Pacific Oceans to the south and east. This resulted in sauropod and other dinosaur faunas that evolved in isolation, producing distinctive forms. The black line indicates modern coasts.

THE GREAT FORESTS: DASHANPU, PROTO–EAST ASIA

Let's start our survey by transporting down to a Middle Jurassic locality called **Dashanpu**, now in **Zigong, southwestern China (Sichuan)**. Because fossil findings are indicating a high sauropod diversity at this geochron point, we've selected a Bathonian/Callovian time portal of 165 MYA, between the Lower and Upper Shaximaio Formations. As we descend through thinning rain clouds, we catch glimpses of a vast, lush forest stretching away in all directions. In some ways, the moderate to very warm and humid climate, flat, well-watered terrain and rich diversity of life-forms are similar to a modern lowland tropical rainforest like that of the Congo Basin in Central

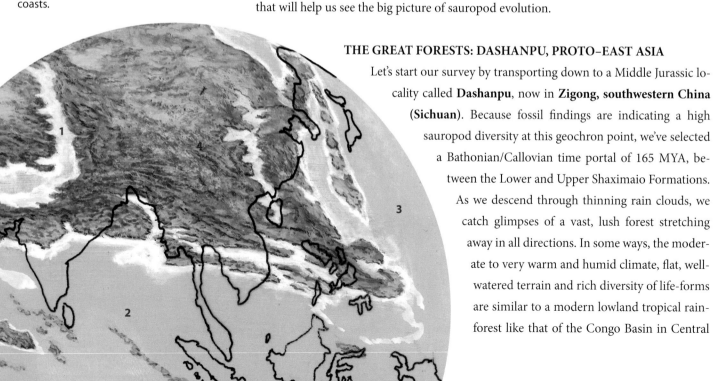

Africa, but this isn't like the modern, neotropical rain-forests we see in films like *Jurassic Park*. The rainfall, however, does come heavily and often, and the rich but shallow soils are highly acidic from leaf decay and other biological turnover. A large lake gradually comes into view, apparently fed by a wide, slow-moving river that broadens into a marshy delta, not far from where we'll land. On emerging into this wet, green world we're simultaneously stifled by the intense humidity but invigorated by the extra oxygen in the air. The open glade/riparian habitat on this early morning is densely covered by a variety of medium to tall cycads and seed ferns glistening from the recent rain, and these give way to thick beds of *Equisetum* that camouflage a small pond. Along the banks a few turtles, *Chenguchelys*, flank a large, temnospondyl amphibian, *Sinobrachyops placentiphalus*, a relict holdover from Triassic times. The drone of insects is everywhere. Besides flies, there are plant-sucking true bugs like *Corollpachymeridium heteroneurus* and a delicate early lacewing, *Bellinympha dancei*. Even this early in time, it has evolved a wing shape that mimics the pointed leaves of a cycad, *Eophyllium*, allowing it to escape the attention of predators. An occasional rustle and flash of cinnamon-brown fur through the carpet of ferns and herbs betray a small mammal, a *multituberculate*, foraging for seeds. These, with their specialized, comb-like lower molars and curved front incisors, are the "rodents" of the Middle to Late Mesozoic and have gradually taken over the understory from woodchuck-like *Bienotheroides zigongensis*, one of the last surviving tritylodont *synapsids*, or "mammal-like" reptiles. Together with tiny, shrew-like true mammalian omnivores and insectivores such as the docodonts *Acuodulodon sunae*, *Dsungarodon zuoi* and *Tegotherium gubini*, "multis" make up a large percentage of the mammal fauna at Dashanpu. In addition, there are rhynchocephalians, or tuatara-like saurians, and lizard-like reptiles. Along with arthropods, they are prey for small and medium-sized coelurosaurian theropods like *Chuandongcoelurus primitivus* and larger predatory lizards like *Archovaranus klameliensis*.

We are startled by a large, rapidly flying shape above us. It's an *Angustinaripterus longicephalus*, a medium-sized, rhamphorhynchoid pterosaur with a 1.6 m (4′) wingspan, which gracefully dips across the pond for an instant to grab a tiny ceratodontid lungfish from below the water's surface. As we follow its flight to the far end of the glen, we catch sight of the first of many basal stegosaurs foraging in the understory. *Huayangosaurus taibaii* has slender limbs and can amble relatively fast while swinging its spiked and clubbed tail at attacking theropods. For good measure, like some of its relatives, this species is also armed with huge, lethal shoulder spines. *Huay-*

Dashanpu, China

Dashanpu, close to the southern edge of Proto-Asia, was part of a vast lowland climax forest ecosystem (**1**) subject to periodic flooding. These floods (*arrows*) swept sauropod and other carcasses (**2**) into a nearby large lake, where fine silt often preserved entire skeletons intact. Sauropods were a "keystone" herbivore at Dashanpu, and their browsing and other activities heavily influenced the growth of many trees and other plants.

At Dashanpu, a *Shunosaurus lii* scores a direct hit with its spiked tail against the mouth of an attacking *Metriacanthosaurus* (see skull, *above*). Another Proto-Asian eusauropod clade, the omeisaurids, had small, bony knobs at the tips of their tails for defense. With or without these, a sauropod tail was a primary weapon in keeping theropods from closing in for a bite.

angosaurus is a low- to medium-level browser, and unlike more advanced forms, its pointed, beaked jaws still have teeth in the front upper jaw. The beaks and little teeth suggest that its diet, like that of the region's other stegosaurs, may be small, nutritious leaves, seeds and other plant structures as opposed to coarse, bulky vegetation. This stegosaur and other sympatric species like larger *Tuojiangosaurus multispinus* and the powerful *Gigantspinosaurus sichuanensis* occupy separate feeding niches that do not directly compete with one another. Later stegosaurs from other times and localities have relatively short forelimbs and longer necks, adaptations that, along with their sled-shaped chevron bones, enable them to rear tripodally to effectively reach medium-height arboreal browse. In this way they've further refined an adaption superficially similar to sauropods, but for feeding at a lower level. There are also small, less common ornithopods like 3.5 m (12′) *Agilisaurus louderbacki*, a fast-running hypsilophodontid whose grasping forelimbs, small beak and shearing, bladed teeth are suited for harvesting small, highly nutritious but tough plants.

As the last of the grunting *Huayangosaurus* melt into the cycads and tree ferns, a new movement, 18 m (60′) high in the surrounding conifer and ginko forest, catches our attention. First one and then another branch of *Araucarioxylon*-like foliage stretches out and snaps back, revealing amazingly long, chocolate-colored streak-patterned necks and bright-eyed heads that vigorously bite off and swallow the coarse cuticles. It's a small group of feeding *Omeisaurus jungshiensis*, among the most common of the Shaximaio sauropods. The jaws open wide to engulf a clump of cuticles, which then briefly swells the blue, underlying throat sac as it begins its long descent down, to be rapidly followed by the next. The browse goes down more easily than expected owing to the viscous, plentiful saliva in the throat, since without the ability to chew, these and future sauropods rely on this to get the bulky food mass down. Saliva is constantly secreted by large glands under the lower jaw, and the sauropod's body efficiently extracts much of the moisture from the foliage to create it. The slowly weaving, bending necks are surreally moving on their own, but when one suddenly rears back and shoots up even higher, powerful shoulders and a midbody appear. To reach the highest browse, an *Omeisaurus* can rear bipedally, letting its relatively long forelimbs drape to the sides of its massive belly. Occasionally, it slowly and carefully takes steps to the side to reach especially select clumps of new, protein-rich cuticles. Some omeisaurid species (such as *Omeisaurus tianfuensis*) have differentially proportioned upper teeth and deep, robust jaws, which means that the bigger, longer teeth come into contact with the browse first and probably serve to crop through the coarser twigs before the remaining teeth close.

With necks 8–12 m (26'–39') long and weighing in at about 4 tons, the omeisaurid *Omeisaurus* and its Oxfordian (161–156 MYA), even longer-necked relatives, the mamenchisaurids, are two closely related families, extreme high-level browsers that evolved in isolation from other, then existing sauropods (the eusauropods or "true" sauropods) after Asia became separated from the other protocontinents. The shorter forelimbs and hindlimbs, as well as longer trunks, of the mamenchisaurids make it harder and less likely for them to rear compared to the omeisaurids, but then they usually don't need to. Even though other, later sauropods such as *Sauroposeidon* and *Supersaurus* will have necks that, at up to 15 m, will be absolutely longer, the 12 m (35') necks of species like *Mamenchisaurus sinocanadorum*, *M. hochuanensis* and *M. youngi* are, in proportion to the body, the longest of any known sauropod and are well able to reach into the highest conifer foliage. The omeisaurids we're watching share the forest with other sauropods. Among these are the large, 15–20 m (45.7'–61') *Yuanmousaurus jianyiensis* and *Dashanpusaurus dongi* (=*Abrosaurus dongpoi?*), the latter a basal macronarian and a possible early titanosauriform, based on its postcranial characters. These, like the omeisaurids and mamenchisaurids, arose from generalized eusauropods that may have already been present in or migrated to Asia before its separation from the rest of Pangea. *Datousaurus bashensis*'s tall shoulders mark it as a high browser, but its shorter neck suggests a lower-canopy food source than the above species— possibly the conifers *Pagiophyllum*, *Brachyphyllum*, *Podozamites*, *Pityophyllum* or another *Classopteris* pollen species, which weren't as tall as *Araucarioxylon*. Although it has an even shorter neck for browsing on *Ginkgo* and other lower-canopy vegetation and is smaller at an average of 9.5 m (30'), another eusauropod, *Shunosaurus lii*, is distinctive in being uniquely well armed. In addition to their weight and length, the shunosaurids' tails have dorsal and posterior *spikes* and clubs. This makes the tails more effective as weapons than the tiny clubs sported by omeisaurids, and a contemporary African relative, *Spinophorosaurus nigeriensis*, actually has two pairs of lateral spikes like a stegosaur. *Shunosaurus* also has well-developed thumb claws that can slash if the animal rears to defend itself.

These defenses are necessary because of Dashanpu's several theropod predators. In size these start with the coelurosaur *Chuandongocoelurus*, go to medium-sized *Gasosaurus* (named after the petro company that helped fund the excavation of the Dashanpu site) *constructus*, followed by the 6–7 m (19'–22') sinraptoroid (or possibly megalosaurid) *Leshansaurus qianweiensis* and end up with large allosauroids like 11 m (35'), 3-ton *Yangchuanosaurus* (*Sinraptor*) *shangyuen-*

The skull and skeleton of *Mamenchisaurus hochuanensis* at the Zigong Dinosaur Museum, China, display its strong, spoon-shaped cropping teeth and fantastic neck length. Narrow-jawed mamenchisaurids were adapted for selective browsing quadrupedally at great heights.

One of several dinosaur species known both from western North America and Proto-Europe, *Torvosaurus* was a large, archaic megalosaurid theropod, which gradually became replaced by allosauroids like *Saurophaganax* by the end of the Jurassic.

sis. Although the medium-size to large species are sometimes powerful enough to kill an old, injured or sick adult, these mainly take a toll on juvenile sauropods.

Young sauropods have in fact been all around us unnoticed since our arrival, and it's only now that the glen has become quiet once more that an occasional movement or soft whine in the stands of *Equisetoides* and *Equisetum* betrays a perfectly camouflaged juvenile. These belong to more than one genus, but only through careful field observation of their subtly differing head proportions and patterns can we tell them apart. All are ravenous little eating machines, at this very young stage competing with each other to consume the same plant species, metabolize calories and grow as quickly as possible. Besides camouflage, their only refuge at this stage lies in their great numbers: for roughly every 30 adults and subadults per square kilometer, there may be perhaps 270 chicks.

In our time, several things make Dashanpu a remarkable locality. One is the extraordinary 50-million-year length of geologic time recorded in the quarry's stratigraphic layers, ranging from the Sinemurian stage of the Early Jurassic period, about 194 MYA, to the 156 MYA Oxfordian stage, the beginning of the Late Jurassic. The Middle Jurassic is currently the least understood period of dinosaur evolution, and the kinds of dinosaurs found here have given us great insight into how sauropods, as well as stegosaurs and theropods, began to radiate and specialize. The oldest Dashanpu formation, the Upper Lufeng, contained a complete skull and postcranial skeleton of a late-surviving prosauropod, *Jingshanosaurus xinwaensis.* By this time animals like this were being outcompeted by the early true sauropods, whose remains elsewhere in China have yet to be discovered. Almost complete skeletons, a rarity in paleontology, happen more often than not at Dashanpu. The preservation of bone at fossil sites depends mainly on two things. One is the amount of *oxidation* (Eh, oxygen-reduction), or the process of dissolving organic substances with the aid of oxygen, that took place in the paleosols, or original prehistoric soils, in which the bones came to rest. In moister environments, oxygen can help break down the organic parts of bone faster than in drier ones. The other is the balance of acidic and alkaline chemicals (pH level) that existed in the paleosol. Rainfall, for example, can dissolve or leach out organic acids in debris-rich forest soils, creating a lower pH level (high acidity, low alkalinity). Less wet environments, like floodplains and especially deserts, have high pH levels (low acidity, high alkalinity), are much less organic and are relatively rich in dissolved minerals, all factors that contribute to fossilization. In forests, moisture and warm temperatures combine to increase biological and chemical activity, which in turn creates acidity and hastens the dissolution and destruction of bone. Evidence of this comes from *fragipan*, a type of paleosol that occurs commonly in areas that once supported hardwoods or coniferous woodland. Fortunately in the case of Dashanpu, the dinosaurs' humid, heavily forested local environment, whose acidic soils would not in themselves have favored bone preservation, bordered on the deltaic area of a major, slow-moving river that occasionally overspilled during heavy rains into a nearby lake. The fluviatile energy of the flooded river, strong enough to move animal corpses and fallen tree trunks, once out in the lake then sharply slowed, allowing muds and silts to gradually cover the remains. Here the fluviatile energy was so gentle that entire skeletons of even the smallest vertebrates like fish and amphibians remained in association and were covered rapidly in the sediment-rich water. In the case of *Shunosaurus*, not only adult but juvenile skulls and postcranial bones were beautifully preserved, often intact, just like the amphibians, reptiles and mammals.

ARCHITECTS OF THE WOODLANDS

In addition to the dinosaur fauna, Dashanpu, together with other roughly contemporary Chinese Middle to Late Jurassic geochrons like the Haifangu, Daohugou and Shishugou Formations, has also painted a very good picture of the overall forest ecosystem. Like much of this area of Proto-Asia, the *mesic* (moist) woodlands were conifer dominated by a mixture of tall foliage types possibly similar to *Araucarioxylon*, followed by shorter forms like *Ptilophyllum*, *Podozamites*, *Elatides*, several species of *Otozamites* and others. Interspersed with these were ginkgoaleans such as *Ginkgoites* and podocarps like *Carpolithus*, and these mid- to high-canopy trees probably formed the basic food sources for the above sauropod adults. Although seasons like those that affect modern temperate forests didn't exist at Dashanpu, some Jurassic conifers might have seasonally shed their leaves or cuticles, as in the living primitive dawn redwood *Glyptostroboides*. If so, the sauropods might have responded by short local migrations within the forest to newly emergent, protein-rich foliage and fruiting organs as they came into seasonal abundance. They later abandoned depleted stands for fresh browse. In their efforts to get as close as possible to their food-producing trees, both adult and juvenile sauropods probably extensively trampled the undergrowth and soil surfaces, an activity known as *bioturbation* and sometimes (when caused by dinosaur trampling) *dinoturbation*. This is similar to the effect elephants and other modern large herbivores have on their local ecosystems, and it can have a profound effect on local habitats. The comings and goings of sauropods and other large dinosaurs on a regular basis would have kept the dominant plants from growing back quickly, allowing many smaller forms to have a chance at sunlight and providing an additional food source for many shorter vertebrates and invertebrates. In this way the sauropods became an important and probably even vital factor in maintaining the biodiversity of a region. When an animal or plant becomes this important to a given environment, it becomes known as a **keystone species**, without

North America

(1) volcanic uplands
(2) inland sea
(3) Fruita

By the Tithonian stage of the Late Jurassic 150 million years ago, North America had drifted away to become a separate continent. The western interior lay between volcanic uplands of the far west and an inland sea to the north, and here the Morrison Formation's well-documented succession of geochrons resulted in the preservation of several large, iconic sauropod species. Like Africa's Okovango Delta, Fruita was alternately saturated by monsoonal rain and flooding (*above, top*) but later by heat and aridity (*above, bottom*). The black line indicates modern coasts.

whose presence an entire ecosystem may be unable to exist. Like modern elephants and some birds, sauropods may have been "ecosystem engineers," shaping the landscape by moving sediment through dinoturbation, removing vegetation and even sowing new plants. In the Daintree Rainforest of modern Queensland, Australia, for example, the southern cassowary (*Casuarius casuarius*) is the only efficient vehicle for spreading certain tree species' seeds. As a by-product of the cassowary's diet, the seeds of the digested fruits are spread over distances of as much as half a mile each day as the big bird travels and defecates, sometimes up hills and across rivers. The seeds of one tree species with a relatively restricted coastal distribution, *Ryparosa kurrangii*, also depend on being in the cassowary's gut to successfully germinate. Of those *Ryparosa* fruits that are eaten, digested and then eliminated, intact and unharmed, 92% germinate, as opposed to the 4% that have no such help.

Just as the Haifangu Formation gives us an idea of the larger forest plants, the distant but contemporaneous Mid-Jurassic Yorkshire flora of England rounds out the probable diversity of the lower-story vegetation. Like some presumed ecosystems in Dashanpu, this is a plant assemblage reflecting mostly deltaic condition and features many ferns like *Cladophlebis*, *Taeniopteris* and others, as well as cycadiforms like *Williamsonia* and *Dictozamites*. Perhaps most importantly for juvenile sauropods, there is abundant evidence of horsetail species like *Equisetites*, *Neocalamites* and *Schmeissneria*. The variety of small to large species would have been the most important food source for these young from the earliest stages.

This highly diverse plant assemblage is what made Dashanpu's equally species-rich fauna possible and is a sign of a stable, long-lasting type of forest ecosystem in which many habitats and feeding niches for animals have developed. The replacement of conifers with angiosperms in today's mesic to wet tropical zone forests, as well as the scarcity of large vertebrates, makes these very different in character from the megafaunal and conifer-dominated Proto-Asian forests of the Mid-Jurassic. In spite of this, Dashanpu's huge, presumably long-lived conifers and probably diverse habitats suggest that it was a true climax community, the Mesozoic equivalent of the great tropical and temperate rain forests of our time.

FLOODPLAINS, FORESTS AND FERN PRAIRIES: FRUITA, NORTH AMERICA

Descending to 27,000′ over western North America, what comes into view is an apparently limitless, flat plain broken by open to dense woodlands and bordered by a vast blue inland sea far to the north and by volcanic highlands in the west. Here and there sunlight sparkles off slowly meandering, interconnecting rivers and an occasional shallow lake as we approach ground level. Our second geochron, **Fruita**, is in what will one day be **Colorado, USA,** and we've traveled 20 million years forward in time from Dashanpu to the Brushy Basin Member of the Morrison Formation, at a point 150 million years before the present. Along the northernmost areas of the Morrison, bordering the inland sea running through future Canada, conditions are humid and heavily forested. Where we've landed, however, it's floodplain,

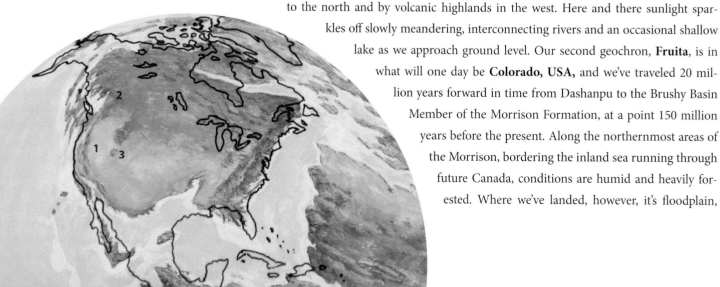
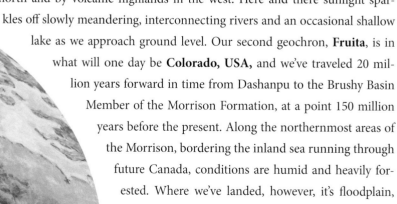

Mesozoic conchostracans, or clam shrimp, were very similar to the modern species *Liostheria*, shown here. Their fossils indicate the presence of ephemeral ponds and a seasonally arid climate.

and the ecosystem is a seasonally wet/dry forested savanna, subject to heavy, almost monsoon-like rains alternating with months of parched aridity. During the rains, the landscape floods to become one gigantic, marshy wetland for miles along the normally sluggish, braided rivers, creating conditions of abundant, lush vegetation. This kind of ecosystem is very similar to the modern Gran Chaco Plain in central South America or the Amboseli Basin of Southern Kenya. We're now a month into the Dry Season, however, and the water has almost totally retreated undergound. In closeup the landscape we now see is an almost continuous, prairie-like community of *xeromorphic*, or dry-adapted, plants that give way to tracts of open to dense conifer woodlands, occasionally taking the form of narrow gallery forests along the margins of rivers and creeks. This "prairie" is mostly made up of fern species similar to the modern *Cheilanthes*, interspersed with the cycads *Cicadolepsis*, *Cicadella* and bennettitaleans like *Nilssonia* and *Zamites*, as well as club mosses like *Selaginella*. During the long dry months, the fern prairie survives on groundwater produced by the heavy rains and drainage in the higher elevations to the west, accumulating in some places as shallow aquifers and artesian wells in subterranean sandstone. The water table is high enough to sustain some permanent marshy places and larger ponds in the river channels, which appear as we trek toward the scattered woodlands and tall gallery forest in the distance. These braided, interconnecting drainage beds are several meters wide, and both they and their banks are slightly *higher* than the surrounding floodplain. During the annual Wet Seasons, successive flooding away from the main river gradually deposits more and more clays and other sediments along the sides of the channels, forming high banks or levees. As these build up, so does the floor of the river channel, but when one of the levees occasionally gives way at a weak point during a particularly high flood, the water fans outward in a new direction and the old watercourse is abandoned. This is where ponds and marshes are likely to occur. Sometimes the escaping water from the breached levee forms a new channel by following "game trails," ruts formed by the passage of large and small dinosaurs.

Fruita, USA

Located far inland, Fruita, North America, experienced seasonal monsoon and dry seasons similar to today's Gran Chaco and Pantanal floodplains. Silt buildup during the Wet Season resulted in distinctively raised river channel levees (1) that occasionally gave way to allow the creation of ephemeral ponds (2) that permitted brief life cycles like those of conchostracons, but these evaporated during the summer (3, section). Deeper, year-round drainage patterns supported dry-adapted conifer forests (4), which were a mainstay of several large diplodocid sauropod species and many other dinosaurs. Vegetation is sectioned to show topography.

During the dry months following the Wet Season at Fruita and other Late Jurassic North American localities, sauropods like *Supersaurus vivianae* were constantly on the move, seeking green fresh conifer browse and water.

As we come close to one of the channel ponds, some small turtles, *Glyptops*, scramble into the water. Poised and staring at us from a sandspit is what at first seems to be a strange, chunky varanid lizard, but which actually turns out to be a small terrestrial crocodylomorph, *Fruitachampsa callisoni*, with a tiny, limp docodontid mammal in its jaws. It sprints away across the wet sand, past the exoskeleton of a crayfish. In the biggest marshes lurk the medium-sized croc *Eutretauranosuchus (Goniopholis) gilmorei*, mainly a predator of fish and other small vertebrates. We know that ponds and marshes like this will survive the Dry Season because of the presence of gastropod snails and unionid clams, both of which require permanent water. Although we haven't yet seen any, the unionids confirm the presence of fish as well, since these clams need the fishes' gills for their larvae to parasitize on their path to adulthood.

The other ponds we see out on the lower areas of the floodplain are different, however—at these the water doesn't stay. Here the litter of tiny, delicate exoskeletons at the edges of the receding water belongs not to crayfish but to *conchostracans,* or clam shrimp. Clam shrimp like *Liostheria* only colonize *ephemeral,* or temporary, ponds: their short life spans and wind-dispersed eggs are both adaptations to places that will soon dry up. The large lungfish lying motionless on the muddy bottom is another indicator of hard times to come, since it will soon burrow into the mud to form a protective, moisture-conserving cocoon from its own mucous. Although we're here to observe the sauropods and other dinosaurs at Fruita, these smaller microfauna are well worth our attention. It's the "micros," the small animals that make their living from and are bound to particular habitats, that actually provide the details that allow us to sketch a vanished environment's portrait, rather than the macrofauna, the bigger animals. What they tell us is that here at Fruita, and throughout much of the vast Morrison area, the climate swings annually between extremes of moisture and aridity. This is backed up by the chemistry and formation of the paleosols, which, unlike the acidic makeup of Dashanpu's soils, are characterized by alkaline conditions, typical of predominantly arid climates. Another sign of this are the tall termite mounds dotting the plain, one of which is being broken into by a small, energetic mammal, *Fruitafossor windscheffeli*, whose strong, clawed forefeet and peg-like teeth recall those of modern aardvarks. Like the latter, *Fruitafossor* specializes in preying on termite colonies.

Let's cut across the floodplain and continue toward the stands of dense woodland. In our own time, fossil pollen sampling and other finds show that when the sediments of the Morrison's Brushy Basin Member were being laid down, several varieties of conifers were present, all suited to the wet/dry seasonality that characterized the region. The conifer forests of the Morrison are extensive, and these include, most importantly for sauropods, cheirolepidiaceans like *Cupressinocladus* and *Pagiophyllum*, as well as other now extinct conifers such as the taxodiacean *Elatides* and others like *Podozamites*, *Pagiophyllum* and *Araucarioxylon*. Today's large fossilized trunk sections indicate that some of these could reach 30 m (100′) or more, creating stands of trees that were the cores of complex animal and plant communities. Modern relatives of these extinct araucarian and podocarp trees still exist on some of Earth's southern continents that used to be Gondwana, reminders of the Late Jurassic woodlands we're now walking toward. Extensive forests of monkey puzzle (*Araucaria araucana*) thrive in poor soil on the slopes of today's Volcan Llaima in southern Chile, while modern podocarps like the Kahikatea (*Dacrycarpus dacrydiodes*) in New Zealand and others in South America and Asia represent the former geographic extent of this family. Here at Fruita genera like *Ginkgo* and *Ginkgoites* are the next most common trees, and their sunlight requirements result in their forming their own communities of more open woodlands. These, in turn, shelter an impressive variety of "seed fern" trees (caytoniales), cycads and cycadeoids, including *Cycadella*, *Nilsonnia*, *Cycadolepsis* and *Williamsonia*. Unlike the araucarians and podocarps, the cycads and bennettitaleans shed their leaves on an annual basis, offering a dinosaur herbivore no food during part of the year. In the Morrison's Kimmeridgian stage *Sequoiadendron* (represented by wood specimens) are the *monospecific*, or dominant, tall-

Diplodocus carnegii was one of the more common diplodocids throughout the Kimmeridgian–Tithonian stage at Fruita and other Late Jurassic western localities.

Proto–Africa/ South America

(1) Tethys Ocean
(2) Tendaguru

At Tendaguru, Proto–Africa/ South America, long-term, alternating marine transgressions and regressions during the Late Jurassic influenced the local forest ecosystems. Although its location was inland like Fruita's and the dinosaur fauna bore some similarities, the climate was only mildly seasonal, and brachiosaurids and dicraeosaurids were more common than diplodocids. The black line indicates modern coasts.

reach winners in the competition for sunlight. These form modest but dense stands, creating so much shade that trees that need more su, like *Pagiopyllum* and *Podozamites*, are forced to grow around marginal clearings, or in the company of ginkgos and cycads. Ginkgophyte trees, now known only from the surviving *Ginkgo biloba*, also shed their leaves, but when in leaf these, along with the nutritious seeds, offer a high-quality energy source to sauropods, especially crude protein. Ginkgoaleans' tolerance for habitat disturbance by dinosaur trampling, along with their resistance to insect damage, make them a successful and common tree type in the Morrison. Growing under the taller vegetation in moist, closed-canopy areas, mesic families of ferns like *Angiopteris*, *Elatides*, *Cladophlebis* and *Osmunda* are common, sharing this habitat with "seed ferns" like *Sagenopteris* and *Czekanowskia*. *Equisetum*, the taller *Equisetoides* and small herbaceous plants provide food and concealment for young sauropods and other small herbivores.

Along the way a patch of trampled fern prairie and the buzzing of flies reveal a kill, the now dismembered and rapidly drying carcass of a small *Camarasaurus lentus*, our first Fruita sauropod and only one of many casualties to come this season. Judging from its size, it was about seven months old. At some point earlier in the day it was brought down by a bite to the throat from a medium-sized theropod, possibly *Marshosaurus bicentissimus*. What's left is now providing protein for two scavenging *Fruitachampsa*, and clouds of flies are laying eggs on the corpse. The juvenile's death occurred away from any of its likely food plants, and a nearby large pile of *Camarasaurus* dung suggests that it was starting to feed from an adult's feces, in order to acquire gut microflora, when it was killed. This would have helped in its transition to eating adult browse. The fibrous excrement is alive with dark, shiny beetles, each of which is industriously carving out and forming its own tiny dung ball, rich in undigested plant remains, for its grubs to eat after being rolled away and buried.

For the last few minutes we've noticed a perceptible, rumbling vibration in the air and on the ground, and suddenly we see them: up ahead, 76 m (750′) away, a small group of 12- to 15-year-old *Camarasaurus lentus*, striding toward a mixed stand of *Podozamites* and *Brachyphyllum*. Big headed for sauropods, their large, observant dark eyes and keen sense of smell and hearing take everything in as they approach the trees, and their disruptive pattern of irregular greenish-brown splotches on light orange, vaguely giraffe-like, visually locks into the foliage and makes them disappear into the embracing branches. The taller ones have already chosen the choicest feeding spots, forcing the smaller juveniles to amble around, moaning loudly in frustration, until they find a place where they can reach. A few small ornithischians, *Drinker nisti* and *Othnielosaurus consors*, scamper out of their way to glean the tender leaflets that fall to the ground as the sauropods feed. The camarasaurs' stout teeth make short work of the tough bracts and leaflets, and although we see a few up-and-down jaw closures, the browse is quickly swallowed, making room for the next mouthful. *Camarasaurus* occurs as at least four species throughout the Kimmeridgian, first appearing as *C. lentus* during the middle of the stage and producing *C. grandis* and *C. lewisi* before apparently evolving into the larger *C. supremus* later on. This advanced macronarian, with differential wear patterns on its most anterior teeth, moderately elongated glenoid fossa and simple orthal bite, has a jaw action suitable for harvesting a broad spectrum of vegetation. The fact that "shoulders," or shelf-like macrowear structures, are commonly found on fossil teeth confirms that a ripping or tearing action takes place during feeding. Because of the potential (but probably limited) ability of its teeth to crush, *Camarasaurus* is a rather generalized, mixed browser, harvesting a variety of conifer and broad-leaved *Ginkgo* foliage. If necessary, it can resort to feeding on tough cycads.

Almost at once the juvenile *Camarasaurus* stop their rumbling and branch tearing, turning their heads toward the west as a series of high-pitched, metallic wails become audible. All now face the sound, and the larger ones quickly rear, presenting large, curved thumb claws. A tall, weaving shape appears in the afternoon haze, moving jerkily as it comes closer. It's a *Brachiosaurus altithorax* male somewhere in its early teens, and a long red stain of blood is gushing from a deep, jagged tear in the right tail base, slowing the sauropod's normally fast amble. As it lurches toward the trees, we also see another, fast-pursuing figure, which turns out to be the theropod *Ceratosaurus magnicornis*. This is the now larger version of the older species, *C. nascornis*, from the earlier Morrison. An adult female, it ambushed the young *Brachiosaurus* male when the sauropod's quest for greener *Sequoiadendron* leaflets took it farther than it realized from its age group. While the young giant was busy feeding, *Ceratosaurus* was there to rush forward and inflict a bite with its massively long, serrated teeth in a critical place, the muscles of the tail base that provided the sauropod with much of its walking power. Instead of turning toward the others in its age group for support, the *Brachiosaurus* panicked and blindly took off.

The tall, stately brachiosaurids are the largest of the Morrison sauropods and carry their massive bodies easily on relatively slender, long legs, moving slowly and effortlessly across the plains from one area of forest to another. They are macronarians and basal titanosauriforms and have *converged*, or evolved in a similar way, to the omeisaurids of Dashanpu. Like the Asian sauropods, their high shoulders, long necks and legs are all specializations for extreme high reach while remaining quadrupedal, and although they are capable of rearing even higher on two legs, it's seldom necessary. Brachiosaurid teeth, although somewhat spoon-shaped like the camarasaurs', are longer, end in chisel-like tips and are set in a mouth that's proportionately wide in relation to its length. These produce a more precise form of cutting or shearing, and even though their teeth are specialized, brachiosaurids can crop huge amounts of their preferred browse. Since *Sequoiadendron* and other high-canopy trees aren't common in Fruita, however, these sauropods aren't either.

As opposed to the greater prevalence in North American localities of diplodocids, in Tendaguru the brachiosaurid *Giraffatitan brancai* appears to have been more common, based on fossil findings.

The Proto-European Archipelago

One of the largest islands in a scattered archipelago that much later would become Europe, Late Jurassic Ibero-Armorica had a more humid climate than Tendaguru, with forests like those of North Australia but also dry scrub. Geographically very close to both North America and Proto–Africa/South America, localities like Lourinha were a crossroads of dinosaur migration (*arrows*), and in addition to having brachiosaurids, diplodocids, dicraeosaurids and possibly camarasaurs, it also featured the huge, distinctive turisaurids, thus far known only from this region. The black line indicates modern coasts.

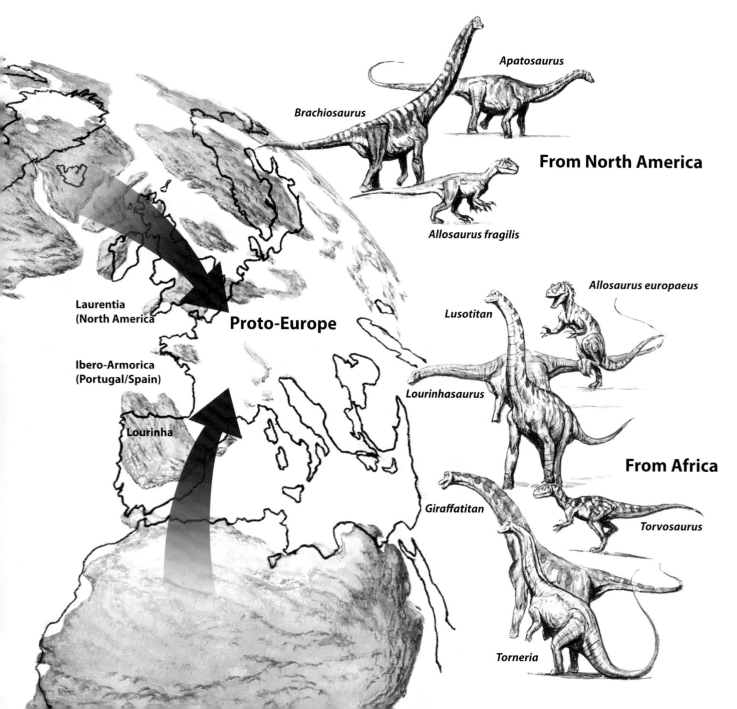

Brachiosaurus

Apatosaurus

From North America

Allosaurus fragilis

Laurentia (North America

Proto-Europe

Ibero-Armorica (Portugal/Spain)

Lourinha

Lusotitan

Allosaurus europaeus

Lourinhasaurus

From Africa

Giraffatitan

Torvosaurus

Torneria

As rare in the Morrison as *Brachiosaurus* is the medium-sized (12 m [38′]), 5-ton, enigmatic *Haplocanthosaurus priscus*. Evolved from the earlier *H. delfsi*, this is a basal diplodocoid that's on its way to extinction, possibly no longer able to compete with advanced diplodocids and perhaps another basal diplodocoid, *Suuwassea emiliae*. *Suuwassea* is both an early flagellicaudatan, or "whip-tailed," diplodocoid and also a basal dicraeosaurid, the other family besides the diplodocids that makes up this group. *Suuwassea*'s remains are found in Montana, farther north than the better-known Morrison exposures that yielded diplodocids like *Apatosaurus* and *Diplodocus*. Because this is located during the Late Jurassic within the even more heavily forested, more humid north near the Canadian inland sea, it could mean that the smaller dicraeosaurids preferred the vegetation found around more densely wooded, coastal habitats. If so, this may have also been true of the Proto–African/South American forms, the basal dicraeosaurids *Amargasaurus*, the more advanced *Dicraeosaurus* and the highly specialized, short-necked *Brachytrachelopan* (see Chapters 5 and 6).

When the *Ceratosaurus* catches up with the injured subadult *Brachiosaurus*, a few massive bites put its cries to an end. Like the brachiosaurids, ceratosaurids aren't as common in the Morrison as other theropods. Their huge teeth suggest a specialization for penetrating the thick hides of older sauropod prey, and they may be solitary hunters. Others like the more common and smaller-toothed *Allosaurus fragilis* take in a much wider prey spectrum. These theropods and others, like the now archaic, long-headed megalosaurid *Torvosaurus tanneri*, often cull the younger sauropod juveniles, as well as basal iguanodontians like the 5 m (16′), 500 kg *Camptosaurus dispar* and much smaller *Dryosaurus altus*. The basal iguanodonts are part of a guild of ground-level and low-canopy browsers, a niche they share with the versatile stegosaurs. Short-bodied and compact, the two Morrison stegosaur forms *Hesperosaurus mjosi* and *Stegosaurus ungulatus* (*stenops*) often browse on short growth, but their long hind legs and short forelegs give them the ability to easily tilt backward to browse on low-canopy vegetation. In this position, their slender heads and pointed beaks let them probe for the choicest, most nutritious plants.

The most accomplished bipedal browsers in Fruita and elsewhere in the Morrison, however, are the diplodocids, typified by *Diplodocus carnegii*. As we walk toward the tall gallery forest fringing one of the depleted rivers, we can hear a series of resonant, echoing moans and rumbles that tell us we're close to them. A small pod of subadults and one or two fully grown adults suddenly betray their subtle camouflage as they momentarily jockey for better feeding positions, keeping an acceptable distance from each other by lashing their thin, whip-like tail ends that bear sharp, keratinized spikes. Able to slowly and carefully walk bipedally if they need to, some *Diplodocus* settle back on their hind legs and proximal tail sections into a tripodal pose when they see a productive amount of foliage. Although their necks are less flexible than the camarasaurs', the anterior portions and steeply angled heads are supple and active. When they encounter a leafy branchlet, *Diplodocus*'s jaws open wide, clamp down and pull backward, efficiently stripping off the leaves with forwardly projecting, slender teeth. As with all sauropods, mouthful after mouthful is easily sluiced down the long esophagus, and a small pod like this will feed for several hours straight before moving on to the next stand of trees. The mature adults turn their heads and hiss as an enormous new sauropod ambles into the feeding area. Massive-bodied and

Found at Tendaguru, *Dicraeosaurus hansemanni* was a low- to mid-level browser that possibly shared the same habitat with stegosaur *Kentrosaurus aethiopus* but did not directly compete because the former was a wide-spectrum feeder, while the latter used its narrow beak to pluck selectively.

robust, with tall vertebral spines and a deep, very wide neck, *Apatosaurus ajax* ignores them as it strolls toward its preferred food tree, a spiky, formidable-looking *Araucaria*. The dead, sharp cuticles on the outside and base are no problem for *Apatosaurus*, which uses its wide, boxy neck to push the mature, less edible branches aside so that it can reach the tender, protein-rich cuticles inside and farther up. Widest at midpoint, its neck at the outside edges is protected against injury by tough, horny calluses. If it needs to, this species can use its robust neck almost like a wrecking ball to break up thicker side branches. As rutting season approaches, well-matched rival male *Apatosaurus* clash with each other in noisy, spectacular displays, rearing upright to batter with their wide necks, as well as roaring loudly and displaying their vividly colored throat sacs. *Apatosaurus*, like *Camarasaurus* and *Diplodocus*, will also become larger and ever more massive as the Kimmeridgian continues, terminating in the enormous *A. ajax*. This genus is less common than *Diplodocus* in the middle to upper levels of the formation, and its more active feeding method may require it to be more solitary. As a fossil it's often found in association with *Stegosaurus*, meaning that the two dinosaurs' food plants may have grown in the same habitats.

Other, even larger diplodocids roam the savannas. This time is the zenith of the diplodocids' evolution in size and diversity, and all have adapted to feeding from different parts of the woodlands' upper to mid-canopies, and at different times, allowing several species to coexist. *Barosaurus lentus*, a close relative of *Diplodocus*, takes over from where its smaller cousin can't reach, browsing 9.25–12.8 m (30′–40′) into the upper canopy level, its longer neck and greater ability to bend the neck laterally allowing it to reach where it wants. *Diplodocus* (*Seismosaurus*) *halli* and *Supersaurus vivianae* are still larger, measuring over 30 m (100′) from nose to tail. Even these, however, may not have been the largest. Based on the discovery made more than a century ago of a single, enormous dorsal vertebra and now more recent material, there *may* have existed during this time a basal but truly gigantic diplodocoid known from possibly two species, *Amphicoelias fragillimus* and *A. altus*. If extrapolations from these bones are correct, *Amphicoelias might* have been 36 m (120′) long and could have extended its head to feed at the staggering height of 18.25 m (60′). Clues to the radiation of the diplodocids have come from the discovery of the early Morrison *Kaatedocus siberi* at Howe Quarry, whose well-preserved juvenile skull

At Tendaguru, Proto–Africa/ South America, long-term, alternating marine transgressions and regressions during the Late Jurassic influenced the local forest ecosystems. Although its location was inland like Fruita's and the dinosaur fauna bore some similarities, the climate was only mildly seasonal, and brachiosaurids and dicraeosaurids were more common than diplodocids.

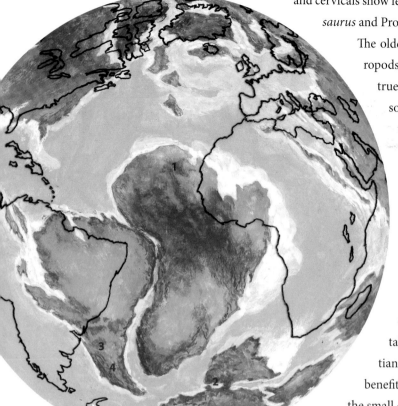

and cervicals show features that suggest a close relationship with the more basal *Supersaurus* and Proto-European *Dinheirosaurus lourinhanensis*.

The older juvenile, subadult and adult diplodocids and some other sauropods of Fruita and elsewhere in the central Morrison's ecosystems are true nomads, forced by their massive hunger and the vast region's seasonal wet/dry conditions to make local to long-distance migrations to find water and browse, possibly to the north and west. Here the sauropods' adaptability spells survival, and although we can't follow the pods, we can imagine what such migrations might be like. If they fail to find the arboreal browse they need, the diplodocids can subsist off the dry but plentiful fern prairies, lowering their long necks to a horizontal position to harvest the nutritionally poor but abundant browse. Under these conditions they're the most vulnerable to theropod ambushes, and at least one older adult forgoes feeding until his turn comes, occasionally rearing up to scan the landscape with sharp eyes that take in the most subtle movements. The smaller basal iguanodontians and other ornithischians travel with the giants in small groups, benefitting from the sauropods' vigilance. It's not all one way, however; the small dinosaurs' keen sense of smell provides an additional early warning to the sauropods if predators are in the area.

In the millennia to come, the Morrison sauropods' fortunes will vary. So diverse in size and species during the Late Jurassic, the diplodocids will dwindle and apparently become extinct, with the exception of at least one South American species, *Leikopal latiocauda*, by the earliest stages of the Cretaceous. The brachiosaurids and some basal macronarians will flourish at least for another 40 million years in the form of species like *Abydosaurus mcintoshi* and gigantic *Sauroposeidon (Paluxysaurus) proteles*. As the Late Jurassic ends and a brief cooling trend ushers in the Cloverly and Cedar Mountain Formations' new ecosystems, the surviving macronarian sauropods will encounter developing angiosperm plant communities, begin to rub shoulders with new dinosaur forms like the basal therizinosauroid *Falcarius utahensis* and occasionally fall prey to the first of the larger predatory dromaeosaurs. Their world is changing.

THE SAME, ONLY DIFFERENT

As we maintain our position over North America, the western part of Laurasia, we can see how close it is to the fragmented island archipelago of Proto-Europe, which almost forms a series of irregularly sized stepping stones between this continent, eastern Laurasia and North Gondwana (Proto-Africa and Proto–South America). What will one day become the North Atlantic Ocean and will eventually separate North America from Africa is still narrow. In addition to Fruita, two other localities, **Tendaguru, Tanzania** (Proto–Africa/South America), and **Lourinha, Portugal** (Ibero-Armorican Island, Proto-Europe), share sauropods that are so similar that they could almost be the same species, while others are ***endemic***, found in one place and no other.

Proto–South America and Proto-Africa

(1) **Northern Gondwana**
(2) **Southern Gondwana:**
India, Madagascar,
Australia, Antarctica
(3) **Patagonia**
(4) **Neuquen Basin**

During the Barremian stage of the Early Cretaceous approximately 125 million years ago, Proto–South America and Proto-Africa were still joined but had begun rifting apart. As a result of their long connection, sauropods such as dicraeosaurids, rebbachisaurids, titanosaurs and a few surviving diplodocids are known from both landmasses. The black line indicates modern coasts.

Tendaguru, a locality very close to the Proto-African east coast, alternates between a series of sea-flooding events (marine transgressions) and periods of being dry enough to exist as a group of tidal estuaries and low-relief coastal plains. During the non-marine periods, these give way farther inland to seasonally wet/dry, subtropical open forests, and like the Fruita ecosystem, they're dominated by conifers like cheirolepidiaceans and araucarians. Here, however, in contrast to Fruita and other Morrison localites, the brachiosaurid *Giraffatitan brancai* is the most common sauropod. It's slightly bigger than its counterpart *Brachiosaurus* in North America, with a longer manus and arm, which makes the shoulders even higher. The high shoulders also anchor unusually deep tendon attachments that dorsally support the neck. This indicates an even more extreme adaptation to maintaining a vertical neck posture for high browsing than *B. altithorax*, and *Giraffatitan*'s abundance here indicates that this brachiosaurid, unlike that of Fruita, is finding enough of the conifer browse it needs. Tendaguru's sauropod diversity doesn't only consist of *Giraffatitan*; other, poorly known basal titanosauriforms, including *Tendaguria tanzaniensis*, *Australodocus bohetti* and *Janenschia robusta,* are also present. *Janenschia* is actually an early or basal **titanosaur** (not to be confused with the more inclusive taxon, **titanosauriforms**) and, along with others of its kind, is destined to be the dominant, and later only, sauropod family of the next great Mesozoic epic, the Cretaceous (see Chapter 11).

Although Tendaguru has at least one diplodocid that's identical or very similar to Fruita's *Barosaurus, Tornieria (Barosaurus?) africanus*, it's also home to the **Dicraeosauridae**, the other diplodocoid family that we met in the form of the Morrison's *Suuwassea*. Like *Suuwassea*, Tendaguru's dicraeosaurids are relatively small and somewhat short necked by sauropod standards, and their cervical architecture is distinctive from that of other diplodocids. Although bifurcated, the neural spines are taller, wider and highest at just forward from the midpoint of the neck. Their height, along with the inclination of the spines caudally from the head and cranially from the neck base, creates especially efficient bracing for the extension and flexion from the midpoint of the neck. It suggests a very different method of feeding from diplodocids like *Diplodocus* or *Apatosaurus*, and together with the more robust mandible, it could mean the ability to exert a stronger, more powerful up-and-down pulling action in cropping. This, in turn, could mean a tougher, less yielding form of conifer browse, either at lower levels or closer to the ground. In Tendaguru, as in Fruita and other Morrison localities, at least one advanced stegosaur, *Kentrosaurus aethiopicus*, is well established and strongly competes for this feeding niche. *Dicraeosaurus* is nevertheless a reasonably common sauropod at Tendaguru, and by the Mid-Tithonian stage the earlier *D. hansemanni* eventually evolves into the later *D. sattleri*. Dicraeosaurids are currently known only from three other genera and species, the basal forms *Suuwassea emeliae* from western North America, *Amargasaurus cazaui* from Early Cretaceous Proto–South America and *Brachytrachelopan mesai* from the same continent. The smaller size and specialized neck morphology of dicraeosaurids may make them well suited for medium-height to low browsing. This, as well as the poorly known presence of rare stegosaurs in Proto–South America, could mean that in this landmass dicraeosaurids evolved to dominate a feeding niche that later came to be partially occupied by stegosaurs in North America and Proto-Europe, with which they found themselves in competition when they expanded into Proto-Africa.

Octavio Mateus, a Portuguese specialist in the dinosaurs of Late Jurassic Ibero-Armorica, is professor of paleontology and biology at the Facultedade de Ciencias e Tecnologia da Universidade Nova de Lisboa and an associate of the Museu da Lourinhã. He has organized international excavations, conducted comparative studies of sauropods from Ibero-Armorica and named new species such as *Lusotitan* and *Europasaurus*.

Lourinha is located on the western edge of Ibero-Armorica, one of the biggest islands in the European archipelago. It's a coastal floodplain environment, but with a slightly wetter climate than Fruita or Tendaguru. Like the other two localities, it's rich in sauropod species, some of which are remarkably similar to each other. Lourinha features not only the huge endemic, basal diplodocid *Dinheirosaurus lourinhanensis* but also an *Apatosaurus* species either the same as, or closely similar to, the one from Fruita. A big brachiosaurid, *Lusotitan atalaiensis*, is also present, paralleling the two species we've met in Fruita and Tendaguru. *Lourinhasaurus alenquerensis*, at present a poorly known macronarian, was possibly the Old World counterpart to the North American *Camarasaurus*. Lourinha is only one locality among several others in Late Jurassic Proto-Europe, and if we include sauropods from nearby northern and eastern Ibero–Armorica (Spain) and Wealden, northern Proto-Europe (Britain), we encounter another group, the **Turiasauria**, a late-surviving clade of eusauropods that at present are thought to be endemic to Proto-Europe. These occur as medium-sized to gigantic forms that, in the case of *Turiasaurus riodevensis*, reached approximately 30 m (100′) and 50 tons, making them the largest known sauropods of Proto-Europe. Originally considered to be diplodocoids, turiasaurs have several unique vertebral features, as well as distinctive, heart-shaped tooth crowns with pointed, asymmetrical and compressed tips. Together with other genera like *Losillasaurus giganteus*, turiasaurs survive at least into the earliest Cretaceous and have relatively shorter necks. For this reason, they may be convergent with the endemic North American and Proto–African/South American dicraeosaurids.

After surveying the geochrons of Fruita, Tendaguru and Lourinha that represent the three regions of North America, Proto–Africa/South America and Proto-Europe, we see that all these areas are very similar in their general environments and in the overall makeup of most sauropod types (except for the endemics) and other dinosaur species. There are two, however, subtly important differences: in Proto-Europe there are apparently *fewer* sauropod taxa (and more theropods) than in either of the others. Also Tendaguru, Proto–Africa/South America, in spite of representing an even bigger landmass than North America, has only *38%* of the dinosaur families found at the latter. Why is this? Since at least intermittent land connections exist from time to time during periods of lowered sea levels to permit animals to migrate, we might expect the populations of the known sauropod species to be pretty much evenly distributed. This, however, isn't the case. So why are certain sauropods (and some other dinosaurs) more diverse in the two larger landmasses, and why are some found in one region and nowhere else?

At least two factors come into play here, and one is geography. Proto-Europe once shared similar dinosaur types with the rest of Pangaea before the giant supercontinent broke up, and although now separated, the archipelago is still located close to its huge neighbors North America in the west and Proto–Africa/South America to the south. As a result, it's a dispersal route or pathway for the migration of dinosaur populations between the other, bigger landmasses. Such areas often acquire animal species from both areas. How do we know that Proto-Europe wasn't the *origin* of dispersals to the other two regions, instead of the other way around? Although it's hard to be certain about the absolute direction of prehistoric animal dispersals, we can reason that bigger landmasses have a higher numerical potential to populate smaller areas. Owing to their vast geographic areas, North America and Proto–Africa/South America contained larger and more diverse dinosaur populations and therefore had more potential to colonize other territories. On the other hand, it could be that because some North American localities in the Morrison probably have up to 100 times the outcrop area of Tendaguru and have been intensively collected in for 150 years (as opposed to a few decades for Tendaguru), the lower species diversity is actually just an artifact of incomplete sampling.

If Proto-Europe is truly a conduit for migrating dinosaurs at this point in the Late Jurassic, why hasn't it actually acquired *more* species, as a result of intermittent migrations, from both of its giant neighbors? Again geography can provide an answer, this time through acting as a *filter*. Some sauropod species may be able to gradually migrate from their homelands in one of the bigger continents to the other by moving across Proto-Europe because the foliage and other conditions are compatible with their food preferences, but others are more limited by less suitable conditions, and others are stopped completely—in other words, they're partially or totally filtered out. This is known as an ***ecological barrier***. At this point we don't know what these are, but it could explain why the sauropods and other dinosaurs of Tendaguru, while sharing many of the forms so far known from the Morrison's Fruita, have only 38% of the latter's families. It could also explain why *Turiasaurus* and its sister species may have been unable to colonize the bigger landmasses and so remained endemic to Proto-Europe. A mystery, however, is that the dicraeosaurids are found in both North America and Proto–Africa/South America but not Proto-Europe, which had to be a dispersal route for this family and should have retained at least some of these specialized diplodocoids. Here the answer could be that there actually were, but these haven't yet been discovered. Biogeography plays a huge role in the evolution of sauropods throughout the Mesozoic, and we'll return to this subject again in this and other chapters.

DESERT UPLANDS, OASES OF LIFE: PATAGONIA, PROTO–SOUTH AMERICA

We speed forward in time and at this point arrive in **Patagonia** (western Argentina), **Proto–South America** at the end of the Barremian stage of the Mid-Cretaceous, about 125 MYA. Proto–South America is unzipping from its still-conjoined Gondwanan twin, Proto-Africa. The landmass's overall climatic conditions are at present like modern Australia's: mostly dominated by aridity, but nevertheless with a humid, moister belt. In the deserts life flourishes but is sparse, supporting only a small number of dinosaur herbivores such as large and small ornithopods and possibly some sauropods, as well as a variety of theropods. The low-lying basins and the slopes surrounding them, however, with their drainage from the dry uplands, are differ-

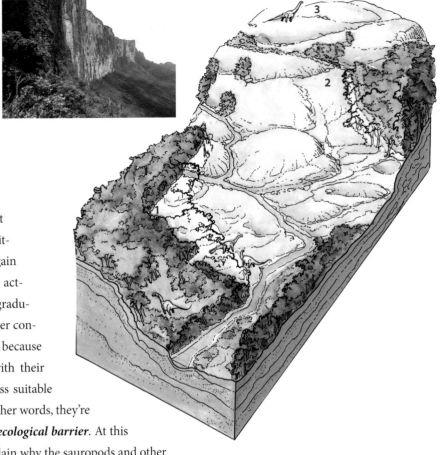

The Neuquen Basin, Patagonia

The Neuquen Basin was one of several huge, widespread lowland basins that sat below forested highlands and arid deserts, which may have been similar to the Roraima Escarpment, Venezuela (*above*). Aridity characterized most of Proto–South America during the Early Cretaceous, but oasis-like basins (**1**) supported species-rich ecosystems. The drier highlands (**2**) and parched deserts (**3**) may have been evolutionary laboratories for evolving rebbachisaurids and titanosaurs. Here the uneven, hilly terrain may have resulted in distinctive locomotion innovations that helped them to radiate after the extinction of diplodocids and eusauropods. Vegetation is sectioned to show topography.

ent, ecologically much more varied environments, and these support far more diverse dinosaur communities. Such basins, remnants of marine transgressions, are hundreds-of-miles-square, Nile delta–sized and larger oases in an arid protocontinent that persist for millions of years. We'll focus on the forest-rimmed Neuquen Basin in the Province of Neuquen, western Argentina, since, in addition to others, this locality contains at least three consecutive formations, the **La Amarga, Lohan Cura** and **Rayoso**. These give us glimpses of key periods of sauropod evolution in Early to Mid-Cretaceous (Barremian–Albian) Proto–South America. During these times, drainage from arid highlands into the flat basins creates extensive systems of broad, meandering rivers and lakes that deposit sands and mudstones across floodplains. These well-watered environments support and preserve a rich variety of small invertebrates like ostracods, clams and insects, as well as aquatic vertebrates like fish, frogs, turtles and crocs. Here and in the surrounding upland hills bordering the deserts, *Classopollis* and other fossil pollen found in the sediments show that cheirolepidiacean conifers and the araucarian *Cyclusphaera radiata* and others are present, as well as the podocarp *Squamastrobus tigrensis*, seed ferns and other understory plants in well-drained areas. Trackways and fragmentary skeletal material indicate that these forests support some ornithischian dinosaur herbivores. Among these are an *Ouranosaurus*-like, large basal iguanodontian, a possible primitive nodosaurid ankylosaur and perhaps one of the last South American stegosaurs, based on finds of cone-shaped to flat dorsal plates.

Our landing portal this bright, sunny midmorning is on a windy, araucarian-dominated mesa with spectacular views of the green Neuquen Basin stretching to the southeast. The trees are tallest and most abundant where the drainage is good, and the shelter an understory of shrubby, cheirolepidiacean conifers, cycadeoids and ferns. Movement in the filtered shade betrays a big animal, and as we stand quietly a sauropod ambles out, foraging among the lower growth but occasionally rearing to reach higher, newer browse. It's the medium-sized (13 m [43'], 4 tons) dicraeosaurid *Amargasaurus cazaui*, and it's amazing. The dorsals, sacrals and paired neural spines of its cervical vertebrae aren't just bizarrely tall—in the case of the cervicals they're also long, pointed and curve backward. At about midpoint on each vertebra there are alternating ring-and-striation patterns where a horny keratin spike begins. These closely resemble scars found on the bony horn cores of living cattle, extending the actual length of the horn by as much as 25%. On *Amargasaurus* the horns are potentially lethal weapons, and in addition to being effective predator deterrents (see Chapter 9), they may come into play to intimidate sexual rivals

Amargasaurus cazui's amazingly long cervical neural spines, probably in life tipped with sharp horns, may have been lethal weapons of defense.

not only by virtue of sheer size but also by both animals actively locking them and neck wrestling. The cervical spines can also be clattered together to create a startling, noisy display to signal aggressive intent. Earlier during the Tithonian stage in the Chubut Province, Argentina, a smaller, even more specialized dicraeosaurid existed, *Brachytrachelopan mesai* (see also earlier in

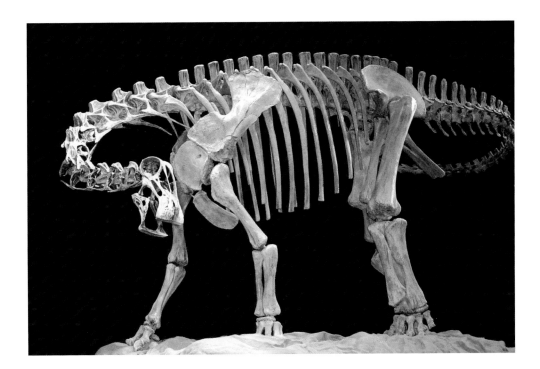

Although some European rebbachisaurids were enormous, most known Proto–South American forms, like *Limaysaurus tessonei*, were medium sized to small, and one Proto-African species, *Nigersaurus taqueti* (*left*), was a very small, short-necked sauropod adapted for low browsing.

this chapter and Chapter 6). *Brachytrachelopan*'s neck, at only a little over 1 m (3.5′) long, is the shortest of any known sauropod, which suggests low browsing. A lack of competition from basal iguanodontians, unknown in this area until much later, and the seemingly rare stegosaurids may have allowed it to occupy this feeding niche.

The basins and surrounding forests, huge green punctuations occurring in a vast, continent-wide desert, preserve a long and spectacular record of sauropod diversity. Although all diplodocoids seem to have become extinct by the earliest Cretacecous in North America, they persist in Proto–South America (and Proto-Africa) during this period. The **La Amarga** geochron is inhabited not only by the dicraeosaurid *Amargasaurus* but also by the basal diplodocoid *Zapalasaurus bonapartei*, and the recent discovery of a diplodocid, *Leikupal laticauda*, confirms that at least relicts of this clade were still able to hang on here. After Barremian times, both dicraeosaurids and stegosaurs will go extinct in Proto–South America/Africa, and this may be due to feeding competition from the basal iguanodontids and other ornithopods that were beginning to disperse into these regions from Laurasian lands.

REBBACHISAURIDS—A NEW WAY OF WALKING?

Moving forward to the Neuquen's next sub-geochron, the **Rayoso/Lohan Cura Formation** of the Aptian stage, we encounter the **Rebbachisauridae**, a third, surviving major family of diplodocoids. These are small to medium-sized forms and will evolve some highly derived species, such as the later Aptian *Nigersaurus taqueti* from the northeastern coastal river deltas of Proto-Africa (see Chapter 6). This, like the earlier dicraeosaurid *Brachytrachelopan*, will be a short-necked, low browser. *Nigersaurus*, with its jaws as wide as its head and batteries of teeth, is a member of a subgroup of rebbachisaurs known as ***rebbachisaurines***, and these will depart from other rebbachisaurs, the ***limaysaurines***. In the Lohan Cura Formation's *Limaysaurus tessonei*, a

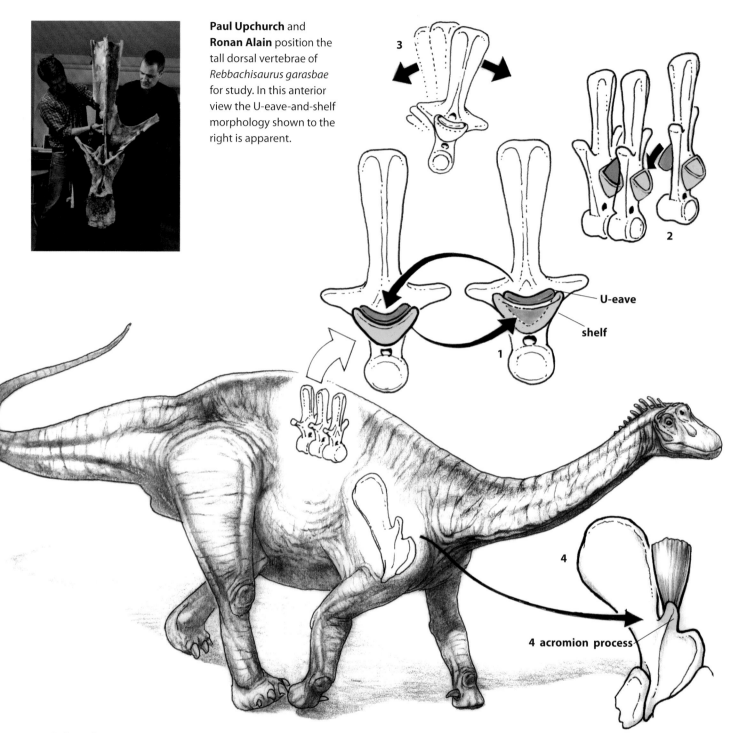

Paul Upchurch and **Ronan Alain** position the tall dorsal vertebrae of *Rebbachisaurus garasbae* for study. In this anterior view the U-eave-and-shelf morphology shown to the right is apparent.

U-eave

shelf

4 acromion process

Rebbachisaurids—Engineered for Suspension

Although limiting the mobility of the dorsal vertebrae gave a sauropod's back more strength in supporting weight, rebbachisaurids such as *Limaysaurus* traded some of this spinal strength for flexibility, possibly to better travel over rough, inclined ground. This was accomplished by having a system of smaller curved shapes (U-eaves, *orange*) projecting from the posterior face of each vertebra, which nested into a larger curve (shelf, *blue*) that projected from the anterior face **(1)**. This gave support between adjacent vertebrae **(2)** but also allowed some torsion, or rotational movement **(3)**, that made the entire back able to constantly adjust to uneven terrain. In addition, rebbachisaurids had a distinctive projection **(4)**, on the scapula, the acromion process (*blue*) that anchored a muscle that gave extra power in lifting the forelimb up and forward. This may have aided in climbing hills.

15 m (50′), 7-ton species, the neural, dorsal and sacral spines are high, but the chevron bones on the underside of the tail are poorly developed, suggesting that tripodal tail support wasn't strongly relied on, as it was in diplodocids. Although limaysaurines and rebbachisaurines didn't have bifurcated cervical and anterior dorsal spines, the height of these suggests that even if they didn't use their tails as a brace, high-level browsing was still, for limaysaurines, a feeding adaptation. Some rebbachisaurines like *Nigersaurus*, however, are very different. Their short necks have low neural spines, which means that upper bracing and high-rise neck control weren't critical for these mid- to low browsers.

While in some anatomical details they seem more primitive than the diplodocids and dicraeosaurines, other rebbachisaurid skeletal features may show innovations. The scapula not only is expanded at its proximal and distal ends but also bears a unique projection, or "hook," that suggests an especially strong dorsal musculature at the withers, or neck base, and shoulders. The dorsal vertebrae, the main supports for the torso, show a remarkable set of features. Whereas most sauropods until now have retained the **hyposphene-hypantrum**, the plug-and-socket accessory neural arch articulation that reinforces the back for strength but still allows dorsal-ventral flexion, in rebbachisaurids this is reduced or absent. The hyposphene-hypantrum was crucial in allowing earlier sauropods to achieve large body size because, while it limited dorsal and ventral flexion, it helped to "lock" and stabilize the central vertebral column—animals of this size and weight can't risk putting their backs out. The centrum is wide and **amphiplatyan** (flat at both ends), but also proportionately smaller to the neural arch than in other sauropods, and this is accompanied by a new kind of intervertebral joint called a **U-eave** (as in the eaves of a roof) **and shelf**. In this joint system, the hyposphene-hypantrum is replaced by two upwardly curved, concave "lips," an upper smaller one at the neural arch's posterior (the "U-eave"), which fits into a lower, larger one on the next neural arch's anterior surface (the "shelf"). Unlike the hyposphene-hypantrum, this may permit limited *torsional*, or rotational, movement, as well as provide support, along the vertebral column's main axis. But why would a sauropod want this, if it compromised stability?

Some paleontologists (Wilson and Alain) disagree with this interpretation of the U-eave and shelf. Others (Apesteguia) think that torsional movement in the dorsal vertebrae, while reducing spinal stability to some extent, would, by giving rebbachisaurids' spines more lateral flexibility, have increased their ability to walk more efficiently and rapidly over *uneven terrain*. During the early Mid-Cretaceous, the rise of the Andes mountain chain is creating uplift over large areas to the east that had once been relatively flat, resulting in extensive highlands and hilly deserts. Sauropod populations living in the higher elevations of this changing ecosystem probably have become marginalized, or perhaps even isolated, and are undergoing selection that favors more efficient locomotion on uneven land surfaces. Lateral spinal flexibility is a plus in negotiating these hills and valleys and is similar to the adaptational challenge that will face large-bodied proboscideans like *Cuvieronius andium* and *Cordillerion humboltii* many millions of years later during the Late Pliocene, when they begin to colonize the Andean highlands. For rebbachisaurids and other Proto–South American sauropods, there's an advantage in efficiently crossing hilly deserts as they migrate to the well-watered oases by shortening their exposure to desiccation and theropod predators. Modern loxodonts, or African elephants, that live in the Namib Desert of southwest Africa are considered by some geneticists to have become reproductively isolated from mainstream savanna loxodonts by about 2 million years ago. Although not considered a separate species, desert loxodonts differ from their relatives in having proportionately longer legs and larger feet, both characteristics that allow them to cross sandy areas quickly and securely by giving them longer strides and more secure footing. The postcranial skeletons of basal rebbachisaurids are poorly known, but if they bear these same characteristics, it may confirm their similar adaptation to at least a partial upland arid environment. As a final piece of indirect evidence for this, there are trackway sites from the much earlier Cretaceous of Brazil's Sousa Formation. These prints are mostly from theropods, with relatively few sauropod tracks, which has led workers like Ismar de Souza Carvalho (Universidade Federal de Rio de Janeiro) to postulate that sauropods inhabited higher areas of the basins near their borders. Here the coarseness of the substrates in these environments made their tracks less likely to be imprinted and preserved.

The rebbachisaurids are a successful, diverse clade. Species like *Rayososaurus agrioensis*, the contemporary Brazilian *Amazonsaurus marahensis* and still others from Proto-Africa and nearby Proto-Europe (*Histriasaurus boscarollii* and a possibly new, unnamed British species) show that by this time the group is widely established, with a probable ghost lineage stretching back into the Late or even Mid-Jurassic. There's a strong likelihood that the rebbachisaurids may have originated in southern Gondwana before Proto–Africa/South America split off, but there also could have been a rapid dispersal into nearby Proto-Europe by the Early Cretaceous.

RISE OF THE TITANOSAURS

The evidence of the evolutionary radiation of the rebbachisaurids in the Neuquen Basin and elsewhere is paralleled by a second major clade of sauropods, the ***titanosaurs***, which as basal forms also inhabited Tendaguru back in Kimmeridgian and Tithonian time. While the rebbachisaurids for the most part follow a diplodocoid feeding strategy with mostly narrow-crowned, pencil-like teeth, however, the titanosaurs show a continuing trend begun earlier by their brachiosaurid and other basal macronarian relatives, that of broader-crowned teeth that produce a precision shear bite. An early form, *Amargatitanis macni*, existed in Neuquen during the Amargan Barremian, and the discovery of possible titanosaur trackways from Mid-Jurassic England indicate that this clade, like the rebbachisaurids, may have had a much earlier origin. Now the group has begun to fully radiate, and we encounter species like *Andesaurus delgadoi*, *Ligabuesaurus leanzi* and massive *Chubutisaurus insignis*. Although as macronarians the titanosaurs are only distantly related to the diplodocoid rebbachisaurids, they show an amazing convergence: like the first clade, they've evolved more flexible spines, *but in a totally different way*. Titanosaur vertebrae, although they, like those of rebbachisaurids, also lose the hyposphene-hypantrum vertebral "lock," differ in that instead of having amphiplatyan centra, these become wide ball-and-socket joints, or *opisthocoelous* ("behind space"). The posterior of each centrum is concave, allowing a convex anterior from the next centrum to fit into it. At the same time, the pre- and postzygapophyseal contacts of the anterior caudal neural arches become broader, cup-like and more widely spaced. This condition, allowing a higher degree of torsional mobility and vertebral column flexibility, is first seen to a limited degree in the basal titanosauriform *Brachiosaurus*. As with the rebbachisaurids, it could also be an adaptation for the need to locomote efficiently over rough terrain (as well as to bear even greater weight), and it is a totally independent conversion of the spine for a new set of needs.

GIANT PREDATORS OF THE SOUTH

As titanosaurs (and rebbachisaurids) evolved and dispersed to other regions, large theropod predators coevolved with them to take advantage of this increasingly abundant and diverse prey base, especially since large ornithopods didn't become more common in Proto–South America and Proto-Africa until the Aptian. In this region, as in other parts of the Mesozoic world, allosauroids kept pace with their sauropod prey and evolved into huge predators, the **carcharodontosaurids**. The oldest known Proto–South American species, the 13 m (42′) Aptian *Tyrannotitan chubutensis*, is known from Argentina's Chubut Province and, along with 14 m (45′), 8-ton *Giganotosaurus* (*Carcharondontosaurus*) *carolinii*, rivals the much later, northern *Tyrannosaurus* in size. These theropods, however, retain the laterally compressed, serrated dentition of their ancestors to deliver quick, slashing bites to weaken their prey by laceration and blood loss. Because the clawed forelimbs in these forms are rather small, the jaws probably do the damage, and it's possible that in some species organized or communal hunting may aid their success (see Chapter 9). As pointed out earlier, the best way to cripple an adult sauropod may have been to hamstring or otherwise wound its powerful hindleg muscles. Against a slashing bite anything to harden the target would have helped, as well as keeping the biter away to begin with. Although its actual sauropod affinity is currently questioned, one animal described from the Aptian of Neuquen, 15 m (50′), 8-ton *Agustinia ligabuei*, could have evolved an effective defense in keeping an attacking carcharodontosaurid away from its body. Although known only from skeletal fragments, its dorsal, sacral and most proximal anterior caudals bear, in dorsal view, winglike transverse processes that, in turn, apparently articulated with paired, meters-long, slightly curved bony extensions. As in *Amargasaurus,* these could have been tipped with pointed, keratinized spikes. A likewise laterally pointed, plate-like and solid bony shield protected the sacrum. The interpretive arrangement of these, although found in pairs and in sequence, is very conjectural, but if actually present, the bizarre ossifications may protect *Agustinia* against these

Elephants of the Namibian Desert in Africa have longer legs than their bush-living cousins on the eastern savannahs. This may be an adaptation to long-distance travel over rough terrain, and if future discoveries of rebbachisaurid leg bones are similar, it could provide a correlation of the hypothesis of their spinal modifications having evolved to handle a similar need.

earliest giant carcharodontosaurids. Along with the carcharondontosaurids, rebbachisaurids and titanosaurs also have to deal with another major theropod group, the broad-skulled **abelisaurids** such as *Ligabueino andesi*, an early Barremian–Aptian form. Adapted to kill by clamping and holding the neck rather than through attrition, abelisaurids will become more common later and, in some forms like *Carnotaurus sastreyi*, evolve relatively long, cursorial hindlimbs and short, powerful jaws.

As the twin continents continue to slowly but inevitably rift apart, the fluvial basins and the surrounding deserts of Proto–South America will continue to be evolutionary laboratories for the rapidly diversifying rebbachisaurids and titanosaurs well into the late Mid-Cretaceous. Long before the growing infant Atlantic finally cuts off more derived forms' dispersals into Africa and the European Isles, their probably differing habitat preferences and feeding ecologies at this time allow them to coexist and avoid competition with each other. Although the rebbachisaurids and basal titanosaurs, together with the carcharodontosaurids and abelisaurids, continue to form an important faunal association of predators and prey, this will eventually change by South America's Late Cenomanian. The above two sauropod clades, so hugely diverse and widespread in the Early Cretaceous, will decline and finally die out, seemingly taking with them into extinction the carcharodontosaurids. As the Cretaceous rolls on, the surviving sauropods, the more advanced titanosaurs, continue to radiate and flourish here and elsewhere around the world.

CARBONATE PLATFORMS AND SALTY LAGOONS: KEM KEM, NORTH AFRICA

This time we're transported not too far ahead into the future to the early Mid-Cenomanian stage, the beginning of the Late Cretaceous. Here our geochron is the **Kem Kem Formation in Morocco, 94 MYA**. Africa is now almost completely separated from its sister continent of South America. From our time and space portal high in the stratosphere, the view coming in over northwest Africa is startling: the modern coastline is now hundreds of miles to the south, where the future Atlas Mountains and Sahara Desert will one day be. To the north, instead of the familiar Mediterranean, there's a much bigger ocean reaching to the horizon, *Tethys*. Beginning in the Early Cretaceous, such conditions as warmer worldwide temperatures, the geologic formation of new, higher-placed oceanic crustal rock and other factors have all come together to displace vast quantities of seawater. This, in turn, has caused ocean levels to rise hundreds of feet, covering great areas of land with shallow **epeiric**, or epicontinental, seas. Below us, an underwater shelf of mottled light blue, fringed by reefs, extends for many miles from the coast and to the east and west. This is our first glimpse of a **carbonate platform.** When epeiric seas in tropical areas occur in combination with low winds and gentle tidal cycles, they favor certain types of warm-water algae. Foraminifera, bryozoans, bivalves and other microscopic to small invertebrates, which feed on the algae, absorb calcium, carbon and other minerals dissolved in seawater to create protective, microscopic exoskeletons and shells. In time the remains contribute to an ever-accumulating layer of marine sediments, and along with minerals like glauconite and sand, these sometimes alternate with clays and marls to become limestone—a carbonate platform. The North African Platform, like the others fringing the shores of Tethys, is part of an enormous, interconnected ecosystem of tidal lagoons, shallow seas and reefs. It's as if, instead of Africa, we were looking down onto the future Great Barrier Reef of northeastern Australia. Here, however, instead of coral as the reef builder, it's often *Exogyra* oysters or other bivalves, known as *rudists*.

INTERTIDAL FORESTS

As we descend closer, we see on the landward side of the platform countless thousands of irregularly shaped, small to large green islands transected by complex mazes of sandy tidal channels. It superficially resembles the great Sundarban swamps at the mouths of the Ganges in Bangladesh or parts of Indonesia's coasts, but instead of modern mangrove vegetation like the angiosperm *Rhizophora* and others, in Kem Kem and elsewhere along the Tethys coasts these are made up mostly of gymnosperms and ferns that have become adapted to high levels of salinity. Species like the cheirolepidiacean conifer *Frenelopsis alata*, the broad-leaved ginkoalean *Eretmopyllum obtusum*, the tree ferns *Weichselia reticulata* and *Acrostichum preaureum,* the narrow-leaved angiosperm *Sapindopsis* and others grow as shrubs and small trees in close association along the tidal flats. When conditions are favorable, these sometimes extend away from the shore to form at first tiny, then larger, offshore extensions or "islands" of crowded growth. These plants don't have anchoring and air-breathing "prop roots" exactly like those of true angiosperm mangroves, but *mangal* or "pre-mangrove" communities like this still develop dense, interconnected root systems that form a trap for organic matter, like decaying leaves and tidally transported marine sediments. The trapped organics become a peaty soil base that enables other plants to grow as well, and eventually these small "islands" may coalesce to become huge mats. As with modern true mangrove ecosystems, this is only possible when there are warm, extremely shallow continental shelves and very low wave turbidities that allow the root systems to trap marine sediments. The end result is that in Cenomanian Kem Kem and elsewhere along the Tethys coasts, the gentle tidal cycles, closeness to the equator and carbonate platforms have created a remarkable intertidal forest system of *halophytic*, or salt-tolerant, plants from inland terrestrial forms we wouldn't expect—mostly conifers, ginkgoaleans and ferns.

Because of the nutrient-rich shelter they provide for small aquatic invertebrates, fish and other species, modern mangroves and tidal marshes are second only to equatorial rainforests in terms of biological productivity. The even greater extent of the Upper Cretaceous tidal swamps at Kem Kem and elsewhere around the globe makes this especially true at this time. As we walk for a few meters along one of the tidal channels between two of the tidal "islands," we flush out a silvery horde of juvenile teleost fish, *Cladocyclus pankowskii*, as well as a *Mawsonia lavocati* coelacanth. Ammonites and turtles like *Bothremys* find their tiny prey among the interconnecting roots, and two of these, as well as a well-camouflaged shark, *Marckgrafia libyca*, yield gracefully at our sloshing approach and veer off into the silty water. Recalling that where there are fish and turtles there also may be large crocs, we turn back along the channel, spotting as we do so a mixed group of several big *Aegisuchus witmeri* and *Elosuchus cherifiensis* sunning themselves along

Africa

(1) Tethys Ocean
(2) North African Carbonate Platform
(3) Kem Kem

By Cenomanian, 70-million-year-old Mid-Cretaceous times the world's seas had dramatically risen. These transgressed into huge low-lying regions like North Africa, resulting in coastal areas that resembled the Sundarbans of Bangladesh (*below*). In North Africa this was bordered to the north by the Tethys Ocean, which would later become the Mediterranean Sea during the Tertiary period following the Mesozoic. The black line indicates modern coasts.

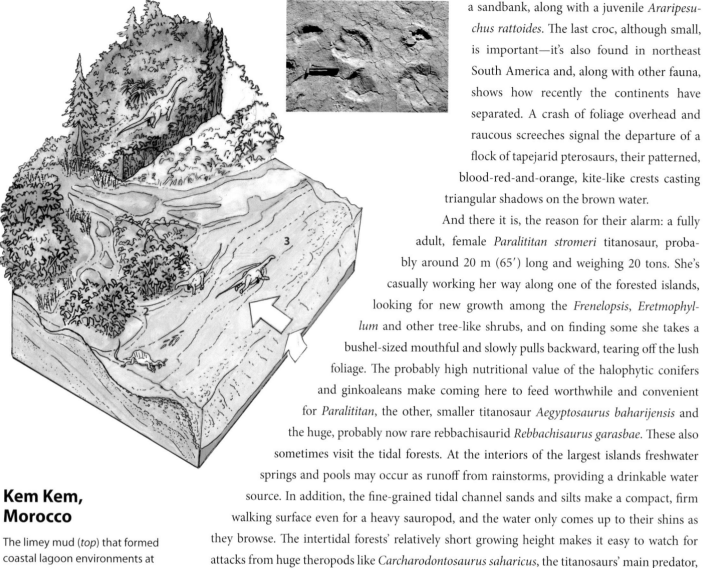

a sandbank, along with a juvenile *Araripesuchus rattoides*. The last croc, although small, is important—it's also found in northeast South America and, along with other fauna, shows how recently the continents have separated. A crash of foliage overhead and raucous screeches signal the departure of a flock of tapejarid pterosaurs, their patterned, blood-red-and-orange, kite-like crests casting triangular shadows on the brown water.

And there it is, the reason for their alarm: a fully adult, female *Paralititan stromeri* titanosaur, probably around 20 m (65′) long and weighing 20 tons. She's casually working her way along one of the forested islands, looking for new growth among the *Frenelopsis, Eretmophyllum* and other tree-like shrubs, and on finding some she takes a bushel-sized mouthful and slowly pulls backward, tearing off the lush foliage. The probably high nutritional value of the halophytic conifers and ginkoaleans make coming here to feed worthwhile and convenient for *Paralititan*, the other, smaller titanosaur *Aegyptosaurus baharijensis* and the huge, probably now rare rebbachisaurid *Rebbachisaurus garasbae*. These also sometimes visit the tidal forests. At the interiors of the largest islands freshwater springs and pools may occur as runoff from rainstorms, providing a drinkable water source. In addition, the fine-grained tidal channel sands and silts make a compact, firm walking surface even for a heavy sauropod, and the water only comes up to their shins as they browse. The intertidal forests' relatively short growing height makes it easy to watch for attacks from huge theropods like *Carcharodontosaurus saharicus*, the titanosaurs' main predator, as well as the smaller *Bahariasaurus (Deltadromaeus) agilis*. Sometimes the theropods are successful, however: at the mouth of one tidal channel is the massive, partially silt-covered skeleton of an adult *Paralititan*. Its presence is evidence that the sauropod died where its bones lie, since the weak tides and long distance from any rivers mean that they couldn't have been transported by water action.

Although the sauropods come here often to feed, their habitat includes inland areas as well. Here the coastal intertidal forests give way to open, tidal lagoons and river estuaries, where less salt-tolerant ginkgoalean species like *Nehvizdyella bipartita* and others take over from the coastally more common forms. Like *Eretmophyllum, Nehvizdyella's* halophytic physiology makes it an evergreen, in contrast to the inland ginkgoaleans like *Ginkgoxylon gruetti* that shed their leaves. Kem Kem's inland forests, like others all over the world, are now coming to be dominated more and more by angiosperms. Although cheirolepidiaceans like *Tomaxellia* and other conifers are still found here, the ferns and cycadeoids are now increasingly shaded by canopies of angiosperms like early magnolias, camphors and laurels.

Kem Kem, Morocco

The limey mud (*top*) that formed coastal lagoon environments at Kem Kem and other geochrons during Early to Late Cretaceous times has provided a remarkable insight into the possible behaviors of sauropods through preserved trackways. *Bottom*: topographic section of Kem Kem showing (**1**) a coastal forest, (**2**) extended, mangrove-like "islands" of salt-tolerant growth and (**3**) a shallow carbonate platform, an offshore shelf created by the accumulations of lime-producing organisms and gentle tides, indicated by arrows (some vegetation deleted to show topography).

HEAVY TRAFFIC

Back at the coast, seasonal fluctuations sometimes dry up the lagoons to expose large tidal flats of limy mud, perfect for recording the daily migrations of the region's dinosaurs, crocs, pterosaurs, turtles and birds. In other areas of the world, spectacular trackways of Aptian diplodocoids and basal titanosauriforms confirm that these sauropods were definitely gregarious, usually moving in pods of 10–22 individuals. The tracks from localities like Glen Rose and Paluxy River, Texas, tend to appear in parallel for long distances without often coming close. This may indicate that individual sauropods generally liked their "space," but some paired trackways at certain places come closer together, and in both the distance between strides decreases, which could mean that the two individuals slowed down and some kind of interaction took place. Depending on locality and horizon, the majority of known sauropod trackways are made up of similar-sized individuals, probably of the same age group. As Lockley has deduced, these trackways also can tell us about sequences of events that took place within a pod. From overlapping footprints we know which animals were at the front (usually the larger individuals) and which followed (the smaller ones). From measurements taken from foot lengths and inferred leg heights at the hip we can also determine approximate speed, and this allows us to calculate that the pod traveled at 1–2 m per second. Based on this, it took one pod at least 30–60 seconds to cross at one tracksite in Texas. Many known sauropod trackways come from paleocoastal localities and often occur in association with those of large theropods. No "chase sequences" have been deduced from these (see Chapter 9), but the fact that in several situations the theropod footprints turn in the same direction, and in the same place, as those of the sauropods suggests that the former may have been actively following their potential prey. This reinforces the probability that carcharodontosaurids and other large allosauroids were principal predators of sauropods in areas where these giants were the dominant herbivores, or where ornithopods were uncommon.

SWIMMING SAUROPODS?

Morocco is one of the very few places, in addition to South Korea and others, where we have definite trackway evidence showing that coastal-dwelling sauropods occasionally may have actually ventured into somewhat deeper water. The current-rippled mudstones of the Iouaridene Formation, now in the central Moroccan High Atlas Mountains, were at the time of deposition during the Late Jurassic located in a semiarid but lagoonal setting near the Tethys coast. One set of trackways studied by Shinobu Ishigaki (Hyashibara Museum of Natural Sciences) indicates two sauropods of about the same size going in the same direction, but at different times. Why different times, and not together? Unlike the well-defined and imprinted manus tracks, in Sauropod A the front of the pes print *always* touches the bottom and makes faint impressions just behind the one left by the manus. In Sauropod B, however, the pes prints show up only intermittently and are totally missing from step numbers 4, 5 and 10. Since both sauropods were roughly the same size and not likely to differ substantially in water displacement, the differences in pes imprints suggest that Sauropod B floated higher in the water, probably when the tide was higher and therefore at another time. In this situation both animals were *not* actively swimming but actually engaging in an activity sometimes called "poling," in which the water was high enough

During the Mid-Cretaceous, Kem Kem, North Africa, and many other coastal localities were part of ecosystems dominated by biologically created carbonate platforms that spread because of the high seas and gentle tides. These were the accumulations of calcium carbonate–rich organisms like rudists (*bottom*) and *Exogyra* oysters (*top*) that formed dense, colonial reefs.

Toward the Maastrichtian stage at the end of the Cretaceous about 70 million years ago, Hateg and other islands were part of a widely scattered archipelago that was Proto-Europe, in some ways resembling Indonesia.

to float the animal's more buoyant hindquarters and tail, but allowed the sauropod to make contact with the lagoon bottom with its fore-limbs. Here it could push back-ward with these to move itself for-ward. Could sauropods actually swim in deeper water? Their fore- and hindlimbs were certainly as well muscled as those of elephants, which are excellent swimmers and are known to be capable of crossing considerable distances with measured, powerful strokes. As discussed in Chapter 7, however, the positive buoyancy indicated by computer modeling presum-ably would have created difficulties for titanosaurs and rebbachisaurids, as well as other sauropods, if they'd tried to swim.

The "poling" capability of these sauropods, however, presents an intriguing idea. If the sau-ropods of Kem Kem and elsewhere along the Tethyan coasts were habitually feeding from the nutritious halophytic plants in intertidal forests and, as all sauropods did, had to continually keep moving in search of new forage, it suggests that some species might have ranged widely for long distances along the North African carbonate platform. It would have been no problem for at least larger subadults and adults to wade and "pole" from island to island as they sought new browse. As in the cases of some other dinosaurs as they dispersed between the rifting continents, sauropods were well suited for long-distance travel. Motivated by their constant need for new foraging areas, they may have used the carbonate platforms as highways to new opportunities. This is a hypothesis we'll explore later on in this chapter.

ISLANDS OF THE DWARF DINOSAURS: THE EUROPEAN ARCHIPELAGO

Our journey spanning thousands of miles and millions of years is now almost at an end. Traveling forward in time north from Kem Kem, we slow to 70 MYA, the beginning of the Maastrichtian stage of the Upper Cretaceous. The scene below us is an extensive chain of huge to small scat-tered islands running from northwest to southeast, and this configuration and their sizes make them strikingly similar to the modern islands of Southeast Asia. We're above **Proto-Europe**, and the geochron we'll be exploring is **Hateg Island, Romania**. Sea levels are even higher now than during the Cenomanian, and the land is still surrounded by carbonate platforms, the largest occurring west and east of Hateg. Our 75,000 km^2 island, like others in Proto-Europe, is partly born of the volcanics resulting from the subduction of the North African Tectonic Plate as it inches northeast and grinds underneath the next. For now, however, the volcanoes are dormant, and as we trek upward from our landing place at the bottom of a hilly terrace near the beach, we find that their green slopes are covered with tropical to subtropical angiosperm-dominated forests largely made up of walnut, laurel, palm and sycamore. Of these, the sycamores (Platana-

By the Maastrichtian stage of the Late Cretaceous, angiosperm trees like sycamore had come to make up an increasingly greater proportion of the arboreal veg-etation in geochrons like Hateg, but there was still enough conifer vegetation to support titanosau-rian sauropods.

ceae) show the greatest diversity and are the most abundant, but growing among them and the other more modern trees, some broad-leaved cheirolepidiacean conifers and tree ferns are still abundant. In the understory, many species of ferns, as well as stands of clubmoss and horsetails, flourish in the wet riparian habitats. The small land snails here confirm the forests' high humidity, but elsewhere in Hateg there are other, drier environments, marshes and small alluvial fans. What will one day become calcium carbonate paleosols show that in some spots the island, in spite of its relatively high humidity, still experiences periods of aridity. These dry spells alternate with ones of heavy rainfall brought in by westwardly prevailing winds, and during the Wet Season, the runoff drains into many fast-moving but shallow streams that deposit their silts and sands onto small, forest-rimmed floodplains.

Hateg's mosaic of habitats shelters a rich variety of small animals. Leaf damage indicates that insects are plentiful here, and by this time in the Late Mesozoic all of the major modern insect families such as Coleoptera (beetles), Lepidoptera (butterflies and moths) and Orthoptera (katydids and others) are present. Some of these are food for the island's small frogs, including *Paralitonia transylvanicus,* and the terrestrial salamander-like *Albanerpeton*, while stealthy lizards, including the mostly swift, long-tailed scincomorph lizards like *Becklesius nopcsai* and at least one of the slower-moving anguimorphs, find abundant invertebrate prey in the decomposing leaf litter. Turtles seem to be everywhere, and one of the most common we encounter in the moist woods is *Kallokibotion bajazidi*, a snapping turtle–sized species with a broad shell. The tropical woods are also alive with tiny mammals, and now, as in Dashanpu long ago, we see scuttling shapes that betray multituberculates, those long-term mammalian survivors that will outlive the dinosaurs far into the Cenozoic. One, however, *Kogaionon ungureanui*, lives only long enough to harvest a nut before a flash of streaked golden brown erupts from the understory foliage and snaps it up. When it pauses to gulp down its prey, we recognize it as a very small theropod, *Elopteryx nopcsai*—our first dinosaur at Hateg.

Proto-Europe (European Archipelago)

When Hateg **(1)** and other Proto-European islands became connected to larger areas during periods of low sea level, originally large-bodied sauropods **(2)** and other dinosaurs **(3, 4)** migrated to these and then became reproductively isolated, evolving into dwarves **(5, 6, 7)**. This cycle occurred many times during the Late Cretaceous. The black line indicates modern coasts.

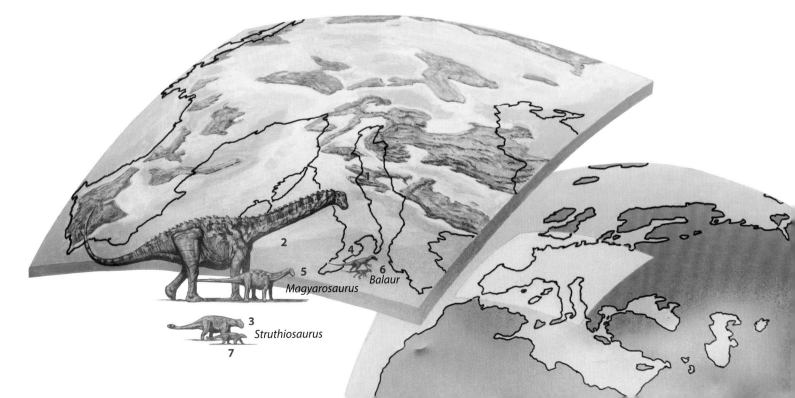

Magyarosaurus
Balaur
Struthiosaurus

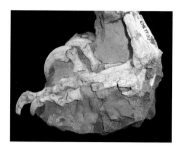

Island dwarfing, in addition to small size, can produce highly distinctive anatomical features. In one species of dromaeosaur, *Balaur bondoc*, the forelimbs bore only two instead of three functional digits, while instead of the usual first digit "killer claw" as in larger mainland forms, there were two.

Not all of Hateg's residents are little, however. In continuing downhill, we almost stumble over a small, lichen-covered sauropod skeleton, apparently from a long-dead older juvenile—the giant adults here can't be far off. What becomes clear by the afternoon, however, is that there *are* no giant sauropod adults on Hateg. The closed sutures on the cervicals, the skull and the condition of the limb joint surfaces of the bones we saw in the forest mean that, in spite of a body length of only 5–6 m (15′–18′), *this animal was a fully grown adult.* As we explore the island, we see other, equally diminutive adult titanosaur sauropods such as *Magyarosaurus dacus* and the somewhat larger but still incongruously named *Paludititan nalatzensis,* all about the same size. Amazing as it seems, this is really as big as sauropods get here, and it's shocking to see one get shouldered out of the way by an aggressive basal hadrosauroid, *Telmatosaurus transsylvanicus,* fully the same size, as it attempts to feed on "cheiro" shrubs. This hadrosauroid, too, is abnormally small compared to the bigger species of these and the basal iguanodontids on the mainland. When we finally watch a sheep-sized ankylosaur, *Struthiosaurus transylvanicus,* amble past us on the way to a watering hole, the realization hits: *we're on an island of dwarf dinosaurs.* How, and why, did they come to be this way?

Among large-bodied herbivores, dwarfism and island living often go hand in hand. Food and water resources in geographically isolated areas like islands are limited compared to those on the mainland, and when a population of big animals immigrates to such a place, competition for these limited resources begins to intensify. If, among the individuals that become established on the island, there are those that are smaller, their nutritional needs will be less, and their chances of long-term survival will be better than the larger animals'. Under these circumstances it no longer pays to be big, and natural selection in such a population starts to favor the ones who can do with less. This situation will repeat itself several times later in the Late Cenozoic. Mammals as diverse as elephants, hippos and buffalo arrive at islands when they're large bodied and then dwarf down to adapt to the available resources.

Insular dwarfing also favors reaching reproduction age early. As we recall from Chapter 8, sauropods and some other dinosaurs had already acquired the trait of early sexual maturity as a way to offset the odds of predation by increasing the numbers of individuals that could reproduce, the earlier the better. This comes in handy on an island, since the smaller animals that sexually mature first can outreproduce the bigger ones and are thus far more likely to pass on their "smallness" gene. In populations of limited size that are geographically separated from larger ones, natural selection intensifies when it comes to favoring and maintaining beneficial characteristics that promote an individual's ability to survive. On Hateg, the result is that several dinosaur clades living here have become radically reduced in size, including the titanosaurs.

Magyarosaurus is the smallest of the titanosaurs here, and this is

Skull of *Europasaurus*

Although not as small as the titanosaurs of Hateg, the Late Jurassic brachiosaurid *Europasaurus holgeri* from Rhenland (now southwest Germany and northeast France) was still a very small sauropod, about the mass of a Holstein cow. Dwarfing was an occurrence that happened many times when high sea levels formed islands.

the species whose bones we found on the hillside. In Chapter 7 we learned that in the case of sauropods more so than in other dinosaurs, the diameter of the humeral and femoral bone shafts grows increasingly greater in proportion to length as body size increases, contributing to their ability to bear weight. Since after arriving at Hateg we've now collected a complete ontogenetic range of these and other bones from juvenile through adult *Magyarosaurus*, we know at what sizes and shapes these began to show adult features, such as the maturity of the joint surfaces. The bones tell us approximately at what size this titanosaur reached full adulthood, and from these, we can separately compare first the juvenile and then adult long-bone widths and lengths by programming sampled measurements onto line graphs. When these are superimposed onto one graph, what we learn is that the *width* of *Magyarosaurus*'s long bones didn't increase very much in proportion to *length* as they grew to be adults. Like *Massospondylus*'s relatively long forelimb versus hindlimb length discussed in Chapter 2, this titanosaur was paedomorphic, or retained some of its juvenile characteristics (in this case bone proportions) into adulthood. It's especially striking when the dwarf's bones are seen against the corresponding ones of big titanosaurs from other times and locations, and they say one thing: *Magyarosaurus*'s limb bones, because of its smaller weight and body size, didn't have to be as strong.

If *Magyarosaurus* and the other herbivores of Hateg have become dwarfs to adapt to the island's decreased food resources, aren't they more vulnerable to predation? After a few days here we haven't seen any big, "regular-sized" theropods, only small, feathered avialan maniraptorans like turkey-sized *Balaur bondoc* and the smaller *Pyroraptor olympius*. *Balaur*, however, at 1.8–2.1 m (5.9′–6.9′) is as roughly proportional to *Magyarosaurus* and some of the basal hadrosaurs here as Fruita's *Allosaurus* would be to *Diplodocus* or *Apatosaurus*, and a group of these

An adult and juvenile *Europasaurus* on Hateg Island (Dinosaurier Park life restoration).

predators would be very capable of taking down one of the dwarf adult titanosaurs. A possible confirmation of this is the avialan's stocky, robust bones and unusual double inner sickle claws, suggesting specialized predatory adaptations. It, too, appears to be an endemic form here, and the titanosaurs were as subject to predation here as their big mainland counterparts. If not, they would have outstripped their food supply and eaten themselves into extinction. Since there were plenty of small predators, including a *Madisoia*-like snake, these would have kept the numbers of hatchlings or juveniles in check.

The dwarf titanosaur sauropods and other dinosaurs of Hateg are unusual in other ways as well. Thought by some earlier paleontologists to be a residual group of species from a much earlier, more diverse dinosaur fauna that occupied Europe when seas were lower and more land existed, some types actually may have arrived later than others and show that some kinds of dinosaurs have changed little while they've been here, while others are evolving relatively fast. One of the hadrosaurs, *Telmatosaurus*, shows basal characteristics in its tooth patterns compared to roughly contemporary but more advanced forms like the North American *Edmontosaurus* and *Lambeosaurus*. Studies based on calculations of the ghost lineage for *Telmatosaurus* suggest that this form evolved at a slow pace, averaging acquisitions of only one new character every 7 million years during its estimated 35 million years of ghost lineage after it split off from the mainstream of hadrosaur evolution. Other big-skulled but selective ornithopod feeders, like the rhabdodontid iguanodontids *Zalmoxes shqiperorum* and *Rhabdodon priscus*, appear to be on a fast track, averaging a new character every 4 million years. These differences in the above forms' rate of evolution are surprising, but it might be an example of Theodosius Dobzhansky's hypothesis of directional versus stabilizing selection. In this corollary to Darwin's original principle of natural selection, a population of organisms is *directed*, or at first subjected to, intense evolutionary pressure that selects in *favor* of any positive adaptation to a set of new environmental conditions in which it finds itself, but once optimal adaptation has been achieved, any deviation from this condition is later selected *against*, resulting in *stabilization*. The evolution of *Zalmoxes* and *Rhabdodon*, which are in the process of morphologically and behaviorally adapting to the food resources of Hateg and elsewhere in the European island archipelago, moves forward when selection favors a trait that makes these dinosaurs better adjusted to this situation. *Telmatosaurus*, on the other hand, is now well adjusted to its environment, and natural selection has begun favoring individuals that simply maintain these optimal traits while discouraging those that diverge from them. The titanosaur species, although also dwarfed like the ornithopods, may have immigrated to Hateg more than once after this area became an island. The originally large size, physical stamina and possible "intertidal island-hopping" behavior of these sauropods could have made this more likely for them than most other dinosaurs, when seas were lower, and increased their potential as colonizers. Although at this point a dwarf, *Paludititan*'s larger size than *Magyarosaurus* may mean that this titanosaur could have immigrated to the island at a later date than the first, and it was still evolving toward its smaller equilibrium size.

Hateg is only one of many islands in the European archipelago, each with its own endemic dwarves, and others with dwarves in the making. In the case of the titanosaurs, some of the bigger ones have larger or smaller species according to when they arrived and how well the land area can sustain them, like Rhenland's (northeast France/southwest Germany) relatively medium-sized, broad-toothed *Ampelosaurus attacis* and Ibero-Armorica's (Portugal–Spain) small *Lirainosaurus astibeae*. Ibero-Armorica harbors one species, *Atsinganosaurus velauciensis*, that's a late-surviving basal lithostrotian titanosaur with affinities to the African *Malawisaurus dixeyi*, and this raises the question, where did the titanosaurs originate? Was it in Gondwana, before it separated into island continents, or did the clade evolve on one of these smaller landmasses, to later populate other areas? We know that earlier in Upper Jurassic Rhenland there was a dwarf brachiosaur, *Europasaurus holgeri*, as well as another in Mid-Cretaceous Croatia. The islands of Proto-Europe are a remarkable laboratory of sauropod evolution that may someday hold the answer. Except for a few possible relict clades dwindling to extinction, the titanosaurs by the Maastrichtian have now become the most common and diverse, if not the only, sauropods in the Upper Mesozoic world, and we'll return to them.

A *Barosaurus lentus* diplodocid lies dead of starvation, a victim of the possible die-off of cheirolepidiaceans.

CHAPTER ELEVEN

END OF EDEN?

During its 4-billion-year history of living things, Earth had never seen anything like them: living cranes that could, if necessary, tower high into the sky to harvest the succulent foliage and fruits of the tall, dominant trees that reached upward in the race to avoid being cropped. From osteoblasts that evolved millions of years earlier to originally give ancient fish protection from sea scorpions, natural selection evolved bioarchitectural marvels of strength and lightness to support vast weight and yet enable flexibility and movement—the mamenchisaurs and diplodocids were among the most amazing and majestic of dinosaurs. But by the beginning of the Cretaceous, they were all gone. What had happened? Would the other clades follow, or would sauropods continue their long and diverse evolution?

G AME OVER. Extinction, like the death of the individual, is the fate of all species, no matter how successful they are. Far from being a sign of ultimate biological failure, it's a normal event, and it is usually the result of a complex interplay of outside circumstances to which a species can find itself unable to adapt. If a set of long-term changes in the environment unfavorable to the species occurs, it's then a race against time: successful new behaviors, sometimes coupled with favorable morphological changes, can give individual populations enough of a chance to reproduce fast enough, and in enough numbers, to outpace the increased attrition from mortality. If adverse conditions occur too fast, however, and over the species' entire range, both mainstream and marginal populations just run out of time. If specialization toward a

A herd of the North American diplodocid *Barosaurus lentus* wanders in search of fresh browse. These, like other closely related species, were perhaps heavily dependent on cheirolepidiacean conifers as a staple food, and when these widespread conifers began to decline, it may have caused massive die-offs and eventual extinction among these sauropods.

particular way of living is too well established, new adaptations generally can't occur fast enough to allow survivors to gain a reproductive edge, and the species—and sometimes an entire clade—flickers out and becomes extinct. This happened to other vertebrate groups long before, as well as after, the dinosaurs came and went. The first known major clade of dinosaurs to disappear from the Mesozoic record were the prosauropods, followed by the stegosaurs, which had a feeding ecology bridging the gap between the smaller bipedal, ornithischian low browsers and the high-browsing, generally non-selective sauropods. As selective feeders, the plated ornithischians could exploit both feeding zones, but, like the core prosauropods (see Chapter 2), they may have been "edged out of a job" by the ever more finely tuned adaptations of the first and second groups that originally bracketed them. Stegosaurs, except for some late-surviving, relict forms like Early Cretaceous *Paranthodon africanus* in South Africa, were gone after the end of the Jurassic period. This is also the case with eusauropods like *Mamenchisaurus* and *Omeisaurus* and diplodocids like *Diplodocus* and *Apatosaurus*. Diplodocids reached their zenith during the Late Jurassic, but, except for the South American *Leikopal laticauda* and a few possible survivors in Africa and Proto-Europe, these had also disappeared by this time. Why, and how, could this happen?

A VANISHED FOOD SOURCE

Like a human crime scenario, we can begin trying to solve the mystery by first looking for patterns. As we've learned, the Late Jurassic (Oxfordian, Kimmeridgian and Tithonian stages) saw a flowering of sauropod feeding specialists that mainly evolved in parts of former Laurasia. Some of these took the form of the Chinese (Proto-Asian), eusauropod mamenchisaurs and omeisaurs, small to average in body size but possessing tall skulls, hyperlong necks and varying from being narrow-muzzled, selective feeders (mamenchisaurs) to broad-headed, non-selective browsers (omeisaurids). Although usually quadrupedal, they also possessed the ability to rear if necessary, and they lived in moist, heavily forested environments. The mostly North American, forest savanna–dwelling diplodocids are characterized by highly evolved bipedal ability, broad heads, long necks and anteriorly located, pencil-like teeth that evolved in combination to reach and efficiently strip or rake quantities of conifer leaflets from stems. Both eusauropods and diplodocids also probably seized and fed on calorie-rich, fruiting cones from araucarians and other conifers. If these foods were scarce, adults of both clades could browse from more close-to-the-ground vegetation they exploited as juveniles, but it wasn't likely a primary food source. For diplodocids leaf stripping was apparently a successful way of feeding, since this clade is represented by a broad diversity of both genera and species. Here these coexisted often as sympatric forms, along with macronarian sauropods. Most of the "classic," well-known flagellicaudatan diplodocids are found in the Late Jurassic (151–145 MYA), when, like the mamechisaurs and omeisaurs, there was an explosive radiation of these forms, followed by an abrupt decline and presumed extinction. While the end of the Jurassic saw the end of the mamenchisaurid/omeisaurid as well as the diplodocid lineages, other diplodocoid families, the dicraeosaurids and rebbachisaurids, survived the extinction. The smallish, mostly African/South American dicraeosaurids are known from the latter continent at least until the Aptian–Albian stages of the Early Cretaceous, while the similar-ranging rebbachisaurids in at least one species, *Cathartesaura anaerobica*, flourished until the South American Middle Cenomanian stage of the Late Cretaceous, about 96 MYA.

A critical factor for the above clades, as with all sauropods, was food abundance and availability. Based on this premise, we could consider the possibility that eusauropods and diplodocid neosauropods might have been heavily (even though not exclusively) dependent on a particular group of closely related plant species that flourished during the Late Jurassic and then declined and became extinct in Proto-Asia and North America as the era closed. Going further, if we assume that these sauropods weren't able to readily adapt to other plant sources, their populations might have become stressed enough to decline and go extinct themselves if this food source failed. Is this likely, and are there any Mesozoic plant groups whose geographic distribution and location in time would fit into this scenario? If we accept the hypothesis (see Chapter 6) that these clades and those of other sauropods were primarily adapted to conifer browse as a main food, we can focus on these plants to see whether there might have been a broad group of Mesozoic conifers, known to herbivore zoologists as **primary browse trees**, on which the extinct forms were dependent. The success of the high-browsing eusauropods and diplodocids may have been based on harvesting a few types of highly abundant conifer browse.

Podozamites represented in living state by modern podocarp-like foliage (*above*), was a foliage type belonging to the family Cheirolepidiaceae. It was, along with many others, an extremely common and widespread conifer form. Like other "cheiros," it was well adapted both to disturbed soil conditions and to a seasonally dry-to-monsoonal climate and was a part of the dense, dry adapted forests that covered the North American West during the Late Jurassic. Cheiros arose from much earlier ancestors to become an important food source for sauropods during this time, and their decline following this period may have meant extinction for some highly specialized clades.

The initial decline of the cheiro-lepidiaceans by the beginning of the Cretaceous may have had a disastrous effect on eusauro-pods and diplodocids in Asia and North America, which may have been highly dependent on this type of conifer as a food source.

Although several of the Mesozoic conifer families are today represented by living species, one, the Cheirolepidiaceae, is totally extinct. This is very unfortunate, not only because we can't do nutrition tests on them as we can with extant conifers, horsetails and other plants to measure digestibility and caloric value, but also because some things that are known about them point to their being especially good candidates for sauropod food. For one thing, fossil finds show that this family arose and became established worldwide by the Late Triassic, just when the prosauropods and early sauropods were evolving. In comparison with other conifer families, the "cheiros" display some of the greatest diversity of growth pattern, habitat and morphology. At least one took the form of tall trees with 1 m (3.5′) trunks that grew at least 23.4 m (76′9″) in height and were long-lived, attested to by the discovery of an in situ stand of trunks from the Late Jurassic of England, with individual trees that may have ranged from 200 to 700 years in age. This meant that they were likely the dominants in a long-lived forest ecosystem. Besides tree-like forms, other types were herbaceous, or shrub-like, growing in low, dense stands in a tidal or coastal marsh setting (see Chapter 10). "Cheiros" were common in many Mesozoic plant communities, ranging from almost monospecific/low-density assemblages in brackish (slightly salty) or hypersaline coastal environments to species-rich communities in mesic (moist) or riparian (streamside) locations. They also apparently thrived under semiarid to completely arid conditions, in strongly seasonal climates and at low paleolatitudes. Fossil cheirolepidiacaean foliage occurs as two types. One, composed of *Brachyphyllum* and *Pagiophyllum*, bears leaves or cuticles that are scale-like and pointing downward and are arranged in a spiral pattern; the other, composed of *Frenelopsis* and *Pseudofrenelopsis*, has leaves or cuticles that form tightly around the stem and have a jointed appearance. Despite these differences, characters that all members of the family have in common are a distinctive type of pollen, separately named *Classopollis*, and less typically thick cuticles than other conifers, with sunken stomata (air exchange pores) and papillae (tiny projections) that extend over these. Other species probably had fleshy leaves, and some were probably deciduous. Abundant, widespread and appearing in many forms, "cheiros" may have been a food source that fueled much of the sauropods' evolutionary radiation, and could have been a particular mainstay of some types in the formerly Laurasian continents of Proto-Asia and North America. The "cheiros" went into a drastic decline in worldwide diversity during the Late Cenomanian and Early Turonian stages of the Cretaceous in the Northern Hemisphere. If this had begun during the earlier Late Jurassic in North America and Proto-Europe and these trees were staple conifer browse for eusauropods and diplodocids, it could have had fatal consequences for these clades. By the Late Cretaceous's Coniacian stage the "cheiros" were almost gone in the Southern Hemisphere continents as well, and with them probably went the last, now relict diplodocoids. Of course, there were lots of other conifer types around then that continued well into the Mesozoic and into our own day and could possibly have also served as food for these sauropods. Modern-day big-bodied mammalian browsers like African loxodonts (elephants, *Loxodonta africana*) actually consume up to almost 50 types of plants depending on seasonal availability, as do rhinos, and aren't dependent on just a certain kind. Others, however, are: Australian koalas (*Phascolarctos cinereus*), although very small bodied by comparison, eat almost exclusively eucalyptus and occasionally some types of gum and tea tree leaves, and they will quickly starve in captivity if offered any other kinds. Even with these they're very picky, eating only a few out of the some 600 known eucalypt species. Koalas' digestive systems are adapted to handle the extreme levels of toxic phytochemicals in eucalyptus leaves, which are actually higher in trees growing on poorer soils. For this reason, they'll sometimes turn down even the preferred species growing under these conditions. If the eusauropods and diplodocids were this specialized in their diets, it's not hard to see why they'd follow their food plants into extinction.

At this point we have only a possible scenario, not any kind of proof, for the die-offs. If we could discover more solid evidence, let's say a predominance of "cheiro" phytoliths among eusauropod and diplodocid teeth, as well as the microwear patterns these plants might have produced, we'd start to have a good case. Another clincher would be to find coprolites, or fossil dung, bearing mostly "cheiro" remains within the body cavities of or in close association with articulated skeletons of the above sauropods. Finally, a database of extensive geochron sampling, extending from the latest Jurassic through the earliest Cretaceous of North America and Proto-Asia and showing a positive decline of "cheiros" in these areas and at this time, would create a very compelling theory to explain the extinctions. For now, we just don't know.

LIFE GOES ON

With the passing of the eusauropods and diplodocids, major feeding niches were left wide open for other potential sauropod browsing specialists, ones that were already well adapted to feeding on conifers. At the very end of the Cenomanian and early Turonian, the other diplodocoid clades, first the dicraeosaurids and then the rebbachisaurids, also perished, making available additional new niches for the remaining neosauropods, the more advanced of the two big original groups. One neosauropod group we met in Chapter 3 are **macronarian titanosauriforms**, which include not only the iconic, generally huge and tall brachiosaurids and the more derived, also enormous somphospondylans but also the **titanosaurs**. Titanosaurs shared the Early Cretaceous with basal macronarians, which were becoming evolutionarily marginalized by their more advanced close cousins. All these clades didn't just suddenly appear full-blown during this time but got their start much earlier during the Mid-Jurassic, when they shared ecosystems with other sauropod groups. As with any evolutionary lineage, our knowledge of sauropod diversity

Initially sharing environments with diplodocoid and basal macronarian sauropods, by the time of later titanosaurs like *Saltasaurus loricatus* the clade could be found in all of the protocontinents and had evolved into both enormous and dwarfed species.

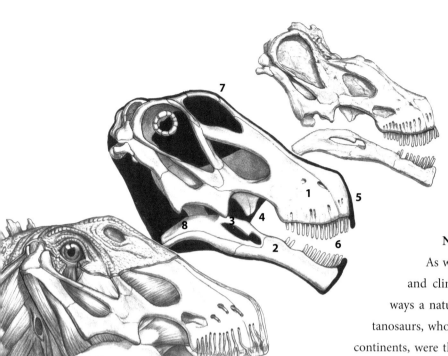

at this time is affected by fossil preservation, and it's probable that we don't have close to the true picture of what was actually unfolding. In the case of the titanosaurs, however, some scrappy bone remains but also some "wide-gauge" trackways show that this group was around in the Kimmeridgian through Tithonian stages of Proto–South America/Africa, and even earlier during the Bathonian–Callovian transition of Proto-Europe.

NEW HEADS, NEW WAYS OF EATING

As we discovered in Chapter 10, the changing geographies and climates of Proto–South America/Africa were in many ways a natural laboratory for new sauropod adaptations, and titanosaurs, whose early representatives were already present on other continents, were the outcome of one of these trends. Characteristic was a continuation of the adaptation shown by their more basal titanosauriform relatives, the brachiosaurs, in developing more narrow-crowned teeth than other macronarians, producing a distinctive *precision-shearing* bite, for orthal jaw action or the cropping/nipping off of vegetation. This was an innovation on both the robust, spoon-shaped dentition of other macronarians and the pencil-like, leaf-stripping morphology of most diplodocoids. Although titanosaur morphology is still very poorly known, this adaptation is present in the better-preserved skulls of later Cretaceous forms, which at present fall into at least three known *morphotypes*, or shapes. One, typified by *Nemegtosaurus mongoliensis*, is very diplodocoid-like, with a long and somewhat broad but not squared-off muzzle, teeth confined to the anterior jaws, a high cranium and retracted nares. The dorsal surfaces of the premaxillary and maxillary bones have foramina and grooves running forward across the bone surfaces, and these indicate that in life the outer coverings of the mouth might have had nerves and blood vessels that supplied an outer skin that was very sensitive, perhaps an aid in selecting tender new leaflets. The second, shown by *Rapetosaurus krausei*, is also diplodocoid-like, but with a lower cranium, and with upper and lower teeth going much farther back. Another main difference is the huge, elongated *antorbital fenestra* in the front of the skull and the high, arched *postdental emargination*. Together with the robust posterior mandible and its high *coronoid process*, this would have made room for particularly powerful jaw-closing muscles, signaling a very strong bite. Finally, there is *Bonitasaura salgadoi*, whose short skull, wide and squared off at the anterior muzzle and with small, anteriorly located teeth, is very reminiscent of the rebbachisaurid diplodocoid *Nigersaurus* and, as in this sauropod, suggests a close-cropping, perhaps low-browsing feeding adaptation. As in the African form, the postdental emargination is high and the antorbital fenestra are large, indicating that this small form had a strong bite for its size, and the narial opening is enormous. Together with the variation in neck lengths (from 13 to 16 cervical vertebrae), the wide diversity in skull forms by the later Cretaceous means that by this

Tapuiasaurus macedoi from Argentina demonstrates some features that typify the feeding specializations of nemegtosaurids and other advanced titanosaurs. These include **(1)** a deep maxilla and **(2)** mandible, **(3)** high coronoid processes for the attachment of strong temporal muscles, **(4)** a postdental emargination for accommodating larger amounts of foliage per bite and **(5)** very wide anterior jaws. All these, along with a pencil-like but cone-chisel dentition **(6)**, were derived from earlier macronarian ancestors. At the same time, titanosaurs convergently evolved some diplodocid features like highly retracted nares **(7)** and a long, forwardly directed infraorbital fenestra and glenoid joint **(8)**, suggesting that some titanosaurs may have combined a more powerful bite with some degree of leaf stripping or raking.

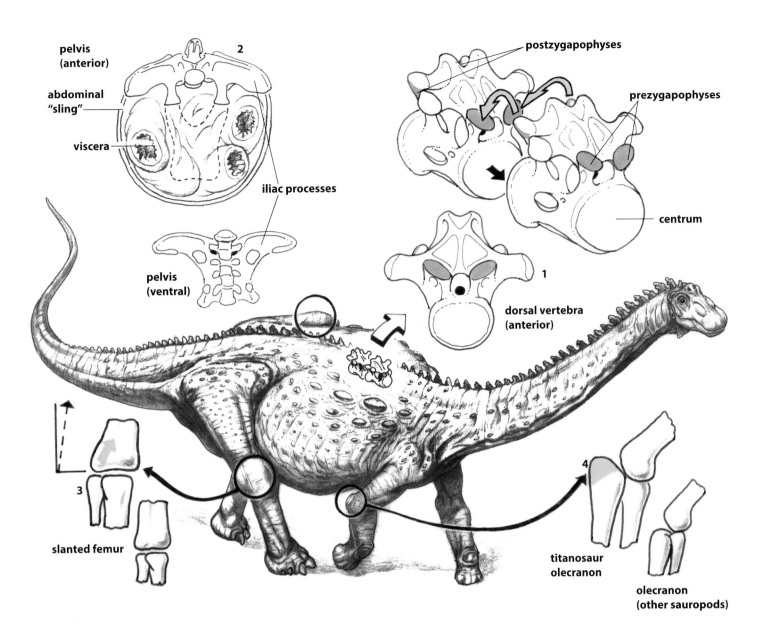

pelvis (anterior)

abdominal "sling"

viscera

iliac processes

pelvis (ventral)

postzygapophyses

prezygapophyses

centrum

dorsal vertebra (anterior)

slanted femur

titanosaur olecranon

olecranon (other sauropods)

Titanosaurs—Built for Wide Loads

Like a wide-bed pickup truck designed to carry extra heavy cargos, the enormously expanded bellies and visceral load of advanced titanosaurs like *Opisthocoelicaudia* had extra support. This was accomplished with **(1)** extrawide pre- and postzygapophyseal contacts and centra (*orange areas*), which extended the amount of intervertebral contact, making the dorsal spine stronger. **(2)** The pelvis's ilium bones jutted far out to the sides, greatly widening the hips and providing anchoring points (iliac processes) for a much stronger abdominal "sling" of strong fascia to supply more lift from below. **(3)** In addition, the femur was slanted inward (*blue arrow*), splaying out to the side to broaden the support from the legs, while the forelimb **(4)** had a bony instead of a cartilaginous elbow, the olecranon process (*blue area*), providing for greater support and strength when the limb was moving. At right, a gigantic pelvis from *Argentinosaurus*.

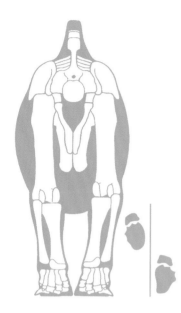

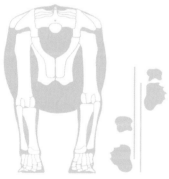

Titanosaurs were in some ways like later, bigger railroad locomotives that needed wider tracks than the older, narrower engines for supporting their larger size. Compared to earlier sauropods (*top*), titanosaurs' much bigger, heavier abdomens were supported by more widely spaced legs (*bottom*) to give them greater stability as they walked, producing "wide-gauge" trackways. Discoveries of these in certain places suggest to some paleontologists that titanosaurs may have preferred more upland terrains than their basal macronarian and diplodcoid cousins.

time titanosaurs had radiated into a number of distinctive types and feeding adaptations. What's fascinating here is that although the skulls have a somewhat diplodocoid look, a macronarian ancestry is betrayed by the strong, deep jaws. While some titanosaurs ("*Campylodoniscus*") retained the wider, macronarian-like tooth shape, some independently acquired a very diplodocoid, pencil-like dental condition (*Rapetosaurus*, *Saltasaurus*, *Nemegtosaurus* and others). These weren't restricted to the very front of the jaws as in true diplodocoids, but it's possible that they were used for a similar leaf-stripping technique in feeding.

As with almost all sauropod remains, there's a frustrating lack of gastrointestinal material and phytoliths that could give us a clear idea as to exactly what kinds of conifers and other plants formed their diets, making it necessary to infer heavily from tooth microwear studies. Another source of potential information, however, is coprolites, or fossilized feces: sauropods had to be very productive, and there should be a lot of these out there. In the case of some titanosaurs we have at least a few clues, with the discovery of plentiful coprolites from the Late Cretaceous Lameta Formation of central India in 1999 by Dahnanjay M. Mohabey (Geological Institute of India), who described the content of the large ("Type A") coprolites from the site. These can be up to 100 cm (39.4″) in diameter and contain a variety of plant remains that in section actually have fragments and phytoliths of conifers, as well as angiosperms like grasses, which had now become more common. These coprolites confirm that certain titanosaurs, such as *Isisaurus colberti*, whose remains are found in the Lameta Formation, were adapted in some degree to the greater prevalence of angiosperms by the end of the Cretaceous. In spite of a general lack of information about titanosaur diets, what's clear is that they innovated in their own directions, which are sure to be better understood as new discoveries are made.

FLEXIBLE SPINES FOR TAKING THE CURVES

As titanosaurs began to fully radiate by the Early to Mid-Cretaceous into the roles vacated by the vanished forms, their postcranial anatomy also changed. At this time we encounter basal species like *Andesaurus delgadoi*, *Ligabuesaurus leanzi* and massive *Chubutisaurus insignis*. Although as derived macronarians these early titanosaurs are only distantly related to their diplodocoid rebacchisaurid relatives, they show an amazing convergence: like the first clade, they've evolved more flexible spines, but in a totally different way. Titanosaur vertebrae, although like those of rebbachisaurs, lose the hyposphene-hypantrum vertebral "lock" (see Chapter 10) and differ in that instead of having **amphiplatyan** ("both flattened") centra, the ends of the centra become wide, ball-and-socket joints, a condition known as **opisthocoelous** ("behind hollow space"). The posterior end of each centrum is a concave or recessed hollow, allowing a convex bulging anterior end from the next adjoining centrum to fit into it. At the same time, the pre- and postzygapophyseal contacts of the anterior caudal neural arches become broader, more cup-like and more widely spaced. So what's all this for? For one thing, the changes in the dorsals allowed (1) a greater amount of joint contact between individual vertebrae, (2) *torsional*, or twisting, action and (3) overall dorsal vertebral flexibility, first seen to a more limited degree in basal titanosauriforms like *Brachiosaurus*. As with the rebbachisaurids, it might be an adaptation for moving more efficiently over the increasingly hilly, uneven terrain of uplifting western Proto–

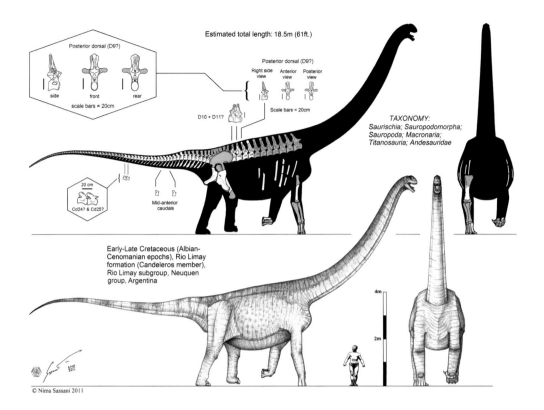

Estimated total length: 18.5m (61ft.)

Posterior dorsal (D9?)

side   front   rear

scale bars = 20cm

D10 + D11?

Posterior dorsal (D9?)

Right side view   Anterior view   Posterior view

Scale bars = 20cm

TAXONOMY:
Saurischia; Sauropodomorpha;
Sauropoda; Macronaria;
Titanosauria; Andesauridae

20 cm

Cd24? & Cd25?

Mid-anterior caudals

Early-Late Cretaceous (Albian-Cenomanian epochs), Rio Limay formation (Candeleros member), Rio Limay subgroup, Neuquen group, Argentina

4m

2m

© Nima Sassani 2011

Nima Sassani, an independent artist-paleontologist, has created well-researched, detailed drawings of several titanosaurs like *Andesaurus delgadoi*, which have cast light on the appearance of this as yet diverse but poorly known major sauropod clade.

South America. Unlike the rebbachisaurids, however, it's a *totally different* anatomical solution in meeting the same environmental change. The procoelous condition of the anterior caudals allowed more tail flexibility at its proximal area, or base, which would have resulted in shorter, more abrupt movements in this spot, possibly for striking at attacking theropods. If this interpretation is correct, it would have paralleled the shorter proximal caudals of ankylosaurids, whose club-tipped tails were effective weapons. Like their rebbachisaurid neighbors, titanosaurs had non-bifurcated, unsplit neural spines, but unlike the diplodocoids, these weren't tall, suggesting a tendency to rely more on quadrupedal feeding instead of bipedal rearing.

WIDE-LOAD SAUROPODS

In addition to greater flexibility in their dorsal vertebrae, advanced titanosaurs begin to show specializations in other parts of their skeletons as well that make them very different from previous sauropods. More derived titanosaurs share with rebbachisaurids a powerful, massive pectoral girdle, but here it's due not only to a robust scapula but also to the combined effects of large, squarish coracoids and big, crescent-shaped sternal plates. These provided more strength, and the shape of the sternal plates may have allowed more fore-and-aft forelimb movement. As they evolve, some titanosaur lineages also develop wider, more massive bodies than ever before known in sauropods, and the preacetabular process of the ilium flares out laterally almost at 90% to the main axis of the pelvis. This meant better belly support, since the process was an important attachment point for broad, sling-shaped abdominal tendons. The sacrum, limited to five vertebra in earlier sauropods, gains a sixth element, contributing strength and support for extra weight.

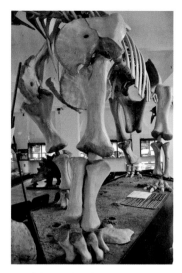

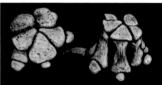

In close-up, the skeleton of *Opis-thocoelicaudia skarzynskii* (top) shows the distinctive forefoot of advanced titanosaurs, in which toes disappeared and any foot pad flexion was entirely sacrificed for support. In addition, the meta-carpal bones, as in species like *Diamantinasaurus*, form a complete circle (bottom) with only one large pad, like an elephant's forefoot.

At the same time, the fore- and hindlimbs are also changing to become even stronger and bear weight, but in a different way. All the articular surfaces are very broad, and the proximal end of the ulna in some forms in some species forms a prominent bump, the **olecranon process**. This probably existed in other sauropods as cartilage and usually wasn't ossified, and it was a major insertion for the powerful, forelimb-straightening *triceps* muscle. In the forefoot, the manual phalanges are becoming more tightly constricted and pillar-like, producing an almost completely cylindrical form compared to the semicircular, horseshoe-shaped foot of earlier sauropods. This gives titanosaur manus prints a more elephant-like look. In titanosaurs support, not toe flexion, is now all that matters, and the phalanges are gradually becoming mere nubbins, in some clades to disappear entirely, even the thumb claw. In the hindlimb, the proximal one-third of the femur is also **medially deflected** and the distal condyles beveled to 10% dorsomedially. As a result of all this, both the fore- and hindlimbs splay noticeably out to the sides, creating a "wide-bodied" stance. This reconstructed posture is supported by finds of trackways from the later Cenomanian of the Neuquen Basin and from the Upper Campanian/Lower Maastrichtian stages of Mendoza Province. Here wide trackways occur that are very different from the typically more narrow ones of earlier diplodocoids and basal macronarians, many of which indicate much bigger animals than some titanosaurs. As with the above spinal modifications for walking on uneven terrain, all these features make sense for supporting greater bulk: the weight-bearing capability of later, more derived titanosaurs compares to that of other sauropods as a wide-bodied pickup truck with heavy-duty suspension might to a regular pickup. In spite of the fact that some titanosaurs, as their name implies, were truly titanic in size, others, as we learned in Chapter 10, were dwarves.

MINERAL RESERVOIRS IN THE SKIN

Yet another distinct feature developed by some titanosaurs, and seen to date in no other sauro-pods, are **osteoderms**, or dermal ossifications. These were embedded in the layers of the outer skin and were covered in keratin; they also occur in other tetrapods like crocodilians, some modern frogs and extinct ground sloths. The osteoderms in known titanosaur species like the Late Cretaceous *Saltasaurus loricatus* range from pebble-sized nodules to thick, keeled discs, and like croc osteoderms, they are made up of an outer layer of dense, compact bone, perforated by vascular cavities. In some species like *Rapetosaurus* from Madagascar, the osteoderms of both juveniles and adults possess a canal that widens into a hollow, internal chamber, whose volume can be as much as 0.9 L (2 pt). This has led Curry Rogers and her other colleagues to suggest that these structures may have acted as stores for the release of the blood-deposited nutrients calcium and phosphorus. In juveniles this would have been vital in fueling rapid skeletal growth, which, as in other sauropods, took place on a high level. In addition, *Rapetosaurus* inhabited an intensely seasonal, semiarid environment on Madagascar that could produce droughts, some-times ending in mass mortality for the juveniles of these and other dinosaurs. The osteoderm reservoirs probably released these minerals during times of food scarcity and other high-stress situations for *Rapetosaurus* all throughout life, as well as helped to supply calcium when needed by egg-producing females. In spite of the many close, pebbled small osteoderms in *Saltasaurus*, so far relatively few osteoderms of any kind have been found in association with other titanosaur skeletons, even when these are relatively complete and articulated, as in *Epachthosaurus*. This makes the argument that in titanosaurs the osteoderms were primarily organs of mineral storage, as described above, and generally weren't extensive or close-fitting enough to form protective armor, as in ankylosaurs. Some workers assign all osteoderm-bearing titanosaurs to a subfamily of advanced forms, the **Lithostrotia**, which lived from the Early Cretaceous until the very end of the dinosaur era. In addition to osteoderms, lithostrotian titanosaurs have even more flexible tail bases, and all the manual phalanges, including the thumb claw, are absent in known species.

UNTANGLING TITANOSAURS

Titanosaurs joined the roster of known sauropod groups when the first, "*Titanosaurus indicus*" (now considered by some workers as an invalid name), was originally described by Hugh Falconer in 1868, and as with sauropods in general, ambiguity reigned for decades as workers attempted to gain an understanding as to how they should be defined and classified. Like some other enigmatic dinosaur clades, titanosaurs became a "catchall" or "ragbag" taxonomic group for fragmentary and poorly known Cretaceous sauropods long after the family was established in 1893. Until the late 20th century, titanosaur remains, unlike the well-preserved and reasonably complete diplodocids and camarasaurs of the Morrison, were frustratingly incomplete. Very little agreement existed until recently as to how to interpret known specimens, with diagnoses, as in other sauropods, being heavily dependent on traditional comparative anatomy studies focused on a limited number of then known characters. Now, beginning in the late 1990s with the increasing attention of sauropod workers toward phylogeny, the scores of new specimens (rarely, extensive skeletons and skulls) from almost all areas of the globe and the help of computer technology in evaluating the flood of new characters based on these discoveries, this is rapidly changing. Any understanding of extinct animal relationships is dependent on comparing bone characters, and in the case of sauropods it frequently focuses on vertebrae, since these

At a bone bed nicknamed "The Graveyard of the Giants," Chubut Province, southern Argentina, members of the Museo Paleontologico Egidio Feruglio research center excavate titanosaur skeletons. As yet unnamed, these add to the growing list of titanosaur taxa from South America.

are among their most distinctive skeletal features. This is especially true with the usually scrappy material for many titanosaurs. As pointed out in Chapter 3, different workers place different interpretations on the same skeletal characters—when the databases for these are limited, so is the knowledge of what can be inferred from them. There is, and continues to be, disagreement regarding titanosaur phylogenies, but let's look at what seems to be gaining acceptance in what we think we know about these remarkable sauropods.

TALENTED NEWCOMERS

Although still very incomplete, the picture that begins to emerge of titanosaurs is that of a small but anatomically advanced sauropod clade that began in the Mid-Jurassic as a minor group but swiftly radiated into many forms during the early Mid-Cretaceous, coming to fill the ecological niches left by the disappearing eusauropods, diplodocoids and basal macronarians. We've met some early representatives like *Janenschia* and *Amargatitanis* back in geochrons such as Tendaguru and Neuquen, and ending with dwarves like *Magyarosaurus* on Hateg Island. Although the reasons for titanosaur origins are still obscure, it's possible that they got their start as populations of marginalized titanosauriforms that were pushed out, at the time of the Mid-Jurassic, by more dominant members of this large clade into marginalized environments and habitats, where they began to evolve somewhat different feeding ecologies. If the former Gondwana continents were truly the birthplace of these sauropods, perhaps places like the uplifting Andean highlands of Proto–South America were what spurred the locomotor adaptations that characterized early titanosaur anatomy. Some of the earliest known species were still fairly broad toothed, but showing a definite trend toward the cylindrical, precision-shear bite mentioned earlier. An early (Aptian–Albian stage) grouping of basal titanosaurs are the **andesaurs**. Though not recognized as a formal clade, these have, in addition to some of the skeletal features already described, more pneumatic vertebral spaces in the trunk and other features that made the spine stronger and lighter. They also had a distal tibia wider than anterior-to-posterior for more support. These may be the first titanosaurs to be embraced by the superfamily **Titanosauria** and include such species as *Andesaurus delgadoi*, *Ruyangosaurus giganteus* and *Sonidosaurus saighangaobiensis*. Some basal titanosaurs even at this early period grew enormous. *Argentinosaurus huinculensis* from western Argentina reached about 30 m (100′) and weighed 50-plus tons, while the later Maastrichtian species *Puertasaurus roulli* was just as big. Because of the claimed presence of osteoderms in its members, somewhat more derived titanosaurs are grouped by some specialists into the clade **Lithostrotia** (see Chapter 3), although it's not recognized by some workers because species included in this group like *Epachthosaurus sciuttoi* don't have them. Other known forms feature *Malawisaurus dixeyi*, *Ampelosaurus atacis*, *Diamantinasaurus matildae* and many others. The group **Lognkosauria** is a subgroup of the lithostrotians based on derived and shared cervical and caudal characters. Some of these, such as *Mendozasaurus negueyelap*, are relatively small and short necked, but others have a very long but deep neck, including gigantic species like *Futalognkosaurus dukei*, one of the largest sauropods known from an extensive skeleton. As basal or relatively primitive as some of these early titanosaurs are, they lasted long into the Cretaceous, in some cases toward the very end.

Malawi, along with Tanzania, has been a source of significant African titanosaur material. **Elizabeth Gomani Chindebvu** (*above*) (Department of Antiquites, Museums of Malawi) was the first researcher to describe a sauropod online with her paper on *Karongasaurus gittelmani* in 2005, and her studies have extended to a reexamination of *Malawisaurus dixei* (*below*).

LATER TITANOSAURS

The Cretaceous continues, and we encounter a new, also more advanced clade: the **Eutano-sauria**, or "true" titanosaurs. These are known mostly from fragmentary, incomplete remains until we get to the provisionally accepted family **Nemegtosauridae**. In contrast to most tita-nosaurs, nemegtosaurids include some of the most completely known of these sauropods, not only from reasonably extensive postcranial skeletons but also from skulls. The most complete of these is the Madagascan *Rapetosaurus*, extensively studied and described by Curry Rogers from an adult skull and a small juvenile. Like the Rosetta Stone, whose discovery opened the way to deciphering ancient Egyptian hieroglyphics, this species has given us so far the most informa-tion about titanosaurs as a group. Along with the equally complete and well-preserved skulls of *Nemegtosaurus mongoliensis* and *Tapuiasaurus macedoi*, nemegtosaurids form a family with a worldwide, Late Cretaceous distribution. Among the nemegtosaurids are wide-mouthed forms like *Antarctosaurus wichmannianus*, the low browser *Bonitasaura* and also some truly strange, highly derived types like *Isisaurus (Titanosaurus) colberti*, whose moderately long neck, very long forelimbs and high shoulders are convergent on some basal macronarians like *Atlasaurus*, and which is the closest known sauropod equivalent to a giraffe in its proportions. Another large, cohesive but unofficial grouping of titanosaurs are **aeolosaurines**. These include species like North American *Alamosaurus sanjuanensis* and *Baurutitan britoi*, which share a biconvex first caudal that's different from the **procoelous** ("forward hollow") condition of the first caudal in more basal titanosaurs. Nested within the aeolosaurines are the **Saltasauridae**, which take in small *Saltasaurus loricatus*, *Neuquensaurus australis*, *Loricosaurus scutatus* and dwarfed species like *Magyarosaurus*. Occurring late in the Late Cretaceous of Mongolia is *Opisthocoelicaudia skarzynskyii*, a species known from an extensive, heavily built skeleton and that gives its name to our last known titanosaur group, the **Opisthocoelicaudinae**.

MORE EXTINCTIONS?

From known body fossils and trackways we know that titanosaurs, although they radiated and flourished mostly during the Mid- to Late Cretaceous, were in existence by or before the Mid-Jurassic, about 165–168 million years ago, and had achieved a worldwide distribution before the protocontinents that formed Pangaea had separated much. This is important, since it's a fact to consider in the debate over whether sauropods as a group actually suffered from more extinction events *after* the one in the Late Jurassic, as well as the final one at the end of the Cretaceous. Those who do support this idea claim that there was a "sauropod hiatus" (gap) in the fossil records of the Mid-Cretaceous of North America and Europe, and that these areas weren't reinhabited by sauropods (titanosaurs, since by now they were the only ones left) until the beginning of the Campanian stage of the Late Cretaceous, about 84 million years ago. According to this hypothe-sis, the above areas only had sauropods once more *after* titanosaurs emigrated from the Southern Hemisphere continents that used to form Gondwana. In tackling this issue, two workers, Phillip Mannion and Paul Upchurch (both University College, London), first constructed a database of known individual sauropod fossil specimens of all types, and from all localities, from the period of the "sauropod hiatus." In assessing sauropod abundance, they took into account the

Along with her associates, **Kristi Curry Rogers** has conducted extensive studies of *Rapetosaurus* and other titanosaurs. The osteo-derms of some titanosaurs were hollow and supplied with nutri-ent canals, indicating that these structures were reservoirs that held stores of calcium and other minerals needed when the ani-mals were undergoing environ-mental stress and during egg laying.

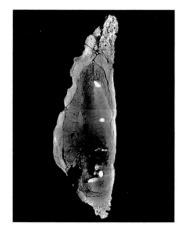

An X-ray of an osteoderm from the Madagascar titanosaur *Rapetosaurus krausei,* showing the large internal cavity that in life probably stored calcium and other dissolved minerals.

possibilities of uneven sampling and importantly divided the finds into those that came from either *inland* or *coastal* localities. Inland areas where sauropods lived generally shrank during times of higher sea levels (marine transgressions), and this is reflected by the fewer numbers of skeletal remains, eggs and trackways that would probably be preserved (this doesn't mean, as we've learned, that sauropods didn't live along the coasts, but it does mean that there were fewer varieties of plants to support a large number of species and bigger populations of these, therefore increasing the chance of becoming fossils). Marine transgressions can result in extinctions, since there's now less habitat to go around—some species can't survive, so there ends up being fewer of them, or none at all. Since a few of these transgressions happened during the "sauropod hiatus," hiatus proponents use these events, as well as an alleged lack of sauropod species, to support the idea of an absence of sauropods in North America and Europe during this time. Mannion and Upchurch, however, based on the findings of their study, criticized this. They found that although there *was* a decline in European and North American inland environments during these times, with a corresponding increase in coastal environments and absence of titanosaurs, the "hiatus" didn't occur anywhere else around the world, where titanosaurs were now well established. The team demonstrated that claims of absences in these continents during certain key periods were incorrect. Titanosaurs *were* present, but in seemingly limited numbers, in Europe and North America during the Cenomanian, in the Late Turonian–Early Coniacian (Europe) and during the Late Campanian (North America)—areas and times during which the "hiatus" said they shouldn't have existed.

South American *Mendozasaurus nguyelap* was a medium-sized member of the titanosaur clade Lognkosauria that lived between 85 and 89 million years ago. It may have been typical of earlier, more basal titanosaurs.

Southern Hemisphere

(1) Africa
(2) India
(3) South America
(4) Sunda-Australia
(5) New Zealand
(6) Antarctica
(7) Proto-Asia

The black line indicates modern coasts.

So titanosaurs were there, but in limited numbers compared to other areas—why? One discovery of Mannion and Upchurch's study was that in other sampled localities from earlier in the Mesozoic, where inland environments and coastal environments were more even, a greater number of titanosaur body and other kinds of fossils were found in inland localities. Those of diplodocoids, basal macronarians and other non-titanosaurids were more likely to come from coastal environments. This is true of geochrons like Tendaguru, Proto-Africa, where we find an abundance of brachiosaur individuals like *Giraffatitan* and to a lesser extent the diplodocoids *Torniera* and *Dicraeosaurus*, but even fewer of the basal titanosaurs like *Janenschia*, *Tendaguria* and *Australodocus*. It could be that the three latter dinosaurs preferred more inland habitats with different primary browse trees. If this is true, the scarcity of titanosaurs in North America and Europe during the so-called hiatus was the result of a bottleneck rather than large regional extinctions, and then titanosaurs bounced back in these areas to become more abundant when the sea levels dropped. A following hypothesis, or a ***corollary***, to the "sauropod hiatus" concept is that titanosaur stocks from other parts of the world, such as Proto–South America/ Proto-Africa, migrated back into these two northern regions to fill the gap once conditions for their existence improved. Here Mannion and Upchurch respond that this is unlikely because of the fact that titanosaur populations were already present in Proto-Europe and North America, and there was no vacuum to be filled. If this is true, the "sauropod hiatus" of these regions is best interpreted as an artifact of uneven sampling in previous studies, based on the scarcity of Mid-Cretaceous inland deposits preserved in the two areas.

SOUTHERN SANCTUARIES

By the Late Cretaceous, the titanosaurs were the only clade of sauropods still surviving, apart from a possible few late-surviving macronarians, in the entire world. In spite of this, they not only had achieved remarkable diversity but also had set a record for sheer numbers of genera and species by this time, more than any of the entire range of sauropod clades that had preceded them. Many are known from Africa, India, Australia, Madagascar and at least one from Antarctica, all former Gondwana landmasses, but thus far an overwhelming number of

After South America and Africa separated from the original main landmass of Gondwana, South America and the remaining protocontinents and large islands of Sunda-Australia, Zealand and Antarctica continued to support a diverse number of conifer species that may have favored the radiation of the titanosaurs, whose genera and species richness is considerable in these regions. *Wollemia* (*below*), a relict from the Mesozoic, survives today in New Zealand.

The advanced Asian *Opisthocoeli-caudia skarzynskii* (*below*) and North American *Alamosaurus sanjuanensis* (*above*) were both huge, in contrast to island dwarf species like *Magyarosaurus dacus* from Hateg. The neck is totally unknown in the Asian form, but if it was proportioned like the neck of *Alamosaurus*, it suggests high reach from a quadrupedal stance, since the dorsal and sacral neural spines were very low.

genera and species are from South America. This could be due to a preservation bias, but it also could reflect two possibilities. One is that, even though they had long since become worldwide in distribution, during the Mid-Cretaceous the titanosaurs may have begun to specialize and evolve derived forms following the decline and extinction of other sauropod clades on this continent first. Eusauropods, diplodocoids and most macronarians were almost entirely gone by the Cenomanian stage in former Laurasia, possibly because of a decline in important conifer food sources (see beginning of chapter). Sampling from such a widespread area isn't complete enough to determine whether conifers as a whole declined in North America, Proto-Europe and Proto-Asia enough to affect overall sauropod diversity in these regions, but many types of conifers and other gymnosperms, all of which were probably important, long-standing titanosaur (and other sauropod) food sources, continued to flourish in South America, Australia, Antarctica and to a lesser extent Africa. These included several species of araucarians, taxodiaceans and podocarps, as well as ginkgoaleans and tree ferns. Could the survival and prevalence of these conifer and other primary browse trees have been the reason for the genera and species richness of titanosaurs in Mid- to Late Cretaceous South America and possibly other former Gondwana regions?

Whether the great concentration of titanosaur species in South America is real or due to a preservation bias, the number of species, extreme size disparity and apparent adaptation to different feeding niches of titanosaurs worldwide refute an earlier long-standing claim that sauropods as a whole were a dying group. According to this, the sauropods were unable to adapt to the possibilities of Cretaceous angiosperms as food sources, of which the widely radiating ornithischians like hadrosaurs and ceratopsians now took advantage. These dinosaurs evolved complex chewing dentitions for processing the new plants, to which the outmoded sauropods couldn't adjust. This, however, ignores certain facts. Although some conifer types like cheirolepidiaceans did decline and become extinct, many others were doing just fine, especially in the southern continents, and were still widespread throughout the Cretaceous. This was a prime food source that, in spite of 90 million years of evolution, the ornithischians as low browsers had never been

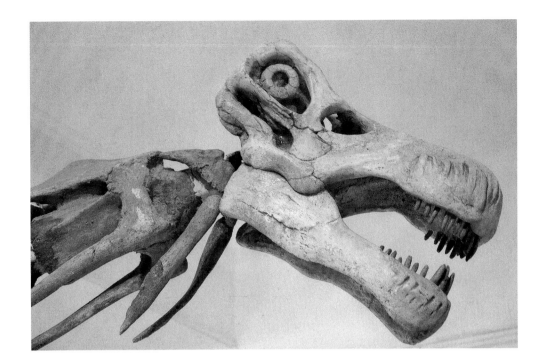

Opisthocoelicaudia (skull, *left*) of Central Proto-Asia was among the last of the titanosaurs to evolve, and these, in turn, represented the final radiation of the sauropods as a clade. The apparently great diversity of titanosaurs in the southern continents of former Gondwana may be linked to the fact that there, more than any other area on Earth, the Mesozoic types of conifers persisted longer than in the Laurasian continents.

able to exploit. As we've already seen, some Indian titanosaurs as non-selective feeders actually branched out into eating grass, and there's no reason why low-browsing forms would have ignored other angiosperm vegetation as well. Titanosaurs not only continued the basic strategy of consuming and retaining large amounts of food, without the need to chew, for efficient energy release but also seemed to have evolved new feeding adaptations. These in some cases retained and actually improved on the macronarians' strong bite, and in others they incorporated the extinct diplodocids' leaf-stripping techniques.

The titanosaurs were the last sauropods to evolve, and they were the last standing when the asteroid hit Earth at the end of the Maastrichtian stage of the Cretaceous. Although with other non-avian dinosaurs they became finally and totally extinct in the catastrophic aftermath, in sauropod evolution they had a very "long run"—100.5 million years—and were not only the most long-lived but also the most diverse, and probably the most widespread, of their kind. By any measure, titanosaurs and all other sauropods were a huge evolutionary success. In the next and final chapter, we'll take a long backward look at sauropods as a whole and summarize what we think we can confidently say about their paleobiology, while focusing on the things we still don't know, offering some speculative but informed possible answers.

A *Diplodocus carnegii* is startled by a fast-moving hypsilophodont, *Othnielosaurus rex*, in a Garden Park forest edge clearing. Small ornithopods like *Othnielosaurus* benefited from the arboreal scraps, normally unavailable to them, dropped by the sauropods.

CHAPTER TWELVE
SUMMING UP SAUROPODS

"Science works with testable proposals. If, after much compilation and scrutiny of data, new information continues to affirm a hypothesis, we may accept it provisionally and gain confidence as further evidence mounts. We can never be sure that a hypothesis is right, though we may be able to show with confidence that it is wrong. The best scientific hypotheses are also generous and expansive: they suggest extensions and implications that enlighten related, even far distant, subjects."

—Stephen Jay Gould, 1984

UNSOLVED PROBLEMS. In this book we've tried to explain the current scientific understanding of how—and—why—sauropods evolved, diversified and thrived. Paleontologists have made huge strides in the past 20 years, both in finding new fossils that open fresh windows into these giants' lives, like discoveries of nests and embryos, and in new techniques like CAT scans and bone histology that allow us to learn as never before from each new find. But there are still plenty of mysteries left to solve. In this final chapter we'll outline at least nine outstanding problems we think will shape the study of sauropods for decades to come—let's now take a look at these.

So what made sauropods ultimately replace prosauropods? In terms of their overall survival strategy, they can be seen from their earliest true forms as evolving a highly advantageous set of key biological adaptations. First, sauropods **lost the complex parental care** that was apparently typical of other dinosaurs, including birds. As with baby sea turtles, a "sweepstakes" strategy was the rule, with more babies produced to offset the mass mortality that occurred—even if most died, at least a minimal number survived to keep the species going. As a result, large clutch sizes (large-scale oviparity), with probably more than one clutch per reproductive season, were developed as a mode of reproduction, meaning that more young could be produced on a more frequent basis. In addition, parental care takes time and energy. By giving up parental care, sauropods could put that energy into growing larger and making more eggs. Second, like many other dinosaurs, sauropods had **a bird-like respiratory system**, which efficiently supplied oxygen to their huge and growing bodies. These bird-like lungs and air sacs also brought additional bonuses mammals don't share: the air sacs were large enough to overcome the respiratory dead space imposed by a long windpipe and allowed for efficient panting to dump heat without hyperventilating. Third, **early sexual maturity** in relatively young animals that hadn't yet reached adulthood meant that breeding could start early, ensuring that sauropods had a reproductive edge to start producing enough individuals to offset predation. Fourth, like other dinosaurs, **sauropods grew very rapidly**, taking individuals to bigger and progressively less vulnerable size stages so they could make it to reproductive age. A simple but effective innovation was **simply swallowing, not chewing food.** Without batteries of heavy cheek teeth for grinding plants, sauropods could have small, light heads that could be lifted high into the trees on long necks. Freedom from chewing was both a consequence and a cause of large size: as body size increases, so does the retention time of food in the gut. Long retention time in the gut increases the possible food particle size that can be efficiently digested. Since they didn't have to spend time chewing, sauropods could harvest and swallow a proportionately much greater amount of food than a mammal in a given amount of time (except for baleen whales), and this greater intake translated to larger and more quickly attained body size.

All these innovations mutually reinforced one another to contribute to the sauropods ultimately replacing prosauropods, but two others also evolved to produce a biological *bauplan* that ensured their long-term success. One was **extensive skeletal pneumaticity**. Sauropods shared a bird-like respiratory system with other saurischian dinosaurs and perhaps with pterosaurs and other non-dinosaurian archosaurs. But not all of these animals benefited from the lighter, stronger skeletons that air-filled bones provide. For example, many prosauropods have shallow excavations on their vertebrae that are the footprints of a bird-like respiratory system, but the amount of bone removed was tiny and had no significant effect on the form or function of the skeleton. But in sauropods extensions of the lungs (**diverticula**) filled the vertebrae and ribs with air, allowing the body to become comparatively light and making 50′ necks possible. This, along with the other previously mentioned key adaptations, enabled sauropods to become gigantic, bulk-feeding herbivores that evolved high reach to exploit and efficiently process a food source of which no other animal has ever fully taken advantage. As we learned in Chapter 4, not only are big animals faced with the challenge of surviving, growing and reproducing, but they also have to succeed at simply being big. And even here sheer size isn't an end in itself; it's a means to an evolutionary end, to survive and reproduce your kind. The above adaptations were all a "package deal," each of whose features depended on the other for everything to work. Like a mechanical ratchet arm that increases leverage the more it's cranked, these coevolving adaptations were boosted to a higher and higher level until sauropods became the biggest and most successful terrestrial herbivores on the planet, either before or since. Although they never developed the highly disparate (different) body shapes and sizes of their contemporaries the theropods and ornithischians, always basically remaining large, long-necked quadrupedal herbivores, this worked for sauropods extremely well—they were one of the most widespread, longest-surviving major lineages in all of dinosaur history.

Early Atmosphere and Sauropod Breathing. The overall makeup of Earth's atmosphere has actually changed over time from what it is today. For the first 2 billion of our planet's 4.5 billion years of history, there was very little free oxygen in the atmosphere. But then about 2.5 billion years ago, photosynthesis by single-celled organisms passed a critical threshold, and there was a rapid increase in the level of atmospheric oxygen. These levels were still much lower than those of today because much of it at first became absorbed into "oxygen sinks"—the oceans and weathering surface minerals like iron. It probably took well over a further billion years before these filled out to the extent that amounts of free oxygen started to climb, not reaching modern levels until after still *another* billion years. By the time of the great coal forests of the Cambrian period (300–350 million years ago), oxygen levels had actually become higher than those of today—maybe as much as 35%, compared to the modern 21% at sea level.

Many models of Earth's atmosphere over time suggest that atmospheric oxygen levels were unusually low during the Triassic and Early Jurassic. That's an intriguing possibility, since it's the same interval during which the dinosaurs—including the very early sauropodomorphs—were experiencing their first evolutionary radiation. Some scientists, most notably Peter Ward (University of Washington), have hypothesized that dinosaurs succeeded in the Triassic *specifically* because their superior respiration could handle low oxygen levels better than those of mammals and other contemporary vertebrates. In this situation the earliest dinosaurs would have had the same advantage as some of today's high-flying birds, such as barred geese, whose similar systems allow them to routinely fly through the oxygen-low air above the 29,000′-plus Himalayas.

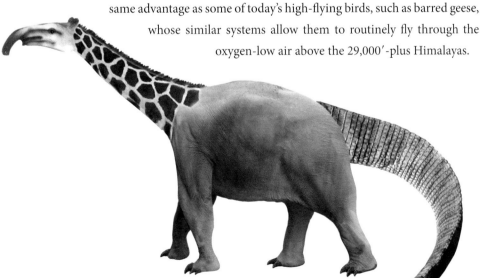

This fanciful creature illustrates an analogy by paleontologist Walter Coombs, that sauropods seem to have been built from parts of many other animals, including reptiles, birds, and mammals. Recalling the old story of the three blind men whose perception of an elephant was based on the parts of its body they touched, it symbolizes how poor our understanding of sauropods once was.

This is a tantalizing possibility, but it's very difficult to test, and there are at least two counter-arguments. The first is that oxygen levels in the Triassic may not have been as low as predicted by some models. Wood can't burn in an atmosphere that's less than 15% oxygen, and the extensive record of fossil charcoal throughout the period shows that there was always enough oxygen to allow forest fires. For comparison, 15% oxygen at sea level is, for modern life-forms, the equivalent of what they'd have to breathe at today's altitude of 9,000′. As a human used to more oxygen-rich lower altitudes, you'd find the air here thin, and in a few days you'd become oxygen deprived, but it's not a real problem for most non-avian animals—even at the level of the Tibetan Plateau (15,000′), mammal species and individuals still outnumber birds, which you'd expect to be more abundant because of their superior respiratory systems.

The second counterargument is that dinosaurs weren't the only terrestrial animals that radiated and flourished during the Triassic–Early Jurassic. Mammals, lizards, turtles and crocodilians all diversified as well, but they just didn't become the dominant large-bodied land animals. If there was an oxygen bottleneck that limited the evolutionary potential of non-dinosaurs, you'd expect that their abundance would be restricted in the fossil record. Having said all this, we know that scientists are still only starting to fill in the blanks regarding the effect of atmosphere on biological evolution, and we can anticipate more breakthroughs on this in the coming years.

Why Aren't There More Juvenile Fossils? We've now learned that sauropods, depending on how long they lived, produced many offspring and, unlike theropods and hadrosaurs, practiced little or no parental care. Given the sheer number of babies that erupted onto the scene during a hatching event and the number of egg clutches laid by a female each reproductive season, predators were confronted with a surfeit of prey, much the way African mammalian carnivores are during the spring calving season of wildebeests and other antelope. Although other dangers awaited them, enough juveniles became old enough to have an increasingly better chance of reproducing on their own. Sauropod juveniles, unlike those of elephants and other large-bodied mammals, would have always outnumbered adults, perhaps by a ratio of 30:1. If this is the case, why don't they turn up more often as fossils?

One answer may lie in the type of habitat that juveniles of a given age group preferred and had a better chance of surviving in, as well as their behavior. As we speculated in Chapter 8, very small individual hatchlings would have been no more than hors d'oeurves to a variety of predators, and their best chance to live was to seek cover—and fast—after emerging from the egg, perhaps under darkness. Dense, low-growing cover like an *Equisetum* marsh or meadow would have probably been ideal, since it would have provided high-quality food as well as shelter. As well as keeping quiet, having concealing patterns and having little or no natural scent, post-hatchlings may also have physically distanced themselves from one another to lower the risk of being detected. As they grew older and outgrew the protection of this habitat, the juveniles would have been forced to find other concealment befitting their size, while still obtaining enough nutrition. We don't know what these environments were, but the question in this case is what potential these had for preserving juvenile sauropod remains. Forested environments, as we learned were present in Dashanpu, China, in Chapter 10, are far less likely to preserve bone because of their relatively high soil acidity, environments that juvenile sauropods may have favored. Rapid burial in a wet, relatively mineral-rich alkaline setting is usually best for bone preservation, but only under rare circumstances might this happen for a juvenile. Another consideration is that predators and scavengers were certainly apt to be as abundant during the Mesozoic as they are now, and the huge mortality rate probably meant that the survival of juvenile skeletal remains was less likely when an entire individual was quickly consumed. On the other hand, the likelihood of preservation probably increased at the point at which certain bones like femora, humerae and vertebral centra started to become too big to be broken down and eaten by predators—this may have been when a juvenile reached the body mass of a small elephant, perhaps by the age of 3–4.

Behavior may have also been an important factor in the prevalence of juvenile fossils. Localities like the Mother's Day Quarry in south central Montana show that small groups of older diplodocoid juveniles of approximately 75% adult size banded together, possibly to improve their survival in the face of predation. Other fossil sites like the Big Bend *Alamosaurus* site near Javelina, Texas, indicate that at some point older juveniles or subadults may have begun grouping with mature adults, not only benefitting from their size and experience in protecting themselves but also to learn by association which food plants to select and to acquire gut microflora by consuming adult feces. Here our speculation must end, since we need more information.

How Did They Keep Warm? Depending on the species, a baby sauropod hatched into the world weighing only several kilograms, about the size of a lion cub. Endothermic, altricial birds and mammals can depend on their parents' warm bodies and a nest for extended periods, but it's different for some types of precocial babies, who must quickly mature and be able to protect themselves. This includes *insulation*, or maintaining body warmth. As we've seen in Chapter 8, a typical sauropod hatchling's size increased thousands of times as it grew to an adult, but for the time being its large surface-to-body ratio meant that, unlike its parents, whose challenge was to rid themselves of excess body heat, one of the baby's main problems was to keep warm. How could they do this? Fossil finds of *filoplume feathers*, evolved specifically for insulation, are now known in a variety of fossil avian theropods from sites like Liaoning, China, showing that this was already an adaptation that allowed this clade of endothermic dinosaurs to conserve warmth. If adults had such a filoplume body covering, it's logical that their chicks, which would have needed it even more, had it also, but there's no direct evidence for "dinofuzz." No prehatchling theropod chick has yet been found to prove this assumption, and we have this problem with other dinosaur babies, including sauropods. The prehatchling nemegtosaurid titanosaur chicks from Auca Mahuevo have a distinctive, tubercle-and-scute pattern, but no feathers; on the other hand, neither do any other known prehatchling theropods. Perhaps the filoplume feathers grew in shortly after hatching, but since precocial birds (like chickens) have these before hatching, this seems unlikely.

Injuries and Disease. The known skeletal remains of sauropods show few of the injuries, and almost none of the diseases, more commonly known in other dinosaurs and modern animals. Although ribs, hips and tails have shown occasional evidence of bone trauma (bruising) in some individuals, the *extoses* and other reactions to the kind of crippling joint damage to the feet of some theropods like "Big Al" the *Allosaurus* are so far unknown. In *Diplodocus* and other di-

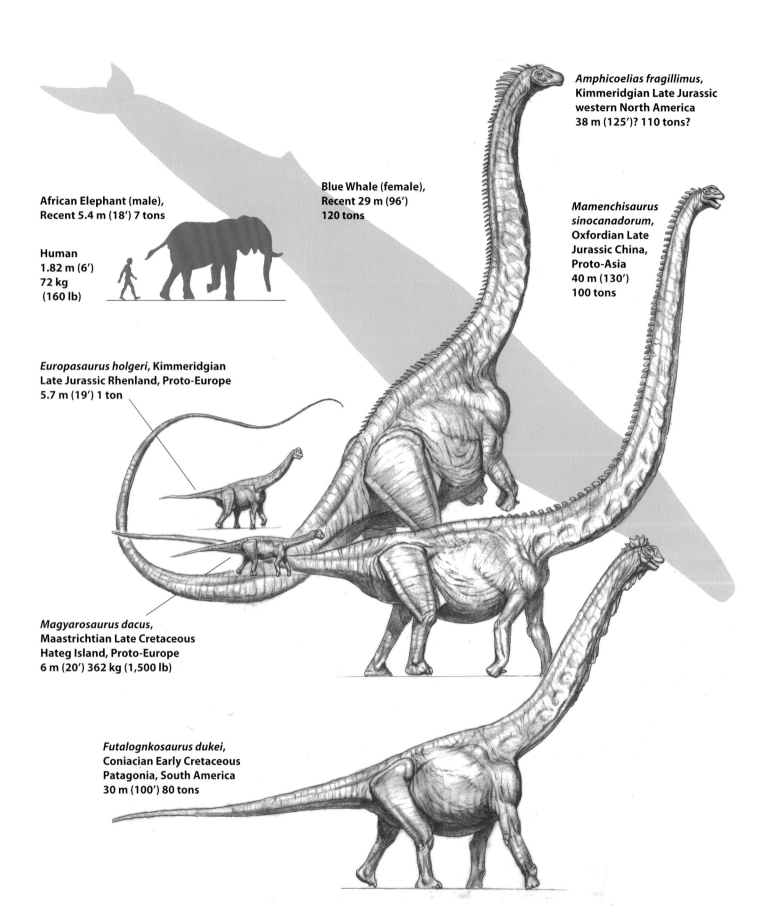

Amphicoelias fragillimus, Kimmeridgian Late Jurassic western North America 38 m (125')? 110 tons?

Blue Whale (female), Recent 29 m (96') 120 tons

Mamenchisaurus sinocanadorum, Oxfordian Late Jurassic China, Proto-Asia 40 m (130') 100 tons

African Elephant (male), Recent 5.4 m (18') 7 tons

Human 1.82 m (6') 72 kg (160 lb)

Europasaurus holgeri, Kimmeridgian Late Jurassic Rhenland, Proto-Europe 5.7 m (19') 1 ton

Magyarosaurus dacus, Maastrichtian Late Cretaceous Hateg Island, Proto-Europe 6 m (20') 362 kg (1,500 lb)

Futalognkosaurus dukei, Coniacian Early Cretaceous Patagonia, South America 30 m (100') 80 tons plodocines what does commonly occur are fused caudal vertebrae due to ligament ossification, ranging from two to sometimes four bones, around the seventh or eighth in the caudal series. This actually occurs in 50% of the individuals with a sufficiently known caudal sequence, and some workers have speculated that it's a response to stress on this area of the tail during tripodal feeding due to weight. The even percentage could imply sexual dimorphism—assuming that this was the typical sex ratio between males and females, maybe one sex was consistently larger than the other and at a certain size needed to rear back more. *Paleopathology* (the study of ancient disease, a term first coined by Marc A. Ruffner in the late 19th century) is a very young science, but it was developed further by Roy L. Moodie in the 1930s and then considerably by the late Larry D. Martin (University of Kansas) and Bruce M. Rothschild (Northeastern Ohio Universities College of Medicine) in the 1980s, who undertook modern studies of dinosaur diseases and injuries. The two workers, although verifying that some dinosaurs did have some forms of modern diseases, found that these were not, as some have claimed, especially prone to osteoarthritis, which as yet has not been found in sauropods. If we could know the kinds of diseases that did occur over a broad spectrum of sauropod species during their geological existence, we might be able to deduce whether some illnesses were more prevalent than others, while a larger database of injuries might hold clues to behavior—were rib and hip bruises due to intraspecific fighting between males, or something else?

Population Structure. Because of their vulnerability to predation until they possibly got big enough to defend themselves in a group (see above), small juvenile sauropods probably didn't go around in what we'd call a herd. At least we haven't yet found any, and we can imagine that a big gaggle of poodle-sized sauropods would have been just a mobile buffet for theropods. Adults are a different question, but even here we shouldn't think of them in the same way we do elephants, where large numbers of full-grown individuals probably reflect a mammalian social system based on close family ties and collective, remembered experience that improve both juvenile and adult survivorship. In addition to the typical number of adult sauropods in a herd (and this, of course, would depend on the species), what was the average size of a mature adult? Were males larger, as in elephants, hippos and other terrestrial, large-bodied mammals, or were females, as in baleen whale species? In mammals we often encounter the condition of *sexual dimorphism* (differences in shape, color and often size between the sexes), and if we determine size differences based on gender in sauropods, we could possibly make inferences as to behavior. If males were appreciably bigger, for example, it could mean that a bull *Sauroposeidon* might have maintained a "harem" of females during the breeding season, as do modern sea lions, because

Size Matters—Biggest of the Big, Smallest of the Small

(*Opposite*) One measure of an animal group's ultimate evolutionary success is its diversity, or species-richness in adapting to different niches. Although the sauropods' key to success was enormous size, digesting huge masses of vegetation and the ability to reach high, some became very small in order to adapt to a limited food source. Here are some of the biggest—and smallest—sauropods known currently to paleontologists.

The buildup of oxygen in the young Earth's atmosphere was a slow, gradual process. Controversy currently exists over whether possible low oxygen levels in the Late Triassic may have spurred the sophisticated respiration of archosaurs and ultimately the system that enabled sauropods to grow huge.

of the advantage of size in intrapecific, male-to-male combat in winning breeding privileges. Alternately, larger females could mean that here bigger size either played a more important role in selecting mates or conferred some biological advantage in reproduction—for example, maximizing the number of eggs produced per season, as in many turtles.

What Kind of Metabolism? An ongoing debate among paleontologists, now decades old, concerns the kind of *thermoregulation*, or temperature control, that sauropods (and other dinosaurs) might have had. This would be a vital part of a sauropod's overall *metabolism*, the physical and chemical processes that occur in an organism. *Ectothermy*, a condition based on an animal's dependence on sources of heat and cooling outside its body to keep its metabolism at an optimum, which is common to fish, amphibians and reptiles, has all but faded as a probable condition for dinosaurs. This conclusion is based on the many currently known facts about dinosaur anatomy and physiology, ranging from upright skeletal stance and the presence of filoplume feathers in adult theropods to the evidence for fast growth in all dinosaurs, unknown in ectotherms, provided by bone histology. *Endothermy*, the maintainance of optimal body conditions based on internal metabolism itself, is seriously considered as the most likely option for theropods (owing to, among other reasons, the growing evidence that probably most members of this clade possessed fluffy, outside structures that helped them retain heat, as in modern birds and mammals), and possibly for many other dinosaurs as well. What about sauropods? Here it gets complicated.

Because they usually grew to such enormous adult sizes, these dinosaurs' volume-to-surface ratio and consequent body heat retention would have given them a de facto endothermic metabolism that would have produced the same activity levels birds and mammals experience. This condition is known variously as *inertial homeothermy*, *megathermy* or *gigantothermy*. Up until this time, however, babies and smaller juveniles would, if they weren't true endotherms to begin with, have had problems with keeping warm enough to maintain the kind of internal metabolism they needed to keep body processes like digestion going (see earlier part of this chapter). Large geochelonid tortoises, for example, became extinct in North America during the Late Miocene period when the average day-to-day winter temperatures dropped to about 15°C. They couldn't digest when they weren't able to keep their bodies at a certain minimum temperature, while crocodiles, which had been present in most of North America for millions of years, disappeared from all but the southernmost areas because their eggs would not develop and hatch at temperatures farther north of these areas. For baby sauropods inertial homothermy doesn't explain how they were able to grow so fast, since this condition doesn't start to have an effect until an animal reaches about 10 kg (22 lb), well above the size of the hatchlings. Ectotherms grow so slowly that if sauropods truly had possessed this kind of metabolism from hatching, they probably would have taken an impossible 70–200 years to reach their mature sizes.

In 2010 Sander and his associates suggested that baby sauropods may have started out as tachymetabolic endotherms, but that as they reached a critical threshold, possibly at about 500–1,000 kg (1,100–2,200 lb), their metabolisms might have gradually shifted toward inertial homeothermy. As adults, this would have then reduced their metabolic rates, in turn curtailing the need for the high caloric intake typical of large-bodied endotherms like elephants. The idea is supported by evidence for extremely rapid growth in the bones of juvenile sauropods, which in turn required a high metabolism—a condition of true endothermy or something equivalent. Although unusual, it's certainly possible that sauropods could have evolved this ontogentic shift, since some modern mammals, like tree sloths, have a comparatively low metabolic rate compared with most mammals, while others temporarily lower their metabolisms while hibernating and then bring it back up again.

The Mysterious Morrison. The Morrison Formation of the North American West has profoundly shaped our understanding of sauropods from the beginning. Paleontologists got their first understanding of what these dinosaurs actually looked like when the first reasonably complete skeletons were discovered in Colorado and Wyoming during the 1870s, and the formation continues to be a rich source of information today, in terms of not only numbers of specimens but also preservation quality. An important question, however, is, actually how representative *is* the Morrison of a Late Jurassic ecosystem? There are some signs that it's actually not a typical sauropod-dominated fauna. For one thing, there's the sheer diversity of sauropods. In a 2006 study of dinosaur species richness, Mike Taylor discussed the "Kimmeridgian sauropod boom," referring to the Kimmeridgian stage of the Late Jurassic. In terms of genera per unit of time, the Kimmeridgian had about *4 times* as many sauropod species as any other stage of the Mesozoic. This apparent burst of diversity is driven in part by sauropod-dominated faunas from Tanzania and Spain, but these are still outnumbered by those of the Morrison. More than one quarry has

Considering how much more common juveniles of sauropods and other dinosaurs were compared to those of adults, it's surprising how rare it is to find their fossils. Except for unusual circumstances, soil conditions may not have favored the preservation of baby sauropods most of the time, and most may have been eaten or scavenged before they could become fossils.

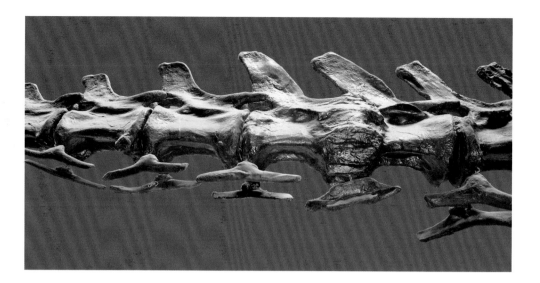

A section of caudal vertebrae from a *Diplodocus* mount at the Museum of Natural History, London, with two that are fused. Rib and vertebral fusion and injury thus far constitutes the only significant type of pathology known in sauropods, but other discoveries may reveal more.

produced remains of five different genera, showing that several species of multi-ton herbivores were living side by side like African antelopes, but on a big scale.

Maybe, though, this was typical of most Mesozoic ecosystems. The Morrison gives us a bigger—and more accurate—window into such a world because the formation itself is so vast, running from Saskatchewan in the north to New Mexico in the south and from Montana in the west to Manitoba in the east. If every dinosaur-bearing formation was that large and well exposed, we might find similar numbers of sauropods in most times and places, but we don't. So the question remains: was the Morrison an unusually spectacular window on an average collection of ecosystems, or was it one that just happened to capture and preserve spectacular fossils? We know that the formation was remarkable in at least one way, in that it was dominated by diplodocids. Even in other Late Jurassic faunas, diplodocids don't seem to be as diverse or abundant as other sauropods. This clade is not only quite specialized but also geologically short-lived compared with others, ranging only from the Kimmeridgian to the Tithonian stages as compared with other diplodocoids, which lasted well into the Cretaceous. As Taylor stated in his study, this clade, widely considered successful, was in fact extremely limited in time and space.

The final Morrison mystery may be the most important for our understanding of sauropod biology: how could these dinosaurs be so diverse and abundant in such seasonally harsh environments? Abundant evidence, including salt lakes and dry soils, indicates that throughout much of its existence the Morrison was periodically arid. Did the giants migrate to sources of better greenery and water, like wildebeests on today's Serengeti? Did they retreat to refuges around the few permanent rivers and non-saline lakes during dry conditions? Adults and subadults could probably migrate hundreds or possibly even thousands of miles, encountering little food along the way but still able to survive. In contrast, this wouldn't have been an option for the fast-growing babies and small juveniles, which would have had even fewer food options and would have been extremely vulnerable to predation. Whatever the answer turns out to be, it will probably be a real surprise and may well shake up our current ideas about sauropod paleobiology.

Mass Extinction—Why the Dinosaurs? After a staggering span of 165 million years, the unthinkable happened. Except for birds, dinosaurs of every type, size and distribution became totally extinct somewhere around 66.5–65 million years ago, at the Cretaceous–Tertiary or "KT" boundary. This, of course, included the sauropods, among the most diverse and successful of all the dinosaurs. The cause for this mass extinction event is usually (and justifiably) pinned on one of two global catastrophes: the bolide (asteroid or comet) that the late Luis and Walter Alvarez (both University of California, Berkeley) and their colleagues theorized struck Earth in the area of the modern southeastern Gulf of Mexico very close to this time, or the consequences of the Deccan Trap volcanic eruptions in India, a colossal spewing of lava intermittently lasting millions of years toward the end of the Cretaceous. According to the Alvarez team, both would have severely impacted Earth's biosphere for at least a few years to perhaps millennia, and popular programs like those on the *Discovery Channel* and BBC's *Walking with Dinosaurs* have latched on to one or both of these events as the obvious cause of dinosaur extinction. In the case of the asteroid collision, an enormous shock wave would have radiated out from the center of impact, immediately killing all dinosaurs and other life-forms within a range of many thousands of square kilometers. Because the impact happened at sea, enormous *tsunamis*, or tidal waves, devastated the low-lying coastal plains, reaching far inland in all directions. Following this, the superheated immediate atmosphere created an inferno of high temperatures, igniting fires that spread across the planet, wiping out forests and other herbivorous dinosaurs' food plants within several weeks. Finally, the pulverized *ejecta*, or dust, falling back to Earth from the aftermath of the impact created two catastrophic conditions. One was the transformation of nitrogen and oxygen within Earth's atmosphere into massive amounts of sulphur dioxide, which, in combination with moisture, later fell as sulphuric acid rain on the land, altering soil and surface water conditions. Secondly, the resulting dust was carried by winds to create a cloud of particles that shut out most sunlight for perhaps as much as several years, severely hindering green plants, the food web's *primary producers*, from effectively photosynthesizing. Along with the global forest fires and acid rain, another corollary to the asteroid impact theory that some meteorologists claimed was that the blocked sunlight also caused worldwide temperatures to plunge to subfreezing conditions lasting from a month to as much as six months. This dealt a double blow to the tropically to subtropically adjusted dinosaurian herbivores, whose extinctions were quickly followed by those of the carnivorous theropods.

The Deccan Trap eruptions of central India, like the much more recent eruptions of Krakatau, Mount Pinatubo and others within historic times, could also have blocked sunlight—and photosynthesis—in the same way by filling the atmosphere with wind-borne ash, which would have stayed for many months to many years. Another consequence might have been the emission of massive amounts of lethal gases. Volcanic activity sometimes releases elements like selenium that are highly toxic to developing embryos, and Hans Hansen (University of Aalborg) reported increased levels of this in the eggshells of the titanosaur *Ampelosaurus* nearest the KT boundary in southern France. Although in this situation the effects would be long lasting, the eruptions themselves would have been intermittent, and dinosaur egg–bearing sediments between some of the volcanic basalt layers testify that life went on in spite of the periodic upheavals. Even so, some think that the massive Deccan Trap eruptions would have disrupted photosynthesis and plant growth so severely and over such a long term that dinosaurian herbivore and then

Approximately 65.6 million years ago, the planet suffered a major bolide impact, which had a devastating effect on both land and marine life-forms, causing widespread extinctions. To date no non-avian dinosaurs are known to have survived, and the mystery remains as to why such nearly complete die-offs occurred among some life-forms and not others.

carnivore food webs would have eventually collapsed. As a result of one or both of these events, dinosaurs, certain other terrestrial animals and other freshwater and marine species worldwide accordingly were doomed and died out at least within a few decades. Among the terrestrial species, only relatively small-bodied life-forms like invertebrates, amphibians, reptiles and orni-thurine birds survived. According to some, these were somehow able to find shelter and avoid the need to depend directly on green plants for food.

Although the direct effects from one or both of these events would have been more than enough to cause widespread extinction among dinosaurs and many other life-forms at the end of the Cretaceous, what remains a mystery is why the extinctions were so selective. On land why were non-avian dinosaurs (and some others archosaurians, like pterosaurs) singled out for elimination, and not the other common terrestrial vertebrates like amphibians, reptiles, crocs, birds and mammals? Let's look at what we actually can determine about the consequences for Late Cretaceous terrestrial vertebrate forms following such events, in light of known biology and the paleontological record. It's true that the bolide hit would have produced a layer of *iridium*, a rare element on Earth but a typical component of asteroids. This substance, deposited in a band all around the planet at about the 65- to 66.5-million-year mark at about the KT boundary, and the presence of "shocked" quartz crystals surrounding the Gulf of Mexico collision site, known as the Chixculub Crater, point to a major asteroid impact at this time. After the Alvarez theory was announced in 1980, however, a chorus of disagreement erupted from some paleontologists and biologists. These workers pointed out that, while the Alvarez team was certainly correct in modeling and describing the primary catastrophic geological and meteorological consequences from such an asteroid impact, their conclusions (and other researchers' corollary hypotheses based on this) were debatable regarding the effects on the overall biological communities of the time. While no one disputes the fact that the bolide impact was certainly linked to the timing of the dinosaurs' extinction, questions still remain as to the exact mechanism of how and when the extinctions occurred from a biological perspective.

For a start, take small animals, which can't easily migrate and are highly dependent on local habitats. These are the most vulnerable to widespread environmental disruption and are conse-quently far more likely to die over a relatively short period of time than large creatures. We know from recent studies that fish, amphibians like frogs and some reptiles are especially sensitive to conditions like acid rain, which for amphibians is lethal to their water-based form of reproduc-tion. Acid rain would also be likely to have a serious effect on the life cycles of the invertebrates they preyed on. If this was severe enough (a very low pH level; see Chapter 10), it could also have killed large animals like dinosaurs through inhalation or contact with their skin or food, but the fact that the North American Late Cretaceous fossil record shows many non-dinosaurian vertebrates like those above surviving the KT transition suggests that the acidity, if it existed at all, wasn't very high following the impact. If these small animals weren't seriously affected, larger ones wouldn't be either.

A second corollary to the Alvarez impact theory, suggested by atmospheric scientist Brian Toon (Colorado University at Boulder) in 1982, was that there was a sudden, severe drop in worldwide temperatures from the typically tropical to subtropical conditions of the Late Cre-

In the BBC production *The Future Is Wild*, gigantic, multi-ton tortoises evolve at a future time, filling an ecological niche occupied by sauropods millions of years earlier. In reality, however, their physiological limitations mean that no ectothermic reptile like a tortoise could evolve into a sauropod-like animal.

taceous to subfreezing conditions, owing to the enormous amount of dust particles blocking sunlight. Sometimes called the "impact winter," it's similar in effect to what might happen if a "nuclear winter" was produced by a concentration of atmospheric dust following an atomic war. This is where, if it did occur, we'd expect to see a real KT decline among small-bodied ectotherms—invertebrates, amphibians and reptiles—which could never have survived the sudden and prolonged drop in temperatures. While many living ectothermic species in temperate-zone areas *can* undergo a period of extended inactivity, or *torpor*, under cold conditions, it happens only when they're given enough time by gradual decreases in outside temperature and light, as well as food supply. As tropical to subtropical animals, they would have been unable to adjust to such a harsh, abrupt thermal drop imposed by an "impact winter." Likewise, as we've seen, some reptiles like tortoises and crocodilians require certain minimal body temperature ranges if they are to digest food and reproduce, and it's unlikely that they would survive this kind of cold for long. What does the KT paleo record say here? In a study of species found in the Lance Formation of northeast Montana during the latest Cretaceous, most ectothermic tetrapods made it successfully across the KT boundary. This finding is bolstered by inference based on another study done by Bill Clemens and Gayle Nelms (both University of California, Berkeley) in 1993. In this they found that among the fauna recorded from Alaska, though less diverse and from an earlier period than that of the northeastern Montana KT fauna, ectothermic species' diversity was low not because of the temperature but because of the three months of darkness annually at latitude 70° north. Here small ectotherm species were scarce because of the smaller amount of primary and secondary food produced during the seasonal darkness, not because of a sudden, catastrophic event's predicted darkness and consequent cold. Although dinosaur endothermy is

The gigantic, hornless rhino *Paraceratherium* of the Miocene period, shown here with an elephant in a fantasy zoo, probably represents the largest mammal that would be likely to evolve considering the limitations of the way these animals eat and process food.

still being debated, most dinosaurs would under these conditions have been able to maintain a high enough body temperature to survive a cold period. If herbivores found enough food during the high-latitude three-month "winter" in Alaska, they might have been able to survive the impact winter following the KT event.

What would happen if the heat from an impact had triggered an apocalyptic global wildfire? If this had occurred, the intensity would have reduced much of the above-ground biomass to ashes, while in freshwater any plants and animals that escaped being boiled would have experienced a choking deluge of charred debris that would have contaminated the water surfaces. There would be almost no way of escaping a rapidly moving inferno like this sweeping across the land, and it would be totally non-selective in the organisms it killed, small as well as large—as J. David Archibald (San Diego State University) stated in his book *Dinosaur Extinction and the End of an Era: What the Fossils Say*, if this happened, everyone would be an equal-opportunity loser. In proof of such a global fire, some workers point to the existence of a carbon and soot layer next to the iridium layer at some sites, arguing that at these places it was caused by the rapid burning of vegetation, while Luis Alvarez and his associates thought that this might locally represent the burning of half the world's modern forests. In spite of this, however, the fossil record at various sites representing the North American KT transition clearly holds contradictory evidence, again that small vertebrates generally survived. What's not taken into account is that if the destruction of the existing plants and the longer-term disruption of photosynthesis occurred on the scale postulated in the above corollary hypothesis, it would have almost eliminated the food web's primary producers—plants—which the smallest herbivores would depend on. These microherbivores made up the overwhelming majority of the planet's animals. Sure, you might have some detritus-eating invertebrates (pill bugs, millipedes, earthworms, some beetles and flies) and mammals that could survive on seed pods and post-KT debris for awhile, but after this was gone, you'd start to get massive microfaunal die-offs from starvation. These would include many (but not necessarily all) invertebrates, which the bird species, small reptiles and mammals of the time, most of which were survivors, must have heavily depended on for food. This weakens the argument that KT terrestrial animals somehow escaped by burrowing or otherwise "hiding out" from the worst of the effects debatable—what would they eat after they emerged from hiding, even if there was a temporary boom in carrion and detritus-feeding invertebrates?

By now we can see that although the immediate main effects of the bolide impact—the shock wave, the geological aftermath and the sunlight-blocking dust cloud that restricted photosynthesis to a significant degree—are certainly likely to have occurred, the paleontological record and modern biological findings don't necessarily support the *corollary* effects: the alleged acid rain, subfreezing temperatures, worldwide forest fires and total cessation of photosynthesis. The selectivity of the animal and plant clades that seemingly *did* survive the KT event just hasn't been convincingly accounted for. Why did most less biologically advanced amphibians reptiles and mammals make it? If birds, the dinosaurs' living representatives that shared so many successful features with saurischian dinosaurs, survived, why not at least some highly adaptable non-avian theropods? If large body size was a fatal flaw, why don't we still have some very small dinosaur species around, other than birds? Finally, was there some distinctive feature of dinosaur paleobiology that made the difference between survival and extinction? If so, it remains unknown.

One important fact that rarely gets mentioned is that practically all of what we surmise about Late Cretaceous dinosaur extinction is based *almost entirely* on a limited stratigraphic area, the Lance (Creek) Formation of North America. From the known species found from this geologic unit of the Maastrichtian Age (the latest of the Cretaceous, and which led up to the KT boundary), some paleontologists, notably Dale Russell (North Carolina State University), think that, based on the number of known species found here, dinosaurs were already declining as an entire class and on their way to extinction toward the end of the last stage of the Cretaceous. The asteroid only performed the final act in what was an inevitable outcome. He and others point to a decline in overall dinosaur species at this geochron, as compared to those of the previous Judithian faunal stage during the Late Campanian. In this situation dinosaurs had (possibly because of more limited ecological conditions related to receding coastlines) become reduced in diversity to a few major types, with *Tyrannosaurus* the only big maniraptoran theropod left, *Triceratops* the only large-bodied ceratopsian, and *Edmontosaurus* the only major hadrosaur, among others. An opposing group of paleontologists disagrees, claiming that this represents a preservation bias, and that the fossil record shows an incomplete record of true dinosaur diversity at this time. The perceived lack of species richness in this presumably well-sampled but relatively narrow stratigraphic unit has to be weighed against the almost complete lack of knowledge we have of dinosaur diversity from other areas of the world at the close of the Cretaceous. In the case of sauropods, for example, the scant but steadily increasing data from findings in Europe show that the titanosaurs were highly diverse well into Maastrichtian times. More species of this clade are being discovered here (as well as in areas like Africa and Asia), on the average of every seven months. From this evidence, titanosaurs were not in decline, and perhaps other European (and world) dinosaur clades and communities continued to be diverse until the very end of their fossil record. Getting back to the KT event, we can't be sure that sauropods and other dinosaurs were automatically snuffed out, in spite of the asteroid theory's claims, at virtually the same time around the globe. As speculative as it sounds, some species *may* have lingered on for a few centuries or possibly a millennium more into the Paleocene period (the first time zone that belongs to the Cenozoic era, or "Age of Mammals") before finally dying out completely around the world. So maybe the sauropods and others went out with a whimper instead of a bang, but at this time we really don't know the precise, actual mechanism.

LOOKING BACK AND LOOKING FORWARD

Having said this, we're still, of course, left with the hard fact that sauropods and all other non-avian dinosaurs—barring the possibility that in the future we actually do find their Paleocene fossils—finally, unarguably became extinct *somewhere* around 66.5-65 million years ago. Unless someday we can actually pull off a "Jurassic Park," we'll certainly never see anything like them on Earth ever again. This is especially so in the case of the sauropods, which no species during the following eons of mammalian evolution ever came close to resembling. In the fascinating, speculative TV documentary series and book *The Future Is Wild*, 100 million years from now tortoises have evolved into 120-ton "toratons," superficially similar to sauropods. But as we've learned, sauropods weren't simply like a tortoise scaled up to gigantic size. Without fast growth driven by a rapid metabolism, these ectothermic reptiles couldn't survive for the decades or centuries it would take them to reach hundreds of tons, and without pneumatization their skeletons would be too heavy to support them on land. Mammals in particular are ill-suited to evolve into future sauropod-mimics, since from a sauropod's point of view mammals do almost everything wrong. Because we chew our food, we have heavy teeth, big heads and short necks. We give birth to live, few young, reducing the chances of greater numbers of individuals making it to maturity (and for mammals, reproduction age), while our extended parental care eats up resources that could be spent on either growing larger or producing more offspring. None of the bones outside of our skulls are air filled, so our skeletons are comparatively heavy. With all these built-in

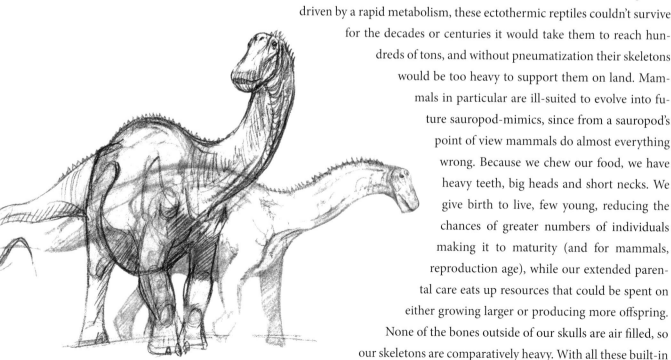

limitations, it's unlikely that something much bigger than a *Paraceratherium (Indricotherium)*, a giant Miocene rhino and the biggest known land mammal, could exist on land.

The fact that sauropods' *bauplan* is so unique was probably a major stumbling block to our understanding of these and other dinosaurs until well into the 20th century. Because they survived and followed the dinosaurs, mammals were considered the ultimate success story: from complex chewing teeth and sophisticated parental care to big brains, tool use and language, mammalian evolution was presented in most books as an ever-rising crescendo of perfection, while dinosaurs were at best a spectacular, bizarre evolutionary sideshow—basically overblown lizards. Their gigantic size, combined with their supposedly reptilian attributes, made them ripe for extinction. The old image of sauropods, wallowing in their swamps, personified this image. Now our perspective is much more balanced, and modern interpretations of success in evolution are based on concepts like geologic longevity, morphological disparity and phylogenetic diversity instead of simply who's left standing at the end of the game. Extinction, pointed to as the badge of failure, is the eventual fate of all life-forms, and sometimes it's just due to bad luck. The dinosaurs were probably one of the most supremely successful of any life-forms, but when something like a 10 km asteroid hits, all the rules change. In this kind of global crisis the normal processes of evolution are overturned, and quirky random factors can decide what becomes extinct and what lives.

Sauropods were the dominant herbivores in most terrestrial ecosystems for almost 160 million years, while our ancestors were trying to avoid being stepped on. If the asteroid had missed, they would probably still be here today. As the current and probably temporary masters of Earth, we would be wise to consider the logical underpinnings of the sauropods' success, as well as the unpredictable causes of their extinction.

GLOSSARY

The following is not intended to be a complete glossary of paleontological terms, but rather a supplemental reference to significant words or concepts that relate to sauropod dinosaurs. Some are omitted here because they have been fully defined in the main text. The definitions presented are based on several sources, but are primarily from *Osteology of the Reptiles* by Alfred Sherwood Romer (1956), *A Dictionary of Scientific Terms* by J. H. Kenneth (1960), *The Illustrated Encyclopedia of Dinosaurs* by David Norman (1985), *Glossary and Manual of the Tetropod Paleoichnology* by Giuseppe Leonardi (1986) and *The New Dinosaur Dictionary* by Donald F. Glut (1982).

A

acetabulum: The hip socket, formed at the junction of the ilium, pubis and ischium of the pelvis.

acromion process: A ventral bony wing extending from the anterior lateral edge of the scapula.

Albian: A stage of the Early Cretaceous period spanning from about 112 to 99 million years ago.

allometric: Pertaining to the differential growth rates within a living organism.

allopatric: When two species do not occur together in the same geographic area.

altricial: Pertaining to vertebrates characterized by a short gestation period and the hatching or birth of relatively helpless, underdeveloped young, as with songbirds and mammalian carnivores.

alveolus (alveoli): Blind sacs in the lungs of mammals for the exchange of oxygen and carbon dioxide.

amniote: A non-taxonomic term for animals that develop away from an external aquatic environment, possess certain embryonic membranes and are protected within an eggshell or the mother's body. Includes reptiles, saurians, birds and mammals.

amphicoelous: Concave on both surfaces, as in *amphicoelous* vertebrae.

amphiplatyan: Condition in which both articular ends of a vertebral centrum are flat.

antorbital: In front of the *orbit* or eye socket.

appendicular skeleton: Parts of the skeleton that include the pectoral and hip girdles, forelimb and hindlimb.

Aptian: A stage of the Early Cretaceous period spanning from about 121 to 112 million years ago.

Archosauria: A class of amniotes, now phylogenetically separated from other reptiles, defined primarily by the possession of an antorbital fenestra and other anatomical and physiological features, which includes dinosaurs, pterosaurs, crocodilians and birds.

articular: The ventral, most posterior mandibular bone that attaches directly to the cranium or skull.

articulated: Joined together.

astragalus: Largest tarsal bone, which directly articulates with the tibia and the metatarsals.

atlas: First cervical vertebrae going posteriorly from the skull.

autapomorphy: In cladistics, a condition in which a species displays a unique character unknown in any other species.

Aves: The monophyletic clade of all living birds; crown-group birds.

avetheropod: Any member of the *Avialae*, the clade that includes living and extinct birds but not dromaeosaurs and troodonts.

avialan theropod: a maniraptoran theropod with bird-like features.

axial skeleton: Parts of the skeleton that include the vertebrae, dorsal ribs, sternal ribs, sternum, gastralia, skull and hyoid.

axis: Second cervical vertebrae going posteriorly from the skull.

B

Barremian: A stage of the Early Cretaceous period spanning from about 130 to 125 million years ago.

basal: For taxa, a taxon whose lineage branched off from the base of the group's family tree. In respect to characters, the lesser or least derived member of a phylogenetic group (replaces term "primitive").

basioccipital: Median, most posterior bone of the skull, forming part of braincase and bearing occipital condyle.

basipterygoid: A process of the basisphenoid bone contacting the pterygoid bone.

basisphenoid: Skull bone between the basioccipital and presphenoid bone.

Bathonian: A stage of the Mid-Jurassic period spanning from about 168 to 165 million years ago.

bed: In geology, a distinct layer of sedimentary rock.

biconcave: Curved inwardly on both ends.

bifurcated: Forked, having two branches or wings (sometimes shortened to "bifid").

bioturbation (dinoturbation): Biological activities at or near ground level that cause the mixing of the sediment, especially from dinosaur trampling.

bipedal: Having the ability to walk on the hindfeet.

browser: A feeder on leafy vegetation, or *folivore* (as opposed to a grazer, or *graminivore*/grass eater).

buccinator: Muscle forming the wall of the cheek, pulling the mandible dorsally.

C

calcaneum: Smaller tarsal bone lateral to the astragalus and distal to the tibia.

Callovian: A stage of the Mid-Jurassic period spanning from approximately 165 to 161 million years ago.

Campanian: A stage of the Late Cretaceous period spanning from about 84 to 71 million years ago.

cancellous: Made up of lamellae and slender fibers, joining to form a network-like structure.

capitulum: The head of a rib, which articulates with the parapophysis of a vertebra.

carina: On some bones and teeth, a narrow, keel-like ridge.

Carnian: The first stage of the Early Triassic period, spanning from about 230 to 217 million years ago.

carnosaur: Technically, a theropod more closely related to *Allosaurus* than to birds. Informally, a term referring to allosauroid theropods.

cartilage: Translucent to fibrous, firm, but elastic tissue usually on a bone's articular surface.

caudal: Pertaining to the tail, or in the direction of the tail.

Cenomanian: A stage of the Early to Mid-Cretaceous period spanning from about 99 to 93.5 million years ago.

centrum: Main body of a vertebra, from which arise the neural and haemal arches.

cervical: Pertaining to the neck.

cervical ribs: Elongate, narrow, paired bones attached to the vertebral diapophyses and parapophyses of cervical vertebrae.

character: a particular feature, anatomical, physiological, genetic or behavioral, displayed by an organism.

chevron: V-shaped bone (haemal arch) in a series ventral to the caudal vertebrae.

choana: Funnel-shaped internal narial opening (internal naris).

clade: In phylogeny, a formally named or ranked group of organisms whose members are all derived from a common ancestor.

cladistics: A scientific method of classifying organisms based on their most recent common ancestor.

cladogram: Diagram representing the distribution of shared/derived characters for organisms.

class: In phylogeny, a group of closely related subclasses sharing a common ancestor.

clavicle: In the pectoral girdle, one of a pair of small bones located dorsal to the coracoid bones. Fused into the *furcula*, or "wishbone," in most theropod dinosaurs, including birds.

cnemial crest: Crest running along the anterior surface of the tibia for muscle attachment.

coelurosaur: Technically, a term covering bird-like theropods that includes tyrannosaurs, ornithomimids, oviraptorosaurs and therizinosaurs in addition to birds. Informally, a term referring to any small theropod.

condyle: A prominence on the articular end of a bone, usually ball- or wheel-shaped.

convergence: A condition of similar anatomy or behavior found in organisms that evolve similar lifestyles but do not share a direct common ancestor.

coprolite: Fossilized feces or dung.

coracoid: In the pectoral girdle, one of a pair of rounded, shield-like bones attached to the scapulae.

coronoid process: A dorsally directed wing of the surangular bone of the mandible or lower jaw, to which jaw-closing muscles attach.

cranium: Collectively, all skull bones except for those of the mandible or lower jaw.

Cretaceous: Third and last period of the Mesozoic era, from about 135 to 65 million years ago.

crown group: The living members of any clade (for example, birds are the *crown group* of dinosaurs).

cursorial: Adapted for fast running.

D

deciduous: In trees and shrubs, pertaining to those that seasonally shed their leaves.

deltopectoral crest: Anterior proximal prominence on humerus for the insertion of deltoid and pectoralis muscles.

dentary: Largest bone of the mandible (lower jaw).

denticle: One of a series of small, bump-like prominences along the edge of a tooth.

derived: More specialized, evolved from a simpler (basal or ancestral) condition. Also, those taxa that branch off later in a group's phylogeny.

diapophysis: The lateral, transverse lamina of the neural arch of a vertebra, which with the parapophysis serves as the attachment for a cervical or dorsal rib.

digitigrade: The condition of walking on the toes instead of the entire foot surface.

dimorphism: The condition of diverging into two forms, usually according to sex.

disarticulated: Pulled apart.

dispersal: In biogeography, radiating or spreading out of a group of organisms.

distal: The end of any body part farthest away from the body's midline, or from the point of attachment.

divergence: Evolutionarily moving away or changing form from a central group.

diverticulum: A sac branching from a larger hollow structure, such as the blind-ended membranous extension branching from an air sac, that invades a pneumatic cavity.

dorsal: Pertaining to the vertebral column or "back," as opposed to *ventral*.

E

ecology: The science of the relationships between organisms and their environments.

ecosystem: A functioning natural system formed by the interactions of organisms with their environment.

ectopterygoid: Ventral bony wing behind the palatine, extending to the quadrate bone.

ectothermic: Relying on external sources of heat to maintain body temperature.

embryo: An organism in a very early, pre-hatching or birth stage of development.

endocast: A replica of the brain area made by filling in the former brain's negative space within the skull.

evolution: The generally gradual change in the characteristics of a population of organisms caused by the inheritance of genetic changes over time, whether resulting from natural selection, genetic drift or other factors.

exoccipital: Skull bone on each side of the foramen magnum.

exposure: In geology, the condition of gradual rock appearance due to weathering.

F

facies: In geology, one of different but contemporarily formed subdeposits, as well as its lithological and paleontological makeup, within an overall sedimentary deposit.

family: In phylogeny, the grouping of related genera with a common ancestor.

fauna: An entire community of animals living in a particular place and time.

femur: The proximal, or upper, bone of the hindlimb, or thigh bone.

fenestra: A relatively large space in a bone or between bones, typically without any continuous structures passing through (in contrast with a *foramen*, which allows continuous structures to pass through).

fibula: The smaller, more narrow outer bone of the distal (lower) leg.

flex: To bend a joint.

flora: An entire community of plants living in a particular time and place.

foramen: Opening through a bone or membrane, to permit the entrance/exit of nerves, blood vessels or diverticula.

foramen magnum: Opening through the occipital part of the skull through which the spinal cord passes.

forensics: The disciplines and techniques used to understand the events surrounding an organism's death and the taphonomic events following this.

fossa: Pit or trench-like depression in a bone.

frontal: A bone of the dorsal skull, anterior to the parietal bone, forming the "forehead."

G

gastralia: Ossified "belly ribs" in the ventral body walls of many amniotes, including saurischian dinosaurs.

gastrolith: Literally "stomach stones," any rocks or pebbles in some animals' digestive tracts swallowed to provide ballast in some aquatic saurians but also in others to help the stomach process food in the gizzard.

genus: In phylogeny, a group of closely related species with a common ancestor.

gizzard: A muscular part of the foregut in some saurians and birds, specialized to grind up food, sometimes with the help of gastroliths, and following the *ventriculus*, or glandular stomach.

Gondwana: Prior to Aptian–Albian time, the Southern Hemisphere supercontinent that incorporated the future landmasses of South America, Africa, Australia, Antarctica, India and Madagascar.

graviportal: Pertaining to very large, slow-moving animals.

grazer: A feeder on grassy vegetation (a *graminivore*, as opposed to a browser or *folivore*, an eater of leafy vegetation).

H

hallux: "Big toe," or first, most medial digit of the *pes*, or hindfoot.

hindlimb dominance: The condition in which the hindlimbs were stronger and bore more of the body's mass than the forelimbs.

histology: The science or study of body tissues.

holotype: A single specimen chosen to designate a species in its first published description.

homodont: Having teeth that are all alike.

homology: The state of having characters that are shared through having a common ancestor.

homoplasy: A condition in which distantly related clades acquire a similar character as the result of a common lifestyle (see *convergence*).

horizon: A stratigraphic layer formed at a particular time, characterized by specific fossil species.

Hox genes: "Master control" genes that coordinate the expression or suppression of various other genes during embryological development.

humerus: The bone of the proximal or upper forelimb.

hyoid: A bone or series of bones located above the larynx and providing partial support/muscular attachment for the tongue and upper throat.

hypantrum: In some saurians, a recessed notch or cavity on the anterior vertebra that articulates with the *hyposphene*.

hyposphene: In some saurians and other amniotes, a posteriorly directed, projecting wedge that articulates with the notch in the *hypantrum*; together these form an accessory joint that provides additional reinforcement to the articulations formed by the pre- and postzygapophyses.

hypothesis: A rational but technically unsubstantiated or unproven idea.

I

ichnology: The science and study of fossil footprints, traces and trackways.

ilium: The paired, largest bone of the dorsal pelvis.

iliofemoralis: A large extender muscle of the thigh, running between the ilium and femur.

infraorder: In phylogeny, a category or rank between a family and an order.

in situ: Latin, "in place," or in its original position on discovery.

integument: An animal's outer covering or skin.

intraspecific: Within the same species.

ischium: The paired ventral and posterior bone of the pelvis.

J

jugal: A cranial bone between the maxilla and quadrate, also known as the "cheekbone."

junior synonym: A taxonomic name no longer considered valid because another name, pertaining to the same fossil materials, was published previously.

Jurassic: The second period of the Mesozoic era, spanning from about 200 to 135 million years ago.

K

Kimmeridgian: A stage of the Late Jurassic period spanning from about 156 to 151 million years ago.

kinetic: Bones that have joints and are capable of movement.

K selection/strategy: Common life-history strategy for populations at or near the carrying capacity of their environments, usually favoring the production of few young that develop slowly and often receive long-term parental care, as with elephants and other large mammals.

L

labial: Pertaining to lips or non-movable, lip-like coverings; "lip-ward" when referring to the sides of teeth or jawbones, as opposed to *lingual*, or on the side of the tongue.

lacrimal: Cranial bone contributing to the anterior border of the orbit.

lamella: Thin strut, or sheet-like bone.

lateral: Away, and at a right angle, from the body midline.

Laurasia: Prior to Albian–Aptian time, the Northern Hemisphere supercontinent that incorporated the future landmasses of North America and Eurasia.

ligament: Strong, fibrous band of tissue that connects one bone to another.

lingual: Pertaining to the tongue; "tongue-ward" in reference to the sides of the teeth or jawbone as oppsed to *labial*.

locality: In geology, a named place where fossil specimens have been found.

M

Maastrichtian: The terminal stage of the Late Cretaceous period, spanning from 71 to 65.5 million years ago.

manus: The forefoot bones or hand, composed of the metacarpals and manual phalanges.

maxilla: The paired, larger tooth-bearing bone in the upper jaw.

medial: Closest to a vertical, anterior-posterior (sagittal) plane through the middle of the body, or a body part closest to this plane.

Mesozoic: An era comprising the Triassic, Jurassic and Cretaceous periods, spanning from about 224 to 65 million years ago, following the Paleozoic and preceding the Cenozoic.

metabolism: The total physical and chemical processes occurring in an organism.

metacarpals: Bones of the manus located between the carpals (wrist) and the manual phalanges (fingers).

metatarsals: Bones of the pes located between the tarsals (ankle) and the pedal phalanges (toes).

monophyletic: In phylogeny, a group derived from one ancestor.

morphology: The science or study of form or shape.

mosaic evolution: A condition in which evolution occurs in a piecemeal pattern.

myelin: A fatty tissue surrounding nerve fibers that maintains their conductivity.

N

naris: One of the paired external nostril openings.

nasal: A paired bone on the anterior dorsal side of the cranium or skull.

natural selection: The evolutionary mechanism by which organisms change based on differential survival and reproduction in fitness to the environment.

neural arch: A bony bridge over the vertebral centrum that protects the spinal cord and serves as a point of attachment for ligaments and muscles.

neural spine: Also called a *neuropophysis*, forms a bony ridge over the neural arch.

neuron: An individual nerve cell.

nomen dubium: Latin, "doubtful name," for a taxon or name founded on fossil material of questionable diagnostic value.

nomen nudum: Latin, "naked name," for a taxon erected with no published description, type designation or figured drawing.

nuchal: Pertaining to the neural spines of the anterior cervical vertebrae.

O

occipital condyle: Condyle at the back of the cranium that articulates the skull with the atlas vertebra.

occlusal: Where the surfaces of the upper and lower teeth meet when biting, cropping or chewing.

olecranon: Process at distal end of ulna where the triceps muscle, a lower forelimb extender, inserts.

olfactory body: Referring to the sense of smell; the neural extension that receives and transmits the detection of odors to the brain.

ontogeny: Growth and development of the individual organism.

opisthocoelous: Condition in which the posterior (caudal) articular surface of a vertebral centrum is concave.

optic lobe: Neural extension that transmits visual stimuli from the optic nerves to the brain.

orbit: The bony cavity or socket enclosing the eye.

order: In phylogeny, a group of closely related families with a common ancestor.

osteoderm: A vascularized, sometimes fluid-filled bony disc embedded in the skin.

oviparous: A reproductive system in which the young develop, obtain nutrients and eliminate wastes within eggs that are spawned or laid outside the mother's body.

Oxfordian: A stage of the Late Jurassic period spanning from about 161 to 156 million years ago.

P

palatine bones: Paired bones that form the *palate*, or roof of the mouth.

paleoecology: The science or study of the relationships between extinct organisms and their environments.

Pangaea: A supercontinent, existing from the Permian until approximately the Early Jurassic period, which encompassed all of Earth's landmasses.

parapophysis: The more ventral of the two attachments on either side of a vertebra for the articulation of a cervical or dorsal rib.

parietal: A paired bone of the dorsal cranium, posterior to the frontals.

pectoral girdle: The paired bones of the shoulders and chest to which the forelimb attaches.

pedomorphism: A condition in which youthful characteristics are retained in adult life.

peduncle: Stalk or stem-like process of a bone.

pelvic girdle: The paired bones of the hips and posterior abdomen to which the hindlimb attaches.

peripheral nerves: All nerves outside of the brain and spinal cord.

pes: The hindfoot bones, composed of the metatarsals and pedal phalanges.

phalanges: The terminal bones of the manus or pes.

phylogeny: The science or study of clades or groups and their relationships to one another.

phylum: In phylogeny, a group of closely related classes that share a basic body plan.

physiology: The science or study of the body processes of organisms.

plantigrade: Walking with the entire ventral surface of the foot touching the ground.

plate tectonics: The science or study of the movements of the large crustal slabs underlying the continents and oceans.

platycoelous: Condition in which both articular surfaces of a vertebral centrum are flat.

plesiomorphy: In cladistics, a condition in which a species possesses an anatomical character similar to or in common with other species.

pleurocoel: A (now-outdated) term for a depression or cavity on the lateral surface of a vertebra.

pneumatic: Containing air, as in pneumatic diverticula and pneumatic bones.

pollex: "Big toe," or first, most medial digit of the pes.

postcranial skeleton: All of the skeleton other than the cranium or skull.

postzygapophysis: One of paired processes on a neural arch's dorsal posterior end that articulates with the prezygapophyses of the following vertebra.

precocial: Pertaining to vertebrates characterized by the hatching or birth of relatively developed, independent young, as with pheasants and hares.

premaxilla: One of paired cranial bones anterior to the maxilla.

prezygapophysis: One of paired processes on a neural arch's dorsal anterior end that articulates with the postzygapophyses of the preceding vertebra.

process: A bony projection, often for articulation with another bone or for the attachment of a muscle.

procoelous: Condition in which the anterior (cranial) articular surface of a vertebral centrum is concave.

pronate (pronation): To orient the surface of the manus, or "hand," to face ventrally or posteriorly (as opposed to *supination*).

proximal: Closest to the center or midline of the body.

pterygoid: One of paired, wing-like bones of the ventral cranium, posterior to palatal bones.

pubis: One of paired, anterior-ventral bones of the pelvis.

Q

quadrate: In amphibians, saurians and archosaurs, one of paired posterior-ventral cranial bones that articulate with the mandible.

quadratojugal: One of paired, ventral cranial bones connecting the quadrate and jugal.

quadrupedality: Having the ability to walk on all four feet.

R

radius: The forelimb bone positioned next to the ulna.

reconstruction: In paleontology, a drawn or modeled skeleton based on the known original fossils, often incorporating extrapolation and/or knowledge based on other, more complete taxa.

r selection/strategy: Common life-history strategy for populations well below the carrying capacity of the environment, usually favoring the early, rapid production of young in large numbers, as with rodents.

restoration: In paleontology, a fleshed-out depiction of how an extinct organism might have appeared in life, usually based on a reconstruction.

S

sacral glycogen body: An expansion of the spinal cord within the sacrum of birds and some non-avian dinosaurs that accommodates the sacral nerve ganglia and may also maintain a reserve of nutrients.

sacral ribs: Tranverse processes of the sacral vertebrae, often fused to the ilium.

sacrum: A structure formed by the fusion of the sacral vertebrae to each other and their associated sacral ribs, providing extra strength in this part of the spine.

sagittal: Pertaining to a vertical, dorsal-to-ventral plane running anterior to posterior along the midline of the body.

sclerotic ring: In many saurians, a series of thin, overlapping small plates supporting the eyeball.

senior synonym: Taxonomic name having priority over another name used to describe the same organism because of its prior publication.

sinus: A space within an area of the body; typically refers to blood-filled venous sinuses or to air-filled cavities in the skull bones.

sister species (sister group): A pair of species or clades derived from a single common ancestor and that are therefore each other's closest relatives.

species: In biology, a formally named population(s) of interbreeding organisms having many shared characters and that are reproductively isolated, through geography or behavior, from breeding with other populations.

squamosal: A paired bone of the posterior ventral or lateral cranium.

stratigraphic: In geology, pertaining to the pattern of rock deposition.

stratum: A layer of sediment.

stride: Pendular movement of the leg, completed when the foot returns to its starting position. In ichnology, the measure of the track segment uniting two corresponding reference points of two consecutive footprints on the same side.

subfamily: In phylogeny, a rank of lesser degree than a family, and which may include one or more genera.

suborder: In phylogeny, a rank of lesser degree than an order but greater than an infraorder, including one or more families.

superorder: In phylogeny, a rank of lesser degree than a class but greater than an order.

supinate (supination): To orient the surface of the manus, or "hand," to face dorsally or anteriorly (as opposed to *pronation*).

suture: A convoluted line where two bones connect, which may later disappear if the bones fuse.

sympatric: When two or more species occur together in the same area.

symphysis: A fibrous joint uniting two paired bones and that mostly allows very little movement, as in the pubic or mandibular symphysis.

synapomorphy: In cladistics, a condition in which two or more species share a derived character because of homology rather than homoplasy.

T

taphonomy: The science or study of the processes following the death of an organism.

tarsals: The bones of the ankle.

taxonomy: The science or study of naming organisms.

tetrapod: A four-legged animal.

tibia: The shin bone, the larger and more medial of the lower leg bones and positioned next to the fibula.

Tithonian: A stage of the Late Jurassic period spanning from about 151 to 145 million years ago.

tooth battery: A set of closely appressed teeth arranged to form a cutting or grinding surface.

transverse processes: Laterally directed wings of the neural arches; the diapophyses are often but not always located on the transverse processes.

Triassic period: First period of the Mesozoic era, spanning from about 225 to 200 million years ago.

trochanter: A prominence on the femur to which muscles are attached.

trophic level: One of a succession of steps in the movement of energy and matter through a food web and an ecosystem.

tubercle: A small, rounded prominence.

tuberculum: One of the two proximal heads of the rib, and which attaches to the diapophysis of a vertebra.

Turonian: A stage of the Late Cretaceous period spanning from about 94 to 89 million years ago.

type specimen: The *holotype*, or the specimen used to describe a new species.

U

ulna: In the lower arm, the larger and more posterior bone positioned next to the radius.

ungual: Pertaining to the terminal digit or phalangeal bone that bears a claw.

V

vascular: Pertaining to the blood vessels or circulatory system.

ventral: The underside of an organism, as opposed to the upper side, *dorsal*.

viviparous: A reproductive system in which the young develop within and obtain nourishment from a placenta within the mother's body before birth, as opposed to remaining within the egg until after being expelled from the mother.

vomer: A paired bone anterior to the palatal bones, contributing to the roof of the mouth.

Z

zygapophysis: A paired, wing-like prominence projecting from the anterior and posterior ends of a neural arch on a vertebra (see *prezygapophysis* and *postzygapophysis*).

SOURCES AND SUGGESTED READING

This list of source materials and references cannot include all of the indispensable, accurate information the authors received from the many general publications, scientific papers, Internet sources and personal communications researched during work on this book. Some principle sources, however, are the following:

Alexander, R. McNeill. *Dynamics of Dinosaurs and Other Extinct Giants*. New York: Columbia University Press, 1989.

Ambrose, J. *Building Structures Primer*. New York: John Wiley & Sons, 1967.

Apesteguía, Sebastián. Bonitasaura salgadoi gen. et sp. nov.: a beaked sauropod from the Late Cretaceous of Patagonia. *Naturwissenschaften* 91, no. 10 (2004): 493–497.

Archibald, J. David. *Dinosaur Extinction and the End of an Era: What the Fossils Say*. New York: Columbia University Press, 1996.

Bader, Kenneth S., Stephen T. Hasiotis, and Larry D. Martin. Application of forensic science techniques to trace fossils on dinosaur bones from a quarry in the Upper Jurassic Morrison Formation, Northeastern Wyoming. *Palaios* 24 (March 2009): 140–158; doi: 10.2110/palo.2008.p08-058r.

Bakker, Robert T. *The Dinosaur Heresies*. New York: William Morrow, 1986.

———. Ecology of the Brontosaurs. *Nature* 229 (15 January 1971): 172–174; doi:10.1038/229172a0.

Balanoff, Amy M., Gabe S. Bever, and Takehito Ikejiri. The braincase of Apatosaurus (Dinosauria, Sauropoda) based on computed tomography of a new specimen, with comments on variation and evolution in sauropod neuroanatomy. *American Museum Novitates*, no. 3677 (March 4, 2010).

Barrett, Paul, and Paul Upchurch. Sauropod feeding mechanisms: their bearing on paleoecology. *6th Symposium on Mesozoic Vertebrates*, Short Papers. Beijing: China Ocean Press, 1995.

Beck, Charles B., ed. *Origin and Evolution of Gymnosperms*. New York: Columbia University Press, 1988.

Behrensmeyer, Anna K., John D. Damuth, William A. DiMichele, Richard Potts, Hans-Dieter Sues, and Scott L. Wing. *Terrestrial Ecosystems through Time: Evolutionary Paleoecology of Terrestrial Plants and Animals*. Chicago: University of Chicago Press, 1992.

Berman, David S., and John S. McIntosh. Skull and relationships of the Upper Jurassic sauropod *Apatosaurus* (Reptilia, Saurischia). *Bulletin of Carnegie Museum of Natural History* 8 (1978): 1–35.

Calvo, Jorge. *Feeding Mechanisms in Some Sauropod Dinosaurs*. Chicago: University of Illinois, 1994.

Carey, M., and Madsen, James. Some observations on the growth, function, and differentiation of sauropod teeth from the Cleveland-Lloyd Quarry. *Utah Academy Proceedings* 49, part 1 (1972): 40–43.

Carpenter, Kenneth. *Eggs, Nests, and Baby Dinosaurs: A Look at Dinosaur Reproduction*. Bloomington: Indiana University Press, 1999.

Carpenter, Kenneth, Karl F. Hirsch, and John R. Horner, eds. *Dinosaur Eggs and Babies*. Cambridge: Cambridge University Press, 1996.

Carr, Archibald. *Africa*. New York: Time-Life Books, 1964.

Chandler, M. Fruiting organs from the Morrison Formation of Utah, USA. *Bulletin of the British Museum (Natural History), Geology* 12, no. 4 (1966): 137–171.

Chiasson, Robert B. *Laboratory Anatomy of the Alligator*. Dubuque, Iowa: William C. Brown, 1962.

Christian, Andreas, Dorothee Koberg, and Holger Preuschoft. Shape of the pelvis and posture of the hindlimbs in *Plateosaurus*. *Paleontology* 70 (November 1996): 591–601.

Christian, Andreas, and Holger Preuschoft. Deducing the body posture of extinct large vertebrates from the shape of the vertebral column. *Paleontology* 39 (1996): 801–812.

Coombs, Walter P. Sauropod habits and habitats. *Palaeogeography, Palaeoclimatology, Palaeoecology* 17 (January 1975): 1–33; doi: 10.1016/0031-0182(75)90027-9.

Currie, Philip J., and Kevin Padian, eds. *Encyclopedia of Dinosaurs*. San Diego: Academic Press, 1997.

Curry Rogers, Kristina, et al. Sauropod dinosaur osteoderms from the Late Cretaceous of Madagascar. *Nature Communications* 2, no. 564 (November 2011); doi:10.1038/ncomms1578.

Curry Rogers, Kristina, and Jeffrey A. Wilson, eds. *The Sauropods: Evolution and Paleobiology*. Berkeley: University of California Press, 2005.

Czerkas, Sylvia. *Dinosaurs: A Global View*. Limpsfield: Dragon's World Ltd., 1990.

D'Emic, Michael D., et al. Evolution of high tooth replacement rates in sauropod dinosaurs, *PLos One* 8, no. 7 (July 17, 2013); http://dx.doi.org/10.1371/journal.pone.0069235.

Dodson, Peter, Anna K. Behrensmeyer, Robert T. Bakker, and John S. McIntosh. Taphonomy and paleoecology of the dinosaur beds of the Jurassic Morrison Formation. *Paleobiology* 6, no. 2 (1980): 208–232.

Endo, H., and R. Frey, eds. *Anatomical Imaging: Towards a New Morphology*, Chapter 6: Using CT to peer into the past: 3D visualization of the brain and ear regions of birds, crocodiles, and non-avian dinosaurs, by Lawrence M. Witmer et al., pp. 67–88. Springer Japan, 2008.

Farb, P. *Ecology*. New York: Time-Life Books, 1963.

Farlow, James O. Speculations about the diet and digestive physiology of herbivorous dinosaurs. *Paleobiology* 13, no. 1 (1987): 60–72.

Farlow, James, Dan I. Coroian, and John R. Foster. Giants on the landscape: modeling the abundance of megaherbivorous dinosaurs of the Morrison Formation (Late Jurassic western USA). *Historical Biology* 22, no. 4 (2010): 403–429.

Fastovsky, David E., and David B. Weishampel. *Dinosaurs: A Concise History*. Cambridge: Cambridge University Press, 2011.

Fiorelli, Lucas E., et al. The geology and palaeoecology of the newly discovered Cretaceous neosauropod hydrothermal nesting site in Sanagasta (Los Llanos Formation), La Rioja, northwest Argentina. *Cretaceous Research* 35 (2012): 94–117.

Fiorillo, Anthony R. Dental micro wear patterns of the sauropod dinosaurs *Camarasaurus* and *Diplodocus*: evidence for resource portioning in the Late Jurassic of North America. *Historical Biology* 13, no. 1 (2009): 1–16; doi: 10.1080/08912969809386568.

Foster, John. *Jurassic West: The Dinosaurs of the Morrison Formation and Their World*. Bloomington: Indiana University Press, 2007.

———. Paleoecological analysis of the vertebrate fauna of the Morrison Formation (Upper Jurassic), Rocky Mountain Region, USA. *New Mexico Museum of Natural History and Science Bulletin* 23 (2003): 1–95.

Gallina, Pablo A., Sebastián Apesteguía, Alejandro Haluza, and Juan I. Canale. A diplodocid sauropod survivor from the Early Cretaceous of South America. *PLos One* (May 14, 2014); doi:10.1371/journal.pone.0097128.

Galobart, Àngel, Maite Suñer, and Begoña Poza, eds. *Dinosaurs of Eastern Iberia*. Bloomington: Indiana University Press, 2011.

Gasparini, Zulma, Leonardo Salgado, and Rudolfo A. Coria, eds. *Patagonian Mesozoic Reptiles*. Bloomington: Indiana University Press, 2007.

Gee, Carole T., ed. *Plants in Mesozoic Time: Morphological Innovations, Phylogeny, Ecosystems*. Bloomington: Indiana University Press, 2010.

Gillette, David D., and Martin G. Lockley, eds. *Dinosaur Tracks and Traces*. Cambridge: Cambridge University Press, 1991.

Gillette, David D., J. Bechtel, and Peggy Bechtel. Gastroliths in the skeleton of a sauropod dinosaur. Albuquerque, New Mexico: Southwest Paleontological Foundation, Inc., 1994.

Gilmore, C. W. A nearly complete articulated skeleton of *Camarasaurus*, a saurischian dinosaur from the Dinosaur National Monument, Utah. *Memoirs of the Carnegie Museum* 10 (1925): 347–360.

———. Osteology of *Apatosaurus* with special reference to specimens in the Carnegie Museum. *Memoirs of the Carnegie Museum* 11 (1936): 175–300.

Glut, Donald F. *Dinosaurs: The Encyclopedia*, Supplements 1–7. Jefferson, North Carolina: McFarland, 1997–2012.

Gomez, Bernard, et al. Plant taphonomy and palaeoecology in the lacustrine Uña delta (Late Barremian, Iberian Ranges, Spain). *Palaeogeography, Palaeoclimatology, Palaeoecology* 170 (2001): 133–148.

Gould, Stephen Jay. *The Panda's Thumb: More Reflections in Natural History*. New York: W.W. Norton, 1980.

Grubb, B. Allometric relations of cardiovascular function in birds. *American Journal of Physiology* 254, no. 4 (1983): H567–572.

Haas, G. A proposed reconstruction of the jaw musculature of Diplodocus. *Annals of the Carnegie Museum* 36 (1962): 139–157.

Hatcher, John Bell. Diplodocus (Marsh): its osteology, taxonomy, and probable habits, with a reconstruction of the skeleton. *Memoirs of the Carnegie Museum* 1, no. 1 (1901): 1–63.

———. Osteology of *Haplocanthosaurus*, with description of a new species, and remarks on the probable habits of the Sauropoda and the age and origin of the Atlantosaurus Beds. *Memoirs of the Carnegie Museum* 2, no. 1 (1903): 1–72.

Henderson, Donald M. Burly gaits: centers of mass, stability, and the trackways of sauropod dinosaurs. *Journal of Vertebrate Paleontology* 26, no. 4 (December 2006): 907–921.

Holland, W. J. The osteology of Diplodocus Marsh. *Memoirs of the Carnegie Museum* 2, no. 6 (1906): 225–264.

Hopson, James A. Relative brain size and behavior in archosaurian reptiles. *Annual Reviews of Ecological Systems* 8 (1977): 429–448.

Janensch, Werner. Material und Formengehalt der Sauropoden in der Tendaguruschichten. *Paleontographica* Supplement VII 1, no. 2 (1929): 3–34.

Jennings, Debra S., and Stephen T. Hasiotis. Taphonomic analysis of a dinosaur feeding site using geographic information systems (GIS), Morrison Formation, Southern Bighorn Basin, Wyoming, USA. *Palaios* 21 (2006): 480–492.

Johnston, Hugh. *The International Book of Trees: A Guide and Tribute to the Trees of Our Forests and Gardens*. New York: Simon and Schuster, 1973.

Kingdon, Jonathan. *East African Mammals: An Atlas of Evolution in Africa*, vol. III, part B. London: Academic Press, 1979.

Klein, Nicole, Kristian Remes, Carole T. Gee, and P. Martin Sander, eds. *Biology of the Sauropod Dinosaurs: Understanding the Life of Giants*. Bloomington: Indiana University Press, 2011.

Lacovara, Kenneth J., et al. The ten thousand islands coast of Florida: a modern analog to low-energy mangrove coasts of Cretaceous Epeiric Seas. *Proceedings* of the 5th International Conference on Coastal Sediments 5 (2003): 1773–1784.

Lockley, Martin S. *Tracking Dinosaurs: A New Look at an Ancient World.* Cambridge: Cambridge University Press, 1991.

Lucas, A., and P. Stettenheim. *Avian Anatomy: Osteology,* Part 2. Agriculture Handbook 362, USDA and Michigan State University, 1972.

Lucas, Spencer G. *Chinese Fossil Vertebrates.* New York: Columbia University Press, 2001.

Maier, Gerhard. *African Dinosaurs Unearthed: The Tendaguru Expeditions.* Bloomington: Indiana University Press, 2003.

Manning, Phillip. *Grave Secrets of the Dinosaurs: Soft Tissues and Hard Science.* Washington, DC: National Geographic, 2008.

Mannion, Philip, et al. Osteology of the Late Jurassic Portuguese sauropod dinosaur Lusotitan atalaiensis (Macronaria) and the evolutionary history of basal titanosauriforms. *Zoological Journal of the Linnean Society* 168 (2013): 98–206.

Martin, John. Mobility and feeding of *Cetiosaurus* (Saurischia: Sauropoda)—why the long neck? In Currie, P.J., and E.H. Koster, eds. Fourth Symposium on Mesozoic Terrestrial Ecosystems, Short Papers, *Occ Pap Tyrrell Mus. Palaeontology* 3 (1987): 154–159.

Martin, John, Valerie Martin-Roland, and Eberhard Frey. Not cranes or masts, but beams: the biomechanics of sauropod necks. *Oryctos* 1(1998): 113–120.

Martin, Thomas, and Bernard Krebs, eds. *Guimarota: A Jurassic Ecosystem.* Munich: Verlag Dr. Pfeil, 2000.

More, Heather, et al. Scaling of sensorimotor control in terrestrial mammals. *Proceedings of the Royal Society B* (2010); doi: 10.1098/rspb.2010.0898.

More, Heather, et al. Sensorimotor responsiveness and resolution in the giraffe. *Journal of Experimental Biology* 216 (2013): 1003–1011.

Myers, Timothy S., and Anthony R. Fiorillo. Evidence for gregarious behavior and age separation in sauropod dinosaurs. *Palaeogeography, Paleoclimatology, Paleoecology* 274, nos. 1–2 (2009): 96–104.

Myers, Timothy, and Glenn W. Storrs. Taphonomy of the Mother's Day Quarry, Upper Jurassic Morrison Formation, South-Central Montana, USA, *Palaios* 22 (2007): 651–666.

Nothdurft, William, and Josh Smith. *The Lost Dinosaurs of Egypt.* New York: Random House, 2002.

Novas, Fernando E. *The Age of Dinosaurs in South America.* Bloomington: Indiana University Press, 2009.

Paul, Gregory S. *Predatory Dinosaurs of the World: A Complete Illustrated Guide.* Baltimore: Johns Hopkins University Press, 1988.

———. *The Princeton Field Guide to Dinosaurs.* Princeton: Princeton University Press, 2010.

———, ed. *The "Scientific American" Book of Dinosaurs.* New York: St. Martin's Press, 2000.

Perrins, Christopher. *Birds: Their Life, Their Ways, Their World.* New York: Reader's Digest Association, 1976.

Platt, Philip R. *The Apatosaurus Notebook.* (self published), 2003.

Poinar, George, Jr., and Roberta Poinar. *What Bugged the Dinosaurs? Insects, Disease, and Death in the Cretaceous.* Princeton: Princeton University Press, 2008.

Prasad, Vandana, et al. Dinosaur coprolites and the early evolution of grasses and grazers. *Science* 310, no. 5751 (2005): 1177–1180; doi: 10.1126/ science.1118806.

Riggs, Elmer. *Brachiosaurus altithorax*: the largest known dinosaur. *American Journal of Science* 15 (1903): 299–306.

Romer, Alfred Sherwood. *Osteology of the Reptiles*, Third Edition. Chicago: University of Chicago Press, 1976.

Rossnagel, W. E. *Handbook of Rigging, for Construction and Industrial Operations*, Third Edition. New York: McGraw-Hill, 1964.

Russell, Dale. *An Odyssey in Time: The Dinosaurs of North America.* Toronto: University of Toronto Press, 1992.

Sander, P. Martin, et al. Adaptive radiation in sauropod dinosaurs: bone histology indicates rapid evolution of giant body size through acceleration. *Organisms Diversity & Evolution* 4, no. 3 (2004): 165–173; doi:10.1016/j.ode.2003.12.002.

Schwartz, Daniela, Eberhard Frey, and Christian A. Meyer. Pneumaticity and soft-tissue reconstructions in the neck of diplodocid and dicraeosaurid sauropods, *Acta Palaeontologica Polonica* 52, no. 1 (2007): 167–188.

Sereno, Paul C., ed. Basal sauropodomorphs and the vertebrate fossil record of the Ischigualasto Formation (Late Triassic: Carnian-Norian) of Argentina. *Journal of Vertebrate Paleontology, Memoir 12* 32 (2013): 1–181.

Seymour, R. S., and H. B. Lillywhite. Hearts, neck posture and metabolic intensity in sauropod dinosaurs. *Proceedings of the Royal Society of London B* 267 (2000): 1883–1887.

Shapiro, H. *Cranes and Derricks.* New York: McGraw-Hill, 1980.

Spotila, James R. *Sea Turtles: A Complete Guide to Their Biology, Behavior, and Conservation.* Baltimore: Johns Hopkins University Press, 2004.

Stevens, Kent. DinoMorph: parametric modeling of skeletal structures. *Senckenburgiana Lethea* 82, no. 1 (2002): 23–24.

Stevens, Kent, and Parrish, J. M. Neck posture and feeding habits of two Jurassic sauropod dinosaurs, *Science* 284 (1999): 798–800.

Stokes, W. L. Fossilized stomach contents of a sauropod dinosaur. *Science* 143, no. 3606 (1964): 576–577.

Sues, Hans-Dieter, ed. *Evolution of Herbivory in Terrestrial Vertebrates: Perspectives from the Fossil Record.* Cambridge: Cambridge University Press, 2000.

Taylor, Michael P., and Mathew J. Wedel. The effect of intervertebral cartilage on neutral posture and range of motion in the necks of sauropod dinosaurs. *PLoS One* 8, no. 10 (2013); http://dx.doi.org /10.1371/journal.pone.0078214.

Tidwell, Virginia, and Kenneth Carpenter, eds. *Thunder-Lizards: The Sauropodomorph Dinosaurs.* Bloomington: Indiana University Press, 2005.

Tosolini, Anne-Marie, et al. Cheirolepidiacean foliage and pollen from Cretaceous high-latitudes of southeastern Australia. *Gondwana Research* 27 (2015): 960–977.

Vakhrameev, Vsevolod A. Jurassic and cretaceous floras and climates of the earth. *Geochimica et Cosmochimica Acta* 57, no. 5 (1993): 1161–1162; doi: 10.1016/0016-7037(93)90053-Y.

Webb, Grahame, and Charlie Manolis. *Australian Crocodiles: A Natural History*. Sydney: Reed New Holland, 2002.

Wedel, Mathew J. Evidence for bird-like air sacs in saurischian dinosaurs. *Journal of Experimental Biology* 311A (2009): 611–628; doi:10.1002/jez.513.

Weishampel, David B., Peter Dodson, and Halszka Osmólska, eds. *The Dinosauria.*

Berkeley: University of California Press, 1990.

Weishampel, David B., and Coralia-Maria Jianu. *Transylvanian Dinosaurs*. Baltimore: Johns Hopkins University Press, 2011.

Whitlock, John A. Inferences of diplodocid (Sauropoda: Dinosauria) feeding behavior from snout shape and microwear analyses. *PLos One* 6, no. 4 (2011); http://dx.doi.org/10.1371/journal.pone.0018304.

Wings, Oliver, and P. Martin Sander. No gastric mill in sauropod dinosaurs: new evidence from analysis of gastrolith mass and function in ostriches. *Proceedings of the Royal Society B* 274 (2007): 635–640; doi: 10.1098/rspb.2006.3763.

Yates, Adam, et al. A new transitional sauropodomorph dinosaur from the Early Jurassic of South Africa and the evolution of sauropod feeding and quadrupedalism. *Proceedings of the Royal Society B* (2009); doi: 10.1098/rspb.2009.1440.

Yates, Adam, and James W. Kitching. The earliest known sauropod dinosaur and the first steps towards sauropod locomotion. *Proceedings Biological Sciences* 270, no. 1525 (2003): 1753–1758; doi: 10.1098/rspb.2003.2417.

Young, Mark T., et al. Cranial biomechanics of *Diplodocus* (Dinosauria, Sauropoda): testing hypotheses of feeding behavior in an extinct megaherbivore. *Naturwissenschaften* 99 (2012): 637–643.

PHOTOGRAPHY AND ILLUSTRATION CREDITS

Unless attributed otherwise, all art is by Mark Hallett. Design concepts by Mark Hallett. Art production/Page layout, Rachel A. Hallett. Vector art, Karyn Servin. Photo acquisitions editors, Michael M. Fredericks, Erik Fredericks.

Page vii: top, Turi Hallett; upper left, Kristi Curry Rogers; upper right, John R. Hutchinson; lower left, Carole T. Gee; lower right, Paul Upchurch. **Page 2**: Gesine Steiner. **Page 3**: Wikipedia. **Page 4**: Wikipedia. **Page 5**: top images, courtesy Michael P. Taylor; bottom image, Wikipedia. **Page 6**: Wikipedia. **Page 7**: top, Museum für Naturkunde; bottom, Phil Hore. **Page 8**: top, American Museum of Natural History; center, John Ryder; bottom, Erwin Christman (courtesy Michael P. Taylor). **Page 9**: top, Winsor McKay; bottom, Robert T. Bakker. **Page 10**: top, Charles R. Knight (courtesy American Museum of Natural History); bottom, Charles R. Knight (courtesy Andrew Milner). **Page 12**: top, middle, Brant Bassam (Brantworks); bottom, Gregory S. Paul. **Page 13**, top left, Phil Platt; top center, courtesy Brant Bassam (Brantworks); middle right, Gregory S. Paul; bottom, Scott Hartman. **Page 14**: Witmer Labs. **Page 15**: top left, top right, Nils Knötschke/Dinosaurier Park. **Page 16**: Mark Hallett. **Page 17**: Phil Hore. **Page 22**: Paul Sereno. **Page 24**: Kieran Davis. **Page 26**: left, Mark Hallett; right, John Collingwood. **Page 27**: Matt Bonnan. **Page 28**: Kieran Davis. **Page 30**: Nils Knötschke/Dinosaurier Park. **Page 31**: Robert Reisz. **Page 32**: top, Hall Train; inset, left, Vlad Konstantinov. **Page 33**: bottom left, right, Phil Platt. **Page 36**: top, bottom, Wikipedia. **Page 37**: bottom left, Mathew J. Wedel (courtesy Dinosaur Isle Museum); top right, Mathew J. Wedel (courtesy Dinosaur Isle Museum). **Page 38**: Joshua Franzoa. **Page 53**: John Collingwood. **Page 54**: model, Tyler Keillor; photo, Mike Hettwer. **Page 57**: Mathew J. Wedel. **Page 60**: Phil Hore. **Page 68**: John R. Hutchinson. **Page 70**: photo, Phil Hore. **Page 71**: top, bottom, Daniela

Schwartz-Wings. **Page 73**: Andrew Huang. **Page 75**: diagram, Daniela Schwartz-Wings; photo, Heinrich Mallison. **Page 76**: photo, Mark Hallett. **Page 78**: top photo, Mathew J. Wedel; middle photo, Bone Clones. **Page 83**: Mathew J. Wedel. **Page 86**: Human Dynamo-Workshop. **Page 88**: Pandawild (Miraslov Nemecek)/Dreamstime. **Page 90**: photo, Heinrich Mallison. **Page 94**: Steve Allen/Dreamstime. **Page 95**: top, Peter Wrege/Elephant Listening Project. **Page 96**: photo, Dreamstime. **Page 97**: Hall Train Studio. **Page 98**: photo, Hannest/Dreamstime. **Page 105**: photo, Gaston Design, Inc. **Page 106**: Susan B. Leibforth/Bastian Voice Institute. **Page 109**: left, Nathan Dahstrom; right, Candace Paulos/Earth's Ancient Gifts; bottom, Karen Chin. **Page 112**: Kts/Dreamstime. **Page120**: Dreamstime. **Page 121**: top, Eugene Zalenko/Wikipedia; middle, Abu Shawka/Wikipedia; bottom, James Field (Jame)/Wikipedia. **Page 122**: top, Tony M. Thomas; bottom, Kathy Rose/Imbala. **Page124**: Grant Wood/Growing Deer. **Page 125**: photo, David C. Freitag. **Page 127**: Olivia/Dreamstime. **Page 135**: John Whitlock/PLOS One. **Page 136**: top, bottom, Dreamstime. **Page 144**: Tomatito/Dreamstime. **Page152**: Mark Hallett. **Page 154**: Larry C. Simpson. **Page 155**: top, Garrido A. Cerda; bottom, Dreamstime. **Page 157**: Mathew J. Wedel/Natural History Museum, London. **Page 158**: top, Kelly Gorham; bottom, P. Martin Sander. **Page 159**: top photo, William Sellers; bottom photo, Michael Fredericks. **Page 160**: top, bottom, Mathew J. Wedel. **Page 162**: Turbosquid Inc. **Page 163**: top, middle, bottom, Sam Noble Oklahoma Museum of Natural History. **Page 164**: photo, Maya Sunpongco. **Page 165**: top, middle, bot-

tom, Mark Young. **Page 168**: top, Marty Daniel/JURA; bottom, Peter Falkingham. **Page 169**: top, Donald M. Henderson; bottom, John Collingwood. **Page 174**: photo, Dreamstime. **Page 178**: top, Dwight Kuhn; bottom, Luis Chiappe/Dinosaur Institute. **Page 180**: top, bottom, Luis Chiappe/Dinosaur Institute. **Page 182**: top, Patricia Maher/Australian Ornithological Services; bottom, Dreamstime. **Page 184**: Luis Chiappe/Dinosaur Institute. **Page 187**: top left, John Collingwood; top right, Dinosaur National Monument; bottom, Troy Weiss/Dinostoreus. **Page 188**: top, Mathew J. Wedel; middle, Ghedoghedo/Wikipedia; inset, John Collingwood. **Page 189**: photo, Dreamstime. **Page 195**: left, Tyler Keillor; right, Ian Cross. **Page 202**: Jennifer A. Coulson/Coulson Harris Hawks. **Page 203**: Hall Train/Hall Train Studios. **Page 207**: John Collingwood. **Page 208**: Cleveland-Lloyd Quarry. **Page 209**: Hannest/Dreamstime. **Page 213**: George Poinar. **Page 218**: top, Malcolm Schuyl/Wikipedia; middle, North Carolina Biological Supply Co.; bottom, Ernie Cooper/Macrocritters. **Page 220**: NASA. **Page 223**: Jeff Shaw. **Page 224**: photo, Zigong Nature Museum. **Page 225**: Zigong Dinosaur Museum. **Page 226**: John Collingwood. **Page 228**: top, bottom, Wikipedia. **Page 229**: top left, North Carolina Biological Supply Co. **Page 231**: John Collingwood. **Page 234**: Dreamstime. **Page 235**: Heinrich Mallison. **Page 236**: photo, Albh/Wikipedia. **Page 239**: Octavio Mateus. **Page 240**: John Collingwood/Melbourne Museum of Natural History. **Page 241**: Jeff Johnson/Wikipedia. **Page 242**: John Collingwood. **Page 243**: Tyler Keillor. **Page 244**: Ronan Alain. **Page 247**: Norbert Schuster. **Page 249**: Wikipedia.

Page 250: Wolfgang A. Hug/JURA. Page 251: top, Wikipedia; bottom, Thomas Steuber. Page 252: top, Leila Hathamore; bottom, Venkatesh K./Wikipedia. Page 254: top, Csiki et al.; bottom, Nils Knötschke/Dinosaurier Park. Page 255: photo, Nils Knötschke/Dinosaurier Park. Page 256: Nils Knötschke/Dinosaurier Park. Page 261: top, Boething Treeland Farms; bottom, Ibaraki Prefectural Museum.

Page 262: Jorget Tanous. Page 265: photo, Jorge Calvo. Page 267: art/photo, Nima Sassani. Page 268: top, Wikipedia; bottom, Scott Hocknull et al., PLOS One. Page 269: Luis Carbillido/Museo Paleontologico Egidio Feruglio. Page 270: top, Sam Banda; bottom, Royal Ontario Museum/Wikipedia. Page 271: Kristi Curry Rogers. Page 272: Kristi Curry Rogers/Jeffrey A. Wilson. Page 273: top,

Diego Alarcon; bottom, Wikipedia. Page 274: top, Mathew J. Wedel; bottom, John Collingwood. Page 275: John Collingwood. Page 279: Mathew J. Wedel. Page 284: Ron Miller/Black Cat Studios. Page 285: Nat Geo Creative. Page 286: John Collingwood. Page 287: Douglas Henderson. Page 290: Masato Hattori.

INDEX

Page numbers in **boldface** refer to illustrations or photos. Scientific genera of sauropodomorphs are initially referenced in brown. Unless specified otherwise, all words and terms refer specifically to sauropods or sauropodomorphs.

caecum, 91, **92**

Calliphora (blow or flesh fly), **218**

Callovian Stage, 222

calories: expenditure in walking, 128

Calvo, Jorge (paleontologist), 14

camarae. *See* vertebrae

camarasauromorphs: camellae of, **101**; camerae of, **101**; characteristics of, 55; location of fossil finds, **55**; phylogeny of, 44, 46

Camarasaurus, 145; *C. grandis*, 205; *C. lentus*, 205; *C. lewisi*, 205; *C. supremus*, 205; defense of calves, **192**; dentition of, **55**, **139**, 145; diet, **143**; discovery of, 7; drinking position, **103**; at Fruita, 232–233; head restoration, **55**; feeding adaptations, **143**; isotopes in teeth, 153–155; location of fossil find, **55**; microwear on teeth of (*see* microwear); neck of, **8**; ontogeny of, 189; orthal bite, 233; phylogeny of, **43**, 46; rearing, **133**; size increase and speciation through time, 205; skeleton of, 8, 55, 72–73; skull of, 12, **55**, **65**, *73*; spinal bifurcations in, 147; in hypothetical taphonomic reconstruction, 216–219; tooth scrapes on bones, **208**

camellae. *See* vertebrae

Camelotia, 26

Camelus (camel): neck extension and flexion of, 83; water conservation in, 106

Camptosaurus (Fruita basal iguanodontian), 235

"*Campylodoniscus*," 266

cancellous bone tissue. *See* bone

Canis (dog genus). *See individual species*

capillaries. *See* cardiovascular system; respiration

carbonate platforms. *See* Kem Kem Formation, North Africa

carbon dioxide (CO_2), 88, 95

carcharodontosaurids, 247

Carcharodontosaurus (carcharodontid theropod), **210**, 249

Cardiodon: description of, 14; discovery of, 4; phylogeny of, 44; tooth of, **5**

cardiovascular system, 93, 96, 109–111; arterial enlargements in sauropods ("auxiliary hearts"), **96**; blood pressure, 93, 109–111; —of average mammals, 110; —of birds, 110; —of giraffe, 109; —of humans, 110; —of sauropods, 109–110; capillaries, 111; heart (*see* heart); *rete mirabile*, **96**, 111

Carey, Mary (paleontologist), 133

Cariama (South American seriema), 199

Carnegie, Andrew, 7

Carnegie Museum of Natural History, 7

carnivores. *See individual species*

carnosaurs, 196

Carnotaurus (abelisaurid theropod), 198

carpals. *See* skeleton, bones of

Carpenter, Kenneth (paleontologist), 83

Carpolithus, at Dashanpu, 227

Carrano, Matt (paleontologist), 68, 137

cartilage: fibroid, 69; hyaline, 69; of manus, **66**; of pes, **67**

caruncle ("egg tooth"). *See* hatching of egg

Casuarius (southern cassowary), 228

CAT scanning. *See* digital imaging

Cathartesaura, 261

Cathetosaurus: pelvic adaptation to rearing, **133**

caudal vertebrae. *See* skeleton, bones of

caudofemoralis muscle, importance of: in defending against theropod attack, **199**, 201; in rearing, 80; in walking, **67**

caytoniales (seed ferns), 231, 232

Cedarosaurus: gastric stones of, 126; location of fossil find, **55**; phylogeny of, 46; predation on, 208

cells: basic requirements, 87–89; diffusion in, 88; needs of, as affected by volume increase, 89

cellulose: difficulty in digesting, 121; digestible energy from, 121

Ceratophaga (tineid moth), 218

ceratopsians, 20; ratio to tyrannosaurids, 211

ceratosaurids, 196

Ceratosaurus: *C. magnirostris*, 206, 233, 235; *C. nasicornis*, 197–198, 206; dentition, 197; at Fruita, 233; in hypothetical taxonomic reconstruction, 217; predation by, 197, 200; size increase and speciation through time, 206

cervical muscles. *See* muscles

cervical ribs. *See* skeleton, bones of

cervical vertebrae. *See* vertebrae, cervical

cetiosaurs ("ceteosaurs"), 4–5, 7; characteristics, 48

Cetiosaurus: discovery of, 4; femur of, 4; location of fossil find, **49**; phylogeny of, 42; vertebra of, **5**, 12

characters (traits), 20, 35; in determining phylogeny, 39–41; parsimony in assigning significance to, 41; terms relating to, 38–39, 41; weighting of, 39. *See also individual characters*

Cheilanthes (Fruita dry-adapted fern), 229

Cheirolepidiaceae, 138–141; extinction of, 261–263

cheirolepidiaceans ("cheiros"), 138, 140–141; pollen, 140, 262; possible pollination by mecopterans, 144; types of foliage, 141, 262

Chenguchelys (Dashanpu chelonian), 223

chevron bones. *See* skeleton, bones of

Chiappe, Luis (paleontologist), 15

Chin, Karen (paleontologist), 108, **109**

Chindesaurus (herrerasaurid basal dinosaur), 20

choroallantois membrane, 175, **177**

Choy, Daniel S. (physiologist), 110

Christian, Andreas (paleontologist), 82

Christman, Erwin (paleosculptor), 8

Chuandongcoelurus (Dashanpu coelurosaurian theropod), 225

Chubutisaurus, 246

Churcher, C. S. (paleontologist), 137

Cicadella (Fruita cycad), 229

Cinnamomum (camphor tree), 250

circulation (of blood). *See* cardiovascular system

clade, 21; definition of, 36. *See also* phylogeny, taxonomy

cladistics: approach in analyzing characters, 13; autapomorphies in, 40, 41; nodes in, 39; outgroups in, 39; parsimony, 41; sister species (groups) in, 39, **40**; steps in, 41; synapomorphies in, 39–40; terms related to, 39–41; weighting of characters, 41

Cladocyclus (Kem Kem teleost fish), 249

cladogram: construction of, 39, **40**; definition of, 39; outgroups, 39, **40**

classes. *See specific classes*; taxonomy

Cladophlebis (Yorkshire Formation mesic fern): at Dashanpu, 228; at Fruita, 232

Classopollis: at Dashanpu, 225; at Patagonia; 242; pollen type of cheirolepidiaceans, 140, 262

Clauss, Marcus (paleontologist), 122

clavicle. *See* skeleton, bones of

claws (pedal): role in walking, **67**, 67–69

cloaca, 172, **174**

Cloverly Formation, Montana, 238

clumped isotope thermometry, 154

clutches. *See* egg: clutch size of

Coelophysis (basal theropod), 20, 197, 200

coelurosaurs, 196

Coelurus (coelurosaurian theropod), **197**, 213

Colbert, Edwin H. (paleontologist), 11

collagen. *See* bone

Coloradosaurus: neck of, 23; phylogeny, 23

communal hunting. *See* predation

communal nesting. *See* nests

Como Bluff, Wyoming, 205, 209

comparative anatomy, 37

competition: between prosauropods and other dinosaurs, 28

compressive loading (force), **64**

computer-aided design (CAD). *See* digital imaging

computer axial tomography (CAT) scanning. *See* digital imaging

Congo Basin, Africa, 222–223

conifers: coevolution with sauropods, 120; diversity of, 119; evolutionary success of, 119; exploitation as sauropod food source, 90, 143, 144, 275; as food source of modern animals, 121–122; general characteristics of, 119; leaflet patterns of, 119; major types of, 119; nutritive value, 121; phytoliths of, 135; pollen of, 120; regenerative capability of, 121; reproduction of, 120; survival in southern continents, 273–275. *See also individual species*

convection (of heat). *See* thermoregulation

convergence (in phylogeny), 38

convex hull estimate, **159**, 162

Coombs, Walter C. (paleontologist), 10

cooperative hunting. *See* predation

Cope, Edward D. (paleontologist), **6**, 6–7, 8, 10, 12, 126

coprolites, **108**; Type "A," 108, 145; of *Isisaurus*, 108

coprophagy, 141, 191

copulation, 172; in male and female *Dicraeosaurus*, **174**

coracoids. *See* skeleton, bones of

"core prosauropods." *See* prosauropods

Corollpachymeridium (Dashanpu true bug), 223

cortical stratifications. *See* bone

countercurrent exchange. *See* respiration

cranium, 61. *See also individual sauropod species*

crocodilians. *See individual species*

Crocodylus, 181, 200, 207

cross-current exchange. *See* respiration

Crurotarsi, 19

cryptic patterning, 183, **189**

Cupressaceae, 122

Cupressinocladus (Morrison cheirolepidiacean), 231

Currie, Phillip (paleontologist), 203, 207

Curry Rogers, Kristina (paleontologist), 15, 156, 159, 268, **271**

Cuvier, Georges (paleontologist), **3**

Cycadella (cycad), 231

Cycadolepsis (Fruita cycad), 231

cycadeoids, 231

cycads, 119, 120. *See also individual species*

Cyclusphaera (Patagonian araucarian), 242

Czekanowskia (mesic fern), 232

Dacrycarpus (podocarp), 231

Darwin, Charles (naturalist), 21,

Dashanpu, China, 222–228; geochron of, 222; paleogeography of, **222**; paleo–Pacific Ocean, **222**; paleo–Tethys Ocean, **222**; sauropods of, 224–225; topographic diagram, **223**; Turgai Strait, **222**

Dashanpusaurus (Abrosaurus): *D. dongi*, 225

data base (data set), 37. *See also subject to which this applies*

Datousaurus: *D. bashensis*, 225

Deccan Trap eruptions, 287–288

defensive behavior, 190, 198–204; from armor by osteoderms, 201; field of vision, **200**, 202; "freeze," 201; group, 190; infrasound detection, 203–204; kicking, **199**; lashing with thumb claws, **198**, **199**, 200; olfactory acuity, 202; stabbing with cervical spines, 202; striking with tail clubs, spikes and whips, 201; striking with tail mass, 200; vocalization, 203

"definitive sauropodomorphs." *See* sauropodomorphs

Deinonychus (dromaeosaurid theropod)*, **196**

Deltadromaeus. See *Bahariasaurus*

Demandasaurus: dentition of, 148

D'Emic, Michael (paleontologist), 180

dentition: bicuspid teeth, 22; canine teeth, 22; compressed-cone-chisel, 133; in diplodocids, **131**, **139**, 148; in eusauropods, 146–147; gnawing function, 134; grasping function, 134; lateral plate in prosauropods, 26; leaf-stripping (raking) function, **131**, 136; in macronarians, 133; molar teeth, 22; morphological comparisons by Barrett and Upchurch, 145; occlusion in, 23; oral processing, 133; orthal bite, 233; pencil-like, 134–135; premolar teeth, 22; robust cropping, 133; in titanosaurs, 150; tooth batteries, **134**; "tooth comb," 133; tooth replacement rate, 134, 135. *See also individual species*

derived (condition of being), 20

Dermestes (dermestid beetle): larva, **218**; role in taphonomy, 218

de Souza Carvalho, Ismar (paleontologist), 246

developmental mass extrapolation (DME), 160

Diamantinasaurus: location of fossil find, **58**; manus, **268**; phylogeny of, 270

Dicraeosauridae: phylogeny of, 41, **43**, 44; at Tendaguru, 239

dicraeosaurids: characteristics, **54**; feeding adaptations of, 148; possible habitat preferences, 235

Dicraeosaurus, **40**; copulation of, **174**; *D. hansemanni*, 148, **235**, 239; *D. sattleri*, 239; dentition of, 145; diet, **143**; feeding adaptations, **143**, 148; location of fossil find, 54; neck, 239; phylogeny, **43**; skeleton, **235**; skull, **235**

Dicroidium (seed fern), 123

Dictozamites (Fruita fern), 228

digital imaging: computer-aided design (CAD), **159**, 163, 165, **169**; computer axial tomography (CAT) scanning, 17, **160**, 162, 165; digital surface scanning, 165; digital 3-D printing, **163**, 165–167; finite element analysis (FEA), **165**, 167–168; of missing elements,163; parametric modeling, **162**, 165; point cloud scan, 163

digital modeling. *See* digital imaging

Dinheirosaurus, 238; gastric stones, 126; occurrence of, at Lourinha, 240

Dinilysia (boid snake), 194

Dinofelis (Dinobastis, machairodont felid), 211

Dinomorph, 165

Dinosauria, 20

Dinosaurier Park, Münchehagen, 15

dinosauriforms, 19

dinosaurimorphs, 19

"Dinosaur Renaissance," 11

dinosaurs, x–xiii, 4–32, **34**, 35–87, 89–118, 121–151, 154–218, **219**, 221–292, **293**. *See also individual species*

"dinoturbation" (bioturbation): at Dashanpu, 227

Diplodocidae: phylogeny of, 45–46

diplodocids, 131; bifurcated (double) neural spines of, **76**, 131–132; body proportions of, 131; chevron bones of, **52**, 131, 146; diets, **143**, 149–150; engineering of neck, **80**, **81**; evolutionary peak, 238, 268; extinction of, 238, 260–261; feeding adaptations, 131, 148–150, 260–261; location of fossil finds, **52**; nares, 105; phylogeny, 45; skulls, **33**, **38**, **52–53**, **64**, **65**; tripodal stance in, 53, 130; vertebral spines of, 76, 80

Diplodocoidea: characteristics, **52**; phylogeny of, **43**, 45

diplodocoids: location of fossil finds, **52**; possible feeding adaptations and diets, **52**, 131–132, 148–149, 261

Diplodocus, 235–237, **276**; bite force of jaws, 90; comparison with *Apatosaurus*, 38; cropping ability, 90–91, **91**; *D. carnegii*, 52, **53**; *D. halli*, 38, **44–45**, 237; dentition of, **52**, 145; discovery of, 7; drinking position, **103**; FEA analysis of skull, 165, 167; feeding ad-

fibula. *See* skeleton, bones of

finite element analysis (FEA). *See* digital imaging

Fiorelli, Lucas (paleontologist), 178

Fiorello, Anthony (paleontologist), 14, 136, 137

Flagellicaudata, 39, **40**

flagellicaudatans, 39, 149

fluvitile energy (in geology), 219

foramina. *See* vertebrae, cervical

forensics: application to paleontology, 216

fossae. *See* vertebrae, cervical

fragipan. *See* paleosols

Frenelopsis (cheirolepidiacean), 249; at Kem Kem, 249

freshwater plants: as food source for sauropods, 128–129

Frey, E. (paleontologist), 127

Fricke, Henry (paleontologist), 154

Fruita, North America, 228–238; geochron of, 228; ecosystems of, 228–233; ephemeral ponds at, 230; flora of, 229, 231–232; inland sea, 228, 235; paleogeography of, **238**; seasonality of, 229–231; topographic diagram, **229**; volcanic uplands of, 238

Fruitachampsa (Fruita terrestrial crocodylomorph), 230

Fruitafossor (Fruita fossorial mammal), 231

fungi: nutritive value of, 124

Futalognkosaurus: location of fossil find, **58**; phylogeny of, 47, 270; size and weight, **282**; skeleton of, **59**

Galveosaurus: location of fossil find, **50**; phylogeny of, 44

gas exchange. *See* respiration

Gasosaurus (Dashanpu theropod), 225

gastralia. *See* skeleton, bones of

gastric mill, 125–126

gastroliths ("gizzard stones"), **124,** 125–126; in *"Seismosaurus"* (*Diplodocus*), 125–126

Gee, Carol T. (paleontologist), 122, 124, 144, 183

geese: foraging habits of, 127; neck flexibility of, 127

genes: expression of, 31–32; Hox, 31; "shortcuts," 31; suppression of, 31–32

genus, 36. *See also* taxonomy; *specific genera*

geochrons, 222. *See also specific geochron*

"Gertie" the dinosaur, **9**

"ghost" lineage, 256

Giganotosaurus (carcharodontid theropod), 247

gigantism: as evolutionary adaptation, 24, 28, 31, 32, 43, 61, **92–93,** 95, 104, 117, 122, 123, 126, 151, 160, 275, 277–279

Gigantospinosaurus (Dashanpu stegosaur), 224

Gill, Bruce D. (paleontologist), 108

Gillette, David (paleontologist), 11

gills (of fish), 96

Ginkgo, 122; at Dashanpu, 224; at Fruita, 232; *G. biloba,* 122, 232

ginkgoaleans (ginkgophytes), 227

Ginkgoites: at Dashanpu, 227

ginkgos, 119, 120, **121**; caloric value of, 122

Ginkgoxylon (Kem Kem ginkoalean), 250

Giraffa (giraffe): blood pressure of, 109; blood volume of, 109; cervical column of, **78**; dentition of, **91**; drinking position of, **98**; heart of, **96**; neck extension of, 83; necks of, 128, 130

Giraffatitan, **15, 118, 233**; air sacs of vertebrae, **75**; comparison with *Brachiosaurus,* 239; feeding adaptations, **143**; jaw muscles, **62**; location of fossil find, **56**; phylogeny, 43; skeleton of, **2, 56, 60**; skull of, **56, 62**; teeth of, **56**; at Tendaguru, 238–239

gizzard (proventriculus), 124–126

Glen Rose, Texas, 208, 251

glycogen body. *See* neural system

Glyptops (Fruita chelonian), 230

Glyptostroboides (conifer): at Dashanpu, 227

Gobiaconodon (triconodontid mammal), 194

Gomani-Chindebvu, Elizabeth (paleontologist), **270**

gonads, 171. *See also* sauropods: reproductive system of

Gondwana, 222; Northern, 221; Southern, 222, 273

Gongxianosaurus: location of fossil find, **48**; pes of, **29**; skeleton, **29, 48**

Gordon, Iain (paleontologist), 137

Gran Chaco Plain, central South America, 229

grasses (graminiforms), **108,** 145

"Graveyard of the Giants," Argentina, **269**

graviportal (limb condition), 68

Great Barrier Reef, Australia, 249

Grellet-Tinner, Gerald (paleontologist), 31, 178

ground-penetrating radar, 11

guilds: of herbivorous mammals, 144; of sauropods, 144; of theropods, 195

gut: bolus in, 91; caecum of, 91, **92**; colon of, **92**; comparison with modern mammals, **92**; energy release within, 94, 122; esophagus of, **93**; floral microbes of, 122; food retention time in, 94, 122; foregut and hindgut systems of, 91, 122; intestines of,

88–89; stomach of, 93; system in sauropods, **92–93,** 122–123

gymnosperms, 119

Haas, G. (paleontologist), 128

habitats: of juvenile sauropods (*see* juveniles); micro, 195. *See also specific habitats*

hadrosaurs, 20, 150; on Hateg Island, 256; ratio to tyrannosaurids in Judith River Formation, 209

Haifangu Formation, China: flora of, 227, 228

Hallett, Mark (coauthor, artist-paleontologist), 11, 47; muscle reconstruction of *Apatosaurus* by, **15**; sauropod drinking hypothesis of, 107

Hall Train Studios, **97**

halophytic plants: as food for juvenile sauropods, 183; at Kem Kem, 249. *See also individual species*

Hansen, Hans (paleontologist), 287

Haplocanthosaurus: camarae of, **101**; camellae of, **101**; extinction of, 235; at Fruita, 235; *H. delfsi,* 235; *H. priscus,* 235; phylogeny of, 235

Hartman, Scott (paleoartist-paleontologist), **13,** 47

Hatcher, John Bell (paleontologist), 114

hatching of egg: of *Ampelosaurus,* 182; caruncle ("egg tooth"), **177,** 182; effect of geothermal conditions on, 178; effect of temperature on, 176, 178–179; predators attracted to, 193–195; synchronous, 183

hatchlings: altricial, 181; of *Argentinosaurus,* **185**; bone growth in, 186; emergence from nest, **170,** 182; growth rate of, 185, 186; of *Massospondylus,* 30–32; in Nile crocodile, 181, **185**; predation on (*see* predation); size and weight, **185,** 186, 195; skull of, **186**; teeth of, 179; tooth grinding in, 179

Hateg Island, Proto-Europe, 252–257; carbonate platforms of, 252; ecosystems of, 252–253; geochron of, 251; insular dwarfing on, 254–257; non-dinosaurian fauna of, 253; paleogeography of, **253**; titanosaurs of, 254–255

Haversian bone. *See* bone

Hay, Oliver P. (paleontologist), 8, 128

Hayden Quarry, New Mexico, 20

hearing. *See* ear

heart: in *Apatosaurus,* 93; of finback whale, 110; heart-to-head distance in sauropods, **96,** 109–110; percentage of sauropod body mass, 110; size in blue whale, 86; size in sauropods, **86,** 110

77–80; extension of, **78**, **80**; flexion of, **78**, **80**, 82; length of (*see individual genus, species*); ligament attachments of, **76**, **79**; mass in, 71, 74; muscles of (*see* muscles); number of vertebrae in, 71; pneumaticity of, 71, 74; sigmoid curve in, **73**, **78**, 83

Nehvizdyella (Kem Kem ginkoalean), 250

Nemegtosauridae. *See* titanosaurs

nemegtosaurids. *See* titanosaurs

Nemegtosaurus, **293**; dentition, **139**; phylogeny of, 47, 271; skull of, **65**

Neocalamites (Yorkshire Formation fern), 228

Neosauropoda, 44–47; characteristics, **51**

neosauropods, **43**

nesting colonies, 180–182; at Auca Mahuevo, 179–182; at Sanagasta Geologic Park, 178

nests: of *Ampelosaurus* (see *Ampelosaurus*); at Auca Mahuevo, 179–182; communal, 180; communal, of alligators and crocs, 180; preparation of, **181**; "prolonged" nesting, 181; "pulse" nesting, 181; at Sanagasta Geologic Park, 178; site fidelity to, 181

Neuquen Basin, Argentina, **241**, 242–243

Neuquensaurus: femur of, **157**; phylogeny of, 271

neural spines. *See* skeleton

neural system, 111–117; afferent nerves, 112; brachial expansion, **116**; brain (see brain); of elephant, **116**; functioning of, 112–117; glycogen body, **116**, 117; interneurons, 112; lumbosacral expansion, 116–117; myelin sheaths, 114; nerves of, 112; neurons, **112**, 114–115; peripheral nerves, 114; plexi, **116**, 117; reflex arc, 112, 115; response time, 114–116; of sauropod, **116**; sensorimotor prediction, 116; spinal cord opening, **116**

niche partitioning. *See* resource partitioning

Nigersaurus, 243; feeding adaptations, **130**, 134, **143**, 148; field of vision, **200**; head restoration, **130**; location of fossil find, **54**; phylogeny of, 46; skeleton of, **54**, **242**; skull of, **54**

Nilssonia (Fruita bennettitalean), 229

North America, 228; dinosaur dispersal across, paleogeographic map (Tithonian stage), **228**

nostrils: in *Macrauchenia*, **105**; in sauropods, **105**

occipital condyle. *See* skull, bones of

occlusion (of teeth). *See* dentition

Ohmdenosaurus: phylogeny of, 42

olecranon process, **265**

omeisaurids: extinction, 260; feeding adaptations, 147, 150; location of fossil finds, **50**; phylogeny, 42, 44

Omeisaurus: at Dashanpu, 224; dentition of, 145; diet, 224–225; feeding adaptations, 146, 150; location of fossil find, **50**; neck, 77, **81**, 225; *O. jungshiensis*, 224; "*O*" *tianfunensis*, 44; phylogeny of, **43**, 44; rearing of, 225; skeleton of, **50**, **81**; skull of, **50**; size/weight, 225

Omosaurus: diagnosis of, 5

ontogeny: of skull in *Tapuiasaurus*, **186**, **187**–**188**, **189**

Opisthocoelicaudia: location of fossil find, **58**; phylogeny of, 47; skeleton of, **59**, *268*, 274; skull of, **275**

Opisthocoelicaudinae. *See* titanosaurs

opisthocoelous (condition of). *See* vertebrae

orders. *See* taxonomy; *specific orders*

organs. *See* viscera; *specific organ systems*

Ornithischia, 20

ornithischians, feeding niches in, 123

Ornithodira, 19

ornithodirans, 71

Ornitholestes (coelurosaurian theropod), **211**, 213

ornithopods, 20

Orthoptera, 253

Osborn, Henry F. (paleontologist, museum director), 7–8, 12, 127

Osmunda (mesic fern), **121**, 124

osteoblast. *See* bone

osteocyte. *See* bone

osteoderms: as character of titanosaurs, 47; as mineral reserves, 180, **272**; as possible defensive armor, 201; in *Rapetosaurus*, **47**, 271, **272**

osteohistology, 14, 155, **156**, 186

osteologically neutral position (ONP), 81–84; "dorsiflection" in, 81; sauropod in, **82**; "ventriflection" in, 82

osteons. *See* bone

ostrich. See *Struthio*

Othnielosaurus (Fruita ornithischian), 232, **276**

Otozamites: at Dashanpu, 227

Otwayia (cheirolepidiacean), 150

outgroups. *See* phylogeny

ovary (ovaries), 171, **174**

oviduct, 171, **174**

oviparity, 173

ovipositor, 181

ovocytes. *See* egg

ovoviparity, 173

Owen, Richard (paleontologist), 3, **4**, 5, 8

oxygen, atmospheric levels of, 279–280. *See also* respiration

Pacinian corpuscles. *See* elephants

paedomorphosis, 31, 255

Pagiophyllum (cheirolepidiacean), 231, 262; at Dashanpu, 225; leaflet pattern of, 145; as staple food for sauropods, **140**, **143**

paleopathology. *See* sauropods

Paleosaniwa (varanid-like lizard), 194

paleosols, 227; at Dashanpu, 226–227; fragipan, 227; oxidization, 227; pH levels (acidity, alkalinity) of, 227

Paludititan, 253

Paluxysaurus. *See Sauroposeidon*

Pampadromaeus. *See Panphagia*

Pangaea: aridity of, 221; dinosaur dispersal from, 222; fragmentation of, 222; supercontinents belonging to, 222

Panorpa (scorpion fly), **144**

Panphagia (*Pampadromaeus*), **18**, 19, 21, 22, **25**

Pantydraco, 22

parabronchi. *See* respiration

Parabuteo (Harris's hawk), 196, **202**; communal hunting of, 207

Paraceratherium (*Indricotherium*, giant Miocene rhino), **290**, 293

Paralititan, 250; diet, 250; location of fossil find, *58*; predation upon, 250

Paralitonia (Hateg frog), 253

parametric modeling. *See* digital imaging

Paranthodon (stegosaur), 260

parasitism (by arthropods), 213

parental care. *See* sauropods

Parrish, Michael J. (paleontologist), 13, 80, 83, 128, 162

parsimony. *See* phylogeny

Patagonia, Proto-South America, 241–247; ecosystems, 241–242; formations of, 242; geochrons of, 241–242; Neuquen Basin, 241; paleogeography of, **238**; topographic diagram, *241*

Patagosaurus: phylogeny, 42; skeleton, *49*

patterning: cryptic, 183; disruptive, **189**, 190

Paul, Gregory S. (artist-paleontologist), 12, **13**, 47, 69

pectoral girdle. *See* skeleton, bones of

pedal claws, 69, 81, 219. *See also* sauropods: locomotion of

pedal phalanges. *See* skeleton, bones of

Pelorosaurus: discovery of, 4; humerus, **5**

pelvis: muscles of (*see* muscles); retroverted condition of, 132. *See also individual species*; skeleton, bones of penis, 172, **174**; flexibility in elephant, 172; in ostriches and waterfowl, 172; shape of, 172, **174**

peripheral nerves. *See* neural system

Perry, Steve (physiologist, paleontologist), 105

pes: caudofemoralis muscle, function of, **67**, 68; claws of, 67; evolution, 67; function, **67**, 69

phalanges. *See* skeleton, bones of

pharyngeal pouch. *See* infrasound

Phascolarctos (koala), 262; dependence on select types of eucalyptus foliage, 262–263

Phillips, John (geologist-paleontologist), 4

pH level. *See* paleosols

Phuwiangosaurus: location of fossil find, **58**

phylogenetic bracketing, 39

phylogenetic trees, 16, **40**, **43**

phylogeny, 20; definition of, 36; terms, 36. *See also specific taxa*

Physalus (finback whale): ventricular thickness of heart, 110

phytochemicals, 120, 122, 263

phytoliths, 108, 135, 144, 266

Pinaceae (pine trees), 122

plantigrade (orientation of pes), 67, 68

plants: problems of preservation in, 145. *See also individual species*

Platanus (sycamore tree), 252

Plateosaurus, 25, 123; as "core prosauropod," 23–24; FEA analysis of, 24–25; feeding adaptations of, 23; manus of, **25**; obligate bipedality in, 24–25; *P. engelhardti*, 23; pes of, **26**; phylogeny, 23; skeleton of, 24; skull of, **28**; virtual modeling of, 24–25

Platt, Phil (paleosculptor-paleontologist), **13**, 69

pleurocoels, **74**

plexi (of nerves). *See* neural system

pneumaticity, 14, 71, **74**, 103, 129, 278; of cervical vertebrae (*see* vertebrae, cervical); role in reducing bone mass, 71, 74; role in thermoregulation, 102

pneumatic spaces. *See* pneumaticity

Podocarpaceae, 119, 122

Podokesaurus (coelurosaurian theropod), 211

Podozamites: at Dashanpu, 225, 227

Poinar, Roberta and George (entomologists), 108

point cloud scan. *See* digital imaging

pollen, 120, 121, 140, 144, 225

posterior iliac process, 68

postzygapophysis. *See* vertebrae

precocial (juvenile condition). *See* juveniles

predation: "attack and retreat" technique, 199; benefit to prey species, 214; communal hunting, 206–207; cooperative hunting, 206–208; defense against by sauropods (see

defense); on hatchlings and pre-hatchlings, 193–195; on sauropods, adult, 199, 200, **203**, **204**; on sauropods, juvenile, 200, **204**; size relationship to prey, 204–206; by theropods (*see* theropods); vulnerability of specific body areas, **199**, 201

predators: amount of flesh consumed by theropods, 215; birds, 194; body size in, 195–197; crocodiles, 200, **204**; mammalian, 193, **194**; in microhabitats, 195; pterosaurs, 195; risks to, 198, **199**; of sauropods,193–198; snakes, 194; theropods, 193, 195–201; trackway evidence of, 208, 250; by varanid-like lizards, 193, 194

predictive regression approach, 160

premolars. *See* dentition

preservation bias, 213

prey. *See individual species*

prezygapophysis. *See* vertebrae

primary browse trees, 261

"probable sauropodomorphs." *See* sauropodo-morphs

procoelous. *See* vertebrae

pronation, 24, 25

propaliny. *See* mandible

prosauropods, 23; "basals," 23, 26, 27; bipedal capability of, **29**; competition with other dinosaurs, 28; "core," 23–24; dentition of, 23, **24**; diets of, 22; extinction, **28**; key adaptations, 27, **28**; "jack of all trades" role, 28; loss of chewing in, **24**; manus of, **26**; pes, 24–26; pronation in, 24; quadrupedality in, 25–26, 29; skeletons of, **29**; skulls, **28**, **29**; supination in, 24. *See also individual species*

Proto-Africa / Proto–South America: paleo-geographic map (Early Cretaceous, Barremian stage), **238**; paleogeographic map (Late Jurassic, Tithonian stage), **232**

Proto-Asia: paleogeographic map (Mid-Jurassic, Bathonian-Callovian stage), **222**

Proto-Europe: dispersal of dinosaur species in, 240; paleogeographic map (Late Jurassic, Tithonian stage), **234;** shared dinosaur faunas with North America and Proto–Africa / South America, 240–241

proton free-precession magnetometry, 11

proventriculus (gizzard), 124

Pseudofrenelopsis (cheirolepidiacean): as staple food for sauropods, 140, **143**, 262

pterosaurs. *See individual species*

Ptilophyllum (conifer): at Dashanpu, 225, 227

pubis (pubes). *See* skeleton, bones of

Puertasaurus: fecal output per day, 108; phylogeny of, 47, 270; predation upon, 206

pulmones. *See* respiration: neopulmon; respiration: paleopulmon

pupation: by dermestids, 219; by dung beetles, 108; by flies, 219; by moths, 219

Pyroraptor (Hateg maniraptoran theropod), 255

quadrupedality, 25–28, **29**

r (reproductive strategy), 187

radius. *See* skeleton, bones of

Rapenomamus (triconodontid mammal), **194**

Rapetosaurus, 134, 135, 264; dentition of, **58**; drinking position, **103**; head restoration, **58**; location of fossil find, **58**; osteoderm(s) of, **59**, 268–269; phylogeny of, 271; predation on, by *Majungasuchus*, **204**; skeleton of, **58**–**59**; skull of, **58**

rauisuchians, 19

Rayleigh waves. *See* elephants; infrasound

Rayoso Formation, Patagonia, South America, 242–243

Rayososaurus, **246**

rearing. *See* sauropods: rearing abilities of; *individual species*

Rebbachisauridae: phylogeny, 41, **43**, 45, 243

rebbachisaurids: characteristics, 54; feeding adaptations of, 149; locomotor adaptations of, 243–245; phylogeny of, **43**, 45, 46; predation upon, 246–247; skeletons of, **54**, **244**

rebbachisaurines, 243, 245; phylogeny of, 46

Rebbachisaurus, 249; location of fossil find, **54**; phylogeny, 46

remodeling (of bone). *See* bone

reproductive isolation, 36, 254–257

reptiles. *See individual species*

resource partitioning, 141–144, **143**; among African ungulates, 142; among herons, 142–143

respiration, 95–102; air capillaries, 97–98; air sacs, 71, **97**, 98–103; alveoli, 95; of archosaurs, 101–104; avian system, 96–101; bronchi, 95; countercurrent/cross-current exchange, 95–96; gas exchange, 95; gills, 95–96; of humans and other mammals, 95, **100**; hyperventilation, 95, 99, 102; lamellae, 95–96; lungs, 95–104; of non-mammalian amniotes, 95; oxygen exchange in, 93–95; paleopulmon, 99–100; parabronchi, 97, **100**; of sauropods, **100**–**101**, 103–104, 278, 280, 284; in thermoregulation, 99–100; tidal, 95–96, **100;** trachea, 99–100, **101**, 105; tracheal dead space, 98–99; ventilation, 95, 97, 98, **100**

respiratory system. *See* respiration

rete mirabile ("wonder net"). *See* cardiovascular system

retroverted pelvises. *See* sauropods

Rhabdodon (iguanodontian ornithopod), 256–257

Rhenland, Proto Europe, **118**, 257

Rhizophora, 249

Rhoetosaurus, location of fossil find, **49**

rhynchocephalians: at Dashanpu, 223

ribs. *See* skeleton, bones of

Ridgely, Ryan C. (paleontologist), 111

Riggs, Elmer S. (paleontologist), 9, 12

Riojasaurus: as "core sauropod," 23

Rogers, Kristina Curry. *See* Curry Rogers, Kristina

rostrum (snout), 179

Rothschild, Bruce M. (paleontologist), 283

rudists, **251**

Ruffner, Marc A. (pathologist), 283

ruminants, 91, 121

Russell, Dale A. (paleontologist), 13

Ruyangosaurus: phylogeny of, 270

Ryder, John (artist, paleontologist), 8

Ryparosa, 228

sacral vertebrae. *See* skeleton, bones of

"sacrificial" ("trial") eggs. *See* egg

sagittal plane, 63

Sagittarius (African secretary bird), 199

Saltasauridae. *See* titanosaurs

Saltasaurus, **263**; camellae of, 74; osteoderms of, **59**, 268; phylogeny of, 271; skeleton of, **59**

Sanagasta Geologic Park, Argentina, 178–179, **182**

Sanajah (boid snake), 194, **195**

Sander, P. Martin (paleontologist), 15, 125, **158**, 159, 285

Sapeornis (avetheropod), 21

Sapindopsis (Kem Kem angiosperm), 249

Sarcophaga (flesh fly), **218**

Sassani, Nima (artist, paleontologist), **267**

Saturnalia, 22, 28

Saurischia, 20; basal condition of, 21

Saurophaganax (allosaurid theropod), 206, 210, 213

Sauropoda: naming and derivation of word, 7

"sauropod hiatus." *See* sauropods: "hiatus" in geological record

Sauropodomorpha: phylogenetic relationships of, 20

sauropodomorphs, 20; advanced, 21, **29**; basal, 21–22; definitive, 22, 23, 24; derived, 21; near, 25; origin of, 20; pes and skeletal evolution, **26**, **29**, 123; "probable," 22, 24

Sauropod Research Biology Team, xiii, 533

sauropods: advanced (derived), 26–28, **29**, 42, 44, 48; aggressive behavior in, 190; assumed aquatic preference, 8–10, 128; assumed decline at Jurassic-Cretaceous boundary, 16; basal, 27, **30**; bioengineering of neck, 128, **129**, 139–141; biology of (*see specific biological process*); biomechanics of (*see biomechanics*); bipedal stance in, 130; bird-like respiratory system, **100–101**, 103–104; blood circulation, 108–111; brains of, **93**, 111–114, **115**, **116**; breathing (*see respiration*); breeding behavior, 172; coprophagy, 142, 191, 232; copulation in, **174**; defensive behavior of (*see defense behavior*); dentition of, **48–58**, 89–90, **91**, **130**, **131**, 133–138, **139**, 146–150, 153–155, 167, 225, 235, 264; digestion, 91, **92–93**, 94–95; drinking position(s), **98**, 102–103; earliest true, 27, 28; effect on Mesozoic ecosystems, 11, 151, 227–228; egg laying of, 176, 178, **181**; elimination in, 108–109; engineering (skeletal) of spine, 77–83, 84–85; extinction of (*see extinction*); eyelids of, 164; fanciful depiction of, **279**; food amount consumed on daily basis, 90; food-gathering ability, 89–90, **91**, 126, 278–279; food plant preferences (possible), 128–129, 140, 144–151; —of diplodocoids, 148–150; —of eusauropods, 147; —of macronarians, 147–148; —of mamenchisaurids, 147; —of titanosaurs, 150; geologic longevity, xii, 287, 292, 293; growth rates of, 156, 158–159; habitat preferences of, 11, 167–168, 273; hearing ability, 202–204; heart (*see cardiovascular system*); "hiatus" in geological record, 271–273; juveniles of (*see juveniles*); juvenile growth rates (*see juveniles*); lack of abundant juvenile fossils, 280–281; locations of early sauropod finds, 5; locomotion in, 68–71; longevity in, 159; loss of chewing, 14, 278; loss of complex parental care, 277; macronarian Titanosauriformes, 263; metabolism (*see thermoregulation*); migratory behavior of, 238, 251; morphological disparity, xiii; Morrison ecosystems (occurrence in), 129, 232–238, 285–287; mortality in juveniles, 183–184, 185, 187, 188–189; "near," 24–26; neural transmission, 114–117; origins, 19–32; paleopathology, 283; parental care in, 182, **189**; phylogenetic diversity, 310; phylogeny of, 37–47; pneumaticity of skeleton (*see pneumaticity*); "poling" locomotion in water, 251; population structure, 283–284; "prob-

able," 24; rearing abilities of, **53**, **81**, **84–85**, 129–132; reproductive strategy in, **185**, 187, 188–189, 278, 284; reproductive system of, **174**; respiratory system of (*see respiration*); retroverted pelvises in, 132; sacral body of, **116**, 117; sensory abilities of, **161**, 163–164; sexual dimorphism of, 159, 283; sexual maturity in, 159; sizes of, 282 (*see also specific genera, species*); skeletal pneumaticity (*see pneumaticity*); skeletons of, 61, **72–73** (*see also individual species*); skull, bones of (*see skulls, bones of*); skull: basic morphotypes of, **65** (*see also specific types, species*); skull: muscles of (*see skull: muscles of*); specialization for medium-high reach, 130–131; speeds of, 68, 69, 201; survival strategies in juveniles (*see juveniles*); temperature regulation in (*see thermoregulation*); Titanosauriformes (*see Titanosauriformes*); titanosaurs (*see titanosaurs*); tooth microwear, 14 (*see also microwear*); tripodal stance in, 130; "true," 27; unsolved problems relating to, 277–281, 283–292; vision (acuity of), 164–165; vision, field of (*see vision: fields of*); weight in, 160–162 (*see also weights of individual species*)

Sauroposeidon (*Paluxysaurus*), xi, 225; camellae of, **101**; location of fossil find, **57**; neck of, 23; phylogeny of, 47; pneumaticity of vertebrae, **101**, 102; predation on, 208; size/weight of, **57**; vertebra of, 37

scapula. *See* skeleton, bones of

scapulacoracoid: action of, in walking, **70**; cartilaginous areas of, **70**. *See also* skeleton, bones of

scavenging: by arthropods, 218; by crocodilians, **215**; by pterosaurs, 215; by theropods, **211**, **212**, **215**; by varanids, 215. *See also individual species*

Schmeissneria (Yorkshire Formation fern), 228

Schweitzer, Mary (paleontologist), **158**, 159

seas: regression/transgression of, 237, 239

seed ferns, 232, 248, 272. *See also individual species*

Seeley, Harry Govier (paleontologist, systematist), 5

"*Seismosaurus*" (*Diplodocus*): as junior synonym of *Diplodocus*, 38; skeleton of, **44–45**

selective/non-selective feeding, 137–140

Sellers, William (paleontologist), 162

semicircular canals (SCCs): in *Camarasaurus*, **161**, 164; in *Diplodocus*, **161**, 164; horizontal, 83, **161**; in *Tyrannosaurus* and other theropods, **161**, 163–164; vertical, 83, **161**